Topics in Applied Physics
Volume 125

For further volumes:
www.springer.com/series/560

Topics in Applied Physics is a well-established series of review books, each of which presents a comprehensive survey of a selected topic within the broad area of applied physics. Edited and written by leading research scientists in the field concerned, each volume contains review contributions covering the various aspects of the topic. Together these provide an overview of the state of the art in the respective field, extending from an introduction to the subject right up to the frontiers of contemporary research.

Topics in Applied Physics is addressed to all scientists at universities and in industry who wish to obtain an overview and to keep abreast of advances in applied physics. The series also provides easy but comprehensive access to the fields for newcomers starting research.

Contributions are specially commissioned. The Managing Editors are open to any suggestions for topics coming from the community of applied physicists no matter what the field and encourage prospective editors to approach them with ideas.

Managing Editor

Dr. Claus E. Ascheron
Springer-Verlag GmbH
Tiergartenstr. 17
69121 Heidelberg
Germany
claus.ascheron@springer.com

Assistant Editor

Adelheid H. Duhm
Springer-Verlag GmbH
Tiergartenstr. 17
69121 Heidelberg
Germany
adelheid.duhm@springer.com

Sergej O. Demokritov · Andrei N. Slavin
Editors

Magnonics

From Fundamentals to Applications

Springer

Editors
Sergej O. Demokritov
Universität Münster
Münster, Germany

Andrei N. Slavin
Oakland University
Rochester, MI, USA

ISSN 0303-4216
ISSN 1437-0859 (electronic)
Topics in Applied Physics
ISBN 978-3-642-30246-6
ISBN 978-3-642-30247-3 (eBook)
DOI 10.1007/978-3-642-30247-3
Springer Heidelberg New York Dordrecht London

Library of Congress Control Number: 2012946930

Printed on acid-free paper

Springer is part of Springer Science+Business Media (www.springer.com)

This book is dedicated
to Prof. Doug L. Mills

Preface

The theoretical and experimental study of ferromagnetic resonance and spin waves – low lying excitations in magnetically ordered substances – has a long history. Experimental methods involving spin waves were used to determine fundamental parameters of magnetic materials, such as static magnetization, exchange constant, anisotropy constants, parameters of magnetic dissipation, etc. Also, spin waves excited at microwave frequencies have group and phase velocities that are much smaller than the velocity of regular electromagnetic waves of the same frequency and, in certain geometries, are non-reciprocal. These properties of spin waves make possible the development of miniaturized microwave signal processing devices, both reciprocal (filters, oscillators, delay lines) and non-reciprocal (isolators and circulators). An additional advantage of microwave signal processing devices based of spin waves is the possibility to tune the operational frequency of these devices by changing the external bias magnetic field applied to this device.

Recently the interest to the spin waves has been revived mostly due to the rapid developments of the novel fields of research, such as nanomagnetism and spintronics, and also due to the realization of the fact that spin waves have very interesting quantum properties. In many situations, spin waves could be (and should be) treated as magnetic quasiparticles – magnons. Magnons obey the Bose–Einstein statistics and can demonstrate interesting and unusual macroscopic quantum phenomena, such as Bose–Einstein condensation of parametrically pumped magnons at room temperature [1]. Magnons are also very interesting in spintronics as they are capable of carrying spin angular momentum (or pure spin current) and, therefore, can be used as possible carriers of spin current in magnetic dielectrics [2].

Another important development made possible by the progress in nanotechnology is the creation of artificial magnetic periodic structures with feature size of the order of the magnon wavelength – the so-called *magnonic crystals*. Similar to the well-known photonic crystals [3], magnonic crystals have designed-to-order band structure of the magnonic spectra and, therefore, novel dynamic properties that cannot be found in natural magnetic materials. The tuneability of magnonic crystals by an external field allows a *dynamic* creation of the magnetic periodicity in an artificial crystal with characteristic times comparable to the time of magnon propagation in the crystal. This possibility results in the existence of the dynamic magnonic crystal with very unusual properties, allowing frequency transformation and reversal of magnon propagation in a purely linear system [4].

Nano-structuring of magnetic materials required significant modifications in the traditional theory of spin waves as the relative influence of the dipole–dipole and exchange interactions on the spectrum of magnons depends on the spatial size of the magnetic structure, the role of the exchange interaction being dominant for the nano-structured magnetic materials.

Experimental investigations of novel dynamic phenomena in magnetism require the development of new methods of probing and manipulation of magnetization dynamics in magnetically ordered substances. Ultrafast optical methods of experimental studies (see, e.g., [5]) made possible probing magnetization dynamics at a time scale that is shorter than the characteristic exchange time. At this time-scale, the length of the magnetization vector is not conserved and the traditional Landau–Lifshitz equation of motion for the magnetization is not valid.

All these factors lead to the formation of a new field of research – magnonics – that includes not only the traditional branches of spin wave research, but also the area of quantum properties of magnons, physics of artificial magnonic crystals, magnetic excitations at nano-scale and ultrafast magnetization dynamics.

The first major conference in the novel field of magnonics entitled "Magnonics: From Fundamentals to Applications" was held in Dresden, in August 2009, sponsored by the visitor program of the Max Plank Institute of Physics of Complex Systems (MPIPKS) [6]. The event had a great success and achieved its main aim – to form a community of magnonics researchers.

The 2nd edition of this conference entitled "International Workshop on Magnonics: From Fundamentals to Applications" was held in a resort city of Porto de Galinhas near Recife, the capital of the state of Pernambuco, Brazil in August 2011 and turned out to be equally successful. It demonstrated that magnonics is rapidly becoming a popular and mature field of research that brings together scientists and engineers interested in a wide variety of research topics ranging from fundamental magnonic properties to their application in the information technologies. The conference also demonstrated that magnonics is a convenient term allowing the unification of several novel branches of magnetization research, such as macrosopic quantum magnetic phenomena, probing and manipulation of magnetism with femtosecond light and polarized electrons, magnon spintronics, and physics of artificial magnonic crystals.

This book is based on the invited papers presented at the second conference on magnonics in Recife, Brazil in 2011 and reflects the current state-of-the-art in the field of magnonics. The volume contains 17 chapters separated into four topical sections.

The first section, devoted to the different magnonic modes and states, starts from the contribution by V.V. Naletov et al. where magnonic modes of magnetic nanopillar are experimentally identified using the method of magnetic resonance force microscopy. P.S. Keatley et al. also study magnonic modes in single nanomagnets and in pairs of dipolar coupled nanomagnets, but use the method of time-resolved scanning Kerr microscopy. Chaotic magnonic states excited in active feedback ring containing macroscopic samples of ferrite magnetic films are studied in the contribution of A.M. Hagerstrom and M. Wu, while S. Rezende provides a theoretical overview of coherent magnonic states, including Bose–Einstein condensates of magnons.

The second section deals with the novel methods of probing and manipulation of magnons using femtosecond light and polarized electrons and starts with the contribution of A. Kiriljuk et al. that discusses the role of angular momentum in ultrafast magnetization dynamics. Probing of magnonic modes in arrays of magnetic antidotes using all-optical femtosecond laser experiments is discussed in the chapter by B. Lenk et al., while Kh. Zakeri and J. Kirschner report on the excitation of high wave-vector magnons in itinerant ferromagnets by means of polarized electron scattering experiments. The role of micromagnetic simulations and numerical solvers based on atomistic spin models in the theoretical understanding of ultrafast magnetization dynamics and interactions of magnons with other fundamental excitations in magnetic solids is discussed in the contribution by M. Dvormik et al.

The third section of the book is devoted to magnon spintronics, or interactions between magnons and spin-polarized electrons. The contribution of H. Adachi and S. Maekawa deals with the spin Seebeck effect that enables thermal injection of spin currents from a ferromagnet into adjacent non-magnetic metal. C. Burrowes and B. Heinrich study the spin pumping at the boundary between the magnetic insulator and non-magnetic metal and compare it with similar process at the boundary between the ferromagnetic and non-magnetic metals. The chapter authored by O. Prokopenko et al. considers the microwave detectors based on spin-torque nano-oscillator structures operating in a passive regime, while V. Demidov et al. report on the control of spin-torque oscillators working in the active generating regime by means of external microwave pumping. The final contribution of this section by S. Bonetti and J. Akerman discusses spin-torque nano-oscillators as building blocks for future active nano-spintronic structures and devices.

The fourth section of the book is devoted to artificial periodic structures supporting magnon propagation – static and dynamic magnonic crystals. The chapter by S. Neusser et al. reports on the experimental study of on two-dimensional static magnonic crystals formed by antidotes (or holes) in a Permalloy film, while the chapter by G. Gubbiotti et al. compares properties of two-dimensional magnonic crystals based on magnetic dots and antidotes. The theory of normal magnonic modes in a waveguide based on a static magnonic crystal is presented in the chapter authored N.Yu. Grigoryeva and B.A. Kalinikos and, finally, the theoretical and experimental properties of dynamic magnonic crystals are discussed in the contribution by A. Chumak et al.

Münster | Sergej O. Demokritov
Rochester, MI | Andrei N. Slavin

References

1. S.O. Demokritov et al., Bose–Einstein condensation of quasi-equilibrium magnons at room temperature under pumping. Nature **443**, 430–433 (2006)
2. Y. Kajiwara et al., Transmission of electrical signals by spin-wave interconversion in a magnetic insulator. Nature **464**, 262–266 (2010)
3. J.D. Joannopoulos, P.R. Villeneuve, S. Fan, Photonic crystals: putting a new twist on light. Nature **386**, 143–149 (1997)
4. A.V. Chumak et al., All-linear time reversal by a dynamic artificial crystal. Nature Commun. **1**, 141 (2010)
5. A. Kirilyuk, A.V. Kimel, T. Rasing, Ultrafast optical manipulation of magnetic order. Rev. Mod. Phys. **82**, 2731 (2010)
6. V.V. Kruglyak, S.O. Demokritov, D. Grundler, Magnonics. J. Phys. D, Appl. Phys. **43**, 264001 (2010)

Contents

In Memoriam Douglas Mills

Our dear friend and colleague Douglas Leon Mills has passed away on March 29, 2012, just few months before this book appears. As a scientist who made seminal contributions to the field of Magnonics even before the term has been invented, Doug Mills was supposed to be one of the key contributors to the book. During the preparation of the book, we had a lot of discussions with Doug about the concept of the book and the ways to make the book more attractive for a wide audience of physicists working in magnetism and solid state physics. Shortly before the deadline for the chapter submission, Doug wrote “I will continue to try. The plan was to have the paper completed at the end of this past week. There was a setback in this damned medical situation so I lost the week. Maybe next week or shortly thereafter, if things stabilize.” Unfortunately, it did not happen. We lost our dear friend and a major theoretician in the field of magnetization dynamics.

Doug received the Ph.D. degree in Physics in 1965 from the University of California, Berkeley under the supervision of Charles Kittel. After having spent a year as a Postdoctoral Fellow at the Université Paris-Sud, in Orsay, France, in the group of Jacques Friedel, Doug joined the Department of Physics at the University of California, Irvine and has worked there for the rest of his life. For most of physicists, both theoreticians and experimentalist, the first visit to Irvine was initiated by Doug.

In his research, Doug made seminal contributions to several areas of condensed matter physics, namely to phonon physics, inelastic light scattering, low energy electron diffraction, infrared absorption, surface excitations, electronic transport properties, and magnetic properties of solids. The majority of this work was done in the context of surface physics. He was an early contributor to the modern theory of surface electromagnetic waves that began in the late 1960s, which laid the theoretical foundations of the contemporary field of plasmonics.

Particularly significant for us was his series of papers devoted to surface waves on magnetic media of several types: ferromagnets, antiferromagnets, and magnetoelastic media. A major interest of his was the scattering of light and of low energy electrons from magnetic solids as a means of studying magnetic excitations. Besides, he produced the first theories of the Raman scattering of light from surface phonon polaritons on semiconductors, from metal surfaces, and from surface spin waves on opaque ferromagnets, as well as of infrared absorption by binary metallic alloys and by multiphonon processes in alkali-halide crystals.

His papers on the scattering of spin-polarized electrons from magnetic surfaces predicted and explained data obtained in such experiments and remain the benchmarks for such calculations to this day.

Although a pen and paper theorist, he collaborated extremely well with experimental and computational groups around the world. He was an enthusiastic person, who loved physics in all its facets, and he was always opened for discussion even with young colleagues. We all will remember lovely and important discussions with Doug.

Among his academic honors, Doug was an Alfred P. Sloan Foundation Fellow (1968–1970), a Fellow of the American Physical Society, a recipient of a Senior U.S. Scientist Award from the Alexander von Humboldt Foundation of Germany (1990), a Yamada Foundation Lecturer in Japan (1984), and a Masse Honor Lecturer of the Kansas State University (1994). The Alumni Association of the University of California, Irvine, honored him with the Distinguished Faculty Research Award in 1984.

Doug served the condensed matter physics community by membership on the organizing committees of several magnetism conferences. In addition, he was a member of the editorial board of the journal Physics Reports, was an editor of Comments on Condensed Matter Physics (Gordon and Breach), and an editor of the book series Contemporary Concepts of Condensed Matter Science (Elsevier).

Doug was a great storyteller and had an excellent collection of them. He had a number of collaborators in Chile, Brazil, Germany, and other countries, which he enjoyed visiting. He was a very generous person, especially to younger colleagues, with his time, ideas, and constant encouragement. He was fighting his illness with amazing courage, approaching it as a problem than needed more and more intricate solution. He was a great optimist until the very end.

Sergej O. Demokritov
Andrei N. Slavin

Part I
Magnonic Modes in Nanomagnets, Chaotic and Coherent Magnonic States

Chapter 1
Spin-Wave Eigen-modes in a Normally Magnetized Nano-pillar

V.V. Naletov, G. de Loubens, S. Borlenghi, and O. Klein

Abstract We report on a spectroscopic study of the spin-wave eigen-modes in a circular spin-valve nano-pillar, perpendicularly magnetized along $\hat{z}$. Spectroscopy is performed by Magnetic Resonance Force Microscopy (MRFM). Distinct spectra are measured depending on whether the nano-pillar is excited by a uniform in-plane radio-frequency (RF) magnetic field or by an RF current flowing perpendicularly through the layers. These results are in agreement with micromagnetic simulations of the time decay response of the local magnetization to excitations with different azimuthal symmetries, $(\hat{x} + i\,\hat{y})e^{-i\ell\phi}$. This demonstrates that the azimuthal ℓ-index is the discriminating parameter for the selection rules, as only $\ell = 0$ modes are excited by the RF magnetic field, whereas only $\ell = +1$ modes are excited by the RF current, owing to the orthoradial symmetry of the induced RF Oersted field.

1.1 Introduction

Technological progress in the fabrication of hybrid nanostructures using magnetic metals has allowed the emergence of a new science aimed at utilizing spin dependent effects in the electronic transport properties [1]. An elementary device of spintronics consists of two magnetic layers separated by a normal layer. It exhibits the well-known giant magneto-resistance (GMR) effect [2, 3] and the converse spin transfer effect [4, 5].

From an experimental point of view, the precise identification of the spin-wave (SW) eigen-modes in hybrid magnetic nanostructures remains to be done [6–11]. Of particular interest is the exact nature of the modes excited by a polarized current in spin transfer nano-oscillators (STNOs). Here, the identification of the associated symmetry behind each mode is essential. It gives a fundamental insight about their selection rules and about the mutual coupling mechanisms that might exist

V.V. Naletov · G. de Loubens (✉) · S. Borlenghi · O. Klein
Service de Physique de l'État Condensé (URA 2464), CEA Saclay, 91191 Gif-sur-Yvette, France
e-mail: gregoire.deloubens@cea.fr

V.V. Naletov
Physics Department, Kazan Federal University, Kazan 420008, Russian Federation

S.O. Demokritov, A.N. Slavin (eds.), *Magnonics*, Topics in Applied Physics 125,
DOI 10.1007/978-3-642-30247-3_1, © Springer-Verlag Berlin Heidelberg 2013

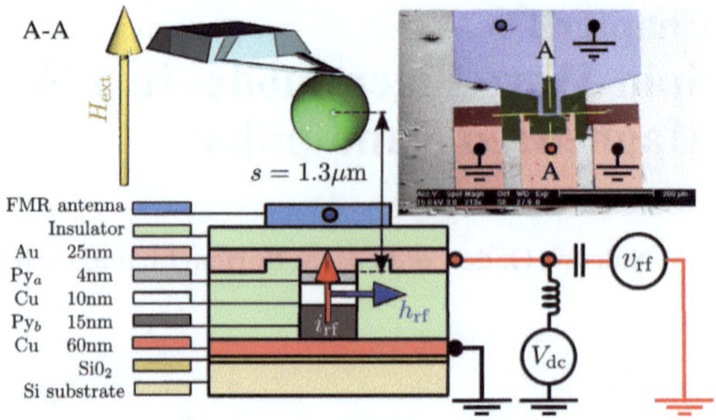

Fig. 1.1 Schematic representation of the experimental setup. Two independent excitation circuits are used: *in red*, the circuit allowing the injection of an RF current perpendicular-to-plane through the nano-pillar (i_{rf}, *red arrow*); *in blue*, the circuit allowing the generation of an RF in-plane magnetic field (h_{rf}, *blue arrow*)

intra or inter STNOs. It also determines the optimum strategy to couple to the auto-oscillating mode observed when the spin transfer torque compensates the damping, a vital knowledge to achieve phase synchronization in arrays of nano-pillars [12]. These SW modes also have a fundamental influence on the high frequency properties of spin-valve devices, and in particular on the noise of magneto-resistive sensors [13, 14].

This work concentrates on a comprehensive identification of the SW eigenmodes in a normally magnetized circular spin-valve nano-pillar, where the axial symmetry is preserved. We shall perform a comparative spectroscopic study of the SW modes excited either by an RF current flowing perpendicularly through the nano-pillar, as used in spin-torque driven ferromagnetic resonance (ST-FMR) [15, 16], or by a homogeneous RF in-plane magnetic field, as used in conventional FMR. The experimental spectra are compared to micromagnetic simulations of the time decay response of the local magnetization to a small perturbation of the magnetic configuration. This allows identifying and labeling the observed SW eigenmodes, according to their symmetry and to the dipolar coupling between the two magnetic layers.

The spin-valve structure used in this study is a standard Permalloy ($Ni_{80}Fe_{20} = Py$) bi-layer structure sandwiching a 10 nm copper (Cu) spacer: the thicknesses of the thin Py_a and the thick Py_b layers are respectively $t_a = 4$ nm and $t_b = 15$ nm. The extended film is patterned by standard e-beam lithography and ion-milling techniques to a nano-pillar of radius 125 nm. The magnetic parameters of this sample have been measured in [17], and are reported in Table 1.1. The top Cu and bottom Au contact electrodes are shown in red in Fig. 1.1. They are impedance matched to allow for high frequency characterization by injecting an RF current i_{rf} through the device. Hereafter, spectra associated to SW excitations by this part of the microwave circuit will be displayed in a red tone.

The originality of our design is the addition of an independent top microwave antenna whose purpose is to produce an in-plane RF magnetic field h_{rf} at the nano-pillar location. In Fig. 1.1, this part of the microwave circuit is shown in blue. Injecting a microwave current from a synthesizer inside the top antenna produces a homogeneous in-plane linearly polarized microwave magnetic field, oriented perpendicularly to the stripe direction. Hereafter, spectra associated to SW excitations by this part of the microwave circuit will be displayed in a blue tone.

We shall use in this study a method independent of transport to detect the magnetic resonance inside a spin-valve nanostructure: a Magnetic Resonance Force Microscope (MRFM) [18–20], hereafter named mechanical-FMR. A first decisive advantage of the mechanical-FMR technique is that the detection scheme does not rely on the SW spatial symmetry because it measures the change in the longitudinal component of the magnetization. It thus probes *all* the excited SW modes, independently of their phase [21, 22]. A second decisive advantage is that mechanical-FMR is a very sensitive technique that can measure the magnetization dynamics in nanostructures buried under metallic electrodes [23]. Indeed, the probe is a magnetic particle attached at the end of a soft cantilever and is coupled to the sample through dipolar interaction.

The mechanical detector consists of a 800 nm diameter sphere of soft amorphous Fe (with 3 % Si) glued to the apex of an Olympus Bio-Lever having a spring constant $k \approx 5$ mN/m. In our setup, the separation between the center of the spherical probe and the nano-pillar is set to 1.3 μm (see Fig. 1.1), which is a large distance considering the lateral size of the sample. The external magnetic field produced by an electromagnet is oriented out-of-plane, exactly along the nano-pillar axis $\hat{z}$. In our study, the strength of the applied magnetic field shall exceed the saturation field (≈ 8 kOe), so that the nano-pillar is studied in the saturated regime. The mechanical sensor measures then the spatial average of the longitudinal component M_z of magnetization:

$$\langle M_z \rangle \equiv \frac{1}{V} \int_V M_z(\boldsymbol{r})\, d^3\boldsymbol{r}, \tag{1.1}$$

where the chevron brackets stand for the spatial average over the volume of the magnetic body. We emphasize that for a bi-layer system, the force signal integrates the contribution of both layers.

The experimental spectra will be shown below in Fig. 1.4(b) and Fig. 1.5(b). The important point is that the two SW spectra excited by h_{rf} and i_{rf} are distinct, implying that they have a different origin. It will be shown in the theoretical Sect. 1.2.1 that the RF field and the RF current excitations probe two different azimuthal symmetries ℓ. Namely, only $\ell = 0$ modes are excited by the uniform RF magnetic field, whereas only $\ell = +1$ modes are excited by the orthoradial RF Oersted field associated to the RF current [24]. The mutually exclusive nature of the responses to the uniform and orthoradial symmetry excitations is a property of the preserved axial symmetry, where the azimuthal index ℓ is a good quantum number, i.e., different ℓ-index modes are not mixed and can be excited separately.

1.2 Identification of the Spin-Wave Modes

In this section, we discuss the boundary-value problem for SW propagation inside normally magnetized disks, where the confinement leads to a discrete SW spectrum.

Neglecting the thickness dependence, only three indices are required to label the resonance peaks: the usual azimuthal and radial indices for a single disk (ℓ, m), plus an additional index referring to the anti-binding or binding (A or B) coupling between the two magnetic layers in mutual dipolar interaction.

1.2.1 Single Magnetic Disk

SW eigen-modes are the solutions of the linearized equation of motion of the magnetization, obtained by decomposing the instantaneous magnetization vector $\boldsymbol{M}(t)$ into a static and dynamic component [25]:

$$\frac{\boldsymbol{M}(t)}{M_s} = \hat{\boldsymbol{u}} + \boldsymbol{m}(t) + \mathcal{O}(\boldsymbol{m}^2), \tag{1.2}$$

where the transverse component $\boldsymbol{m}(t)$ is the small dimensionless deviation ($|\boldsymbol{m}| \ll 1$) of the magnetization from the local equilibrium direction, $\hat{\boldsymbol{u}}$. In ferromagnets, $|\boldsymbol{M}| = M_s$ is a constant of the motion, so that the local orthogonality condition $\hat{\boldsymbol{u}} \cdot \boldsymbol{m} = 0$ is required.

We restrict our study to the case of thin layers so that one can assume that the magnetization dynamics is uniform along the thickness. For a normally magnetized disk, where $\hat{\boldsymbol{u}} = \hat{\boldsymbol{z}}$, the SW modes can be classified according to their behavior under rotations in the x–y plane (2D vector equations with polar coordinates ρ and ϕ):

$$\boldsymbol{m}_\ell = \frac{1}{2}(\hat{\boldsymbol{x}} + i\,\hat{\boldsymbol{y}})e^{-i\ell\phi}\psi_\ell(\rho), \tag{1.3}$$

where the functions $\psi_\ell(\rho)$ describe the dependence of the SW mode on the radial coordinate ρ and have to be determined from the dynamical equations of motion. To satisfy the appropriate boundary conditions at the edges of the magnetic body [26], we shall use as the trial vectors the radial profiles of the form $\psi_\ell(\rho) = J_\ell(k_{\ell,n}\rho)$, where $J_\ell(x)$ is the Bessel function and $k_{\ell,n}$ are SW wave-numbers determined from the pinning conditions at the disk boundary $\rho = R$. For our experimental conditions ($t_a, t_b \ll R$), the pinning is almost complete, and we shall use $k_{\ell,n} = \kappa_{\ell,n}/R$, where $\kappa_{\ell,n}$ is the nth root of the Bessel function of the ℓth order. The set of Bessel function form here a complete basis in the space of vector functions $\boldsymbol{m}$.

Figure 1.2 shows a color representation of the Bessel spatial patterns for different values of the index $\nu = \ell, n$. We restrict the number of panels to two values of the azimuthal mode index, $\ell = 0, +1$, with the radial index varying between $n = 0, 1, 2$. In our color code, the hue indicates the phase (or direction) of the transverse component $\boldsymbol{m}_\nu$, while the brightness indicates the amplitude of $|\boldsymbol{m}_\nu|^2$. The nodal positions are marked in white. A node is a location where the transverse component vanishes, i.e., the magnetization vector is aligned along the equilibrium axis. This coding scheme provides a distinct visualization of the phase and amplitude of the precession profiles. The black arrows are a snapshot of the $\boldsymbol{m}_\nu$ vectors in the disk and are all rotating synchronously in-plane at the SW eigen-frequency.

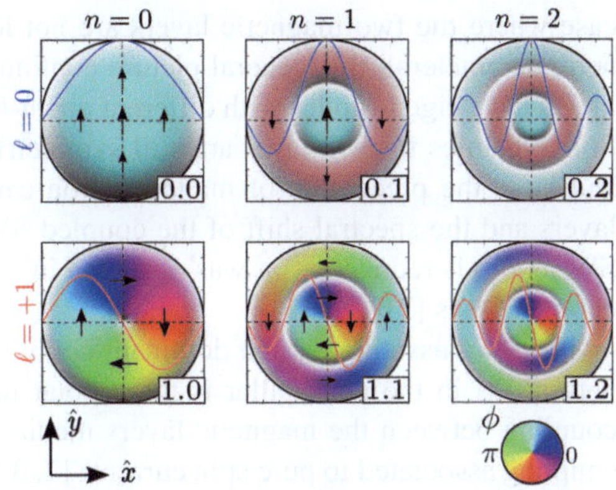

Fig. 1.2 Color representation of the Bessel spatial patterns for different values of the azimuthal mode index ℓ *(by row)* and radial mode index n *(by column)*. The *arrows* are a snapshot of the transverse magnetization $\boldsymbol{m}_\nu$, labeled by the index $\nu = \ell, n$

The top left panel shows the $\nu = 0, 0$ ($\ell = 0$, $n = 0$) mode, also called the uniform mode. It usually corresponds to the lowest energy mode since all the vectors are pointing in the same direction at all time. Below is the $\ell = +1$, $n = 0$ mode. It corresponds to SWs that are rotating around the disk in the same direction as the Larmor precession. The corresponding phase is in quadrature between two orthogonal positions and this mode has a node at the center of the disk. The variation upon the $n = 0, 1, 2$ index (ℓ being fixed) shows higher order modes with an increasing number of nodal rings. Each ring separates regions of opposite phase along the radial direction. All these spatial patterns preserve the rotation invariance symmetry.

The coupling to an external coherent source depends primarily on the ℓ-index [27] as the excitation efficiency is proportional to the overlap integral

$$h_\nu \propto \langle \overline{\boldsymbol{m}}_\nu \cdot \boldsymbol{h}_1 \rangle, \tag{1.4}$$

where $\boldsymbol{h}_1(\boldsymbol{r})$ is the spatial profile of the external excitation field. It can be easily shown that a uniform RF magnetic field, $\boldsymbol{h}_1 = h_{\mathrm{rf}} \boldsymbol{x}$, can only excite $\ell = 0$ SW modes. Obviously, the largest overlap is obtained with the so-called uniform mode ($n = 0$). Higher radial index modes ($n \neq 0$) still couple to the uniform excitation but with a strength that decreases as n increases [19, 28]. The $\ell \neq 0$ normal modes, however, are hidden because they have strictly no overlap with the excitation. In contrast, the RF current-created Oersted field, $\boldsymbol{h}_1 = h_{\mathrm{Oe}}(\rho)(-\sin\phi\hat{\boldsymbol{x}} + \cos\phi\hat{\boldsymbol{y}})$ has an orthoradial symmetry and can only excite $\ell = +1$ SW modes.

1.2.2 Double Magnetic Disks

The interaction between two identical magnetic layers leads to the splitting of the uniform mode in each layer into two collective modes: the binding and anti-binding modes. This has been observed in interlayer-exchange-coupled thin films [29] and in trilayered wires where the two magnetic stripes are dipolarly coupled [30]. In the case where the two magnetic layers are not identical (different geometry or magnetic parameters), this general picture continues to subsist. Although both isolated layers have eigen-modes with different eigen-frequencies, the collective magnetization dynamics in each layer are still symmetrically or anti-symmetrically coupled. But here, the precession of magnetization can be more intense in one of the two layers and the spectral shift of the coupled SW modes with respect to the isolated SW modes is reduced, as it was observed in both the dipolarly- [30] and exchange-coupled cases [31].

Here, we assume that the dominant coupling mechanism between the two magnetic disks in the nano-pillar is the dipolar interaction. We neglect any exchange coupling between the magnetic layers mediated through the normal spacer or any coupling associated to pure spin currents [7, 32] in our all-metallic spin-valve structure. Our nano-pillar consists of two dislike magnetic layers, having different thicknesses, with $t_a < t_b$, and different saturation magnetizations, with $M_a < M_b$. In the normally saturated state which is considered here, this difference of magnetization leads to a difference in the internal demagnetizing field of each layer (self dipolar coupling) and in the stray field of one layer on the other one (cross dipolar coupling). As a result, the eigen-frequency of the uniform mode of the thin layer is larger than the one of the thick layer, $\omega_a > \omega_b$. This is represented in Fig. 1.3(a), on both sides of the bi-layer diagram, where we have reported on an energy scale their relative positions.

In the perpendicular geometry, the strength of the dynamical dipolar coupling Ω between the layers depends on the cross tensor element jj', associated to the in-plane component of the dipolar magnetic field produced inside layer j by the in-plane component of the magnetization of layer j' (see [17, 33]). It is attractive (lower in energy) when both layers precess in anti-phase because the dynamical dipolar charges in each layer are alternate (anti-ferromagnetic coupling). Thus the binding state corresponds to a collective motion where the two layers vibrate anti-symmetrically (B) and the anti-binding state to a collective motion where they vibrate symmetrically (A). In this case, the larger of the frequencies (ω_a) shifts up by

$$\Delta\omega = \frac{\Omega^2}{\omega_a - \omega_b}, \tag{1.5}$$

while the smaller one (ω_b) shifts down by the same amount. This effect is summarized in Fig. 1.3.

A numerical estimate of the coupling strengths between the lowest energy SW modes in each disk can be found in [17]. For the experimental parameters, $\Omega/2\pi \simeq 0.5$ GHz. This coupling is almost an order of magnitude smaller than the frequency splitting $\omega_a - \omega_b$, caused mainly by the difference of effective magnetizations of two disks: $\gamma 4\pi(M_b - M_a) \simeq 2\pi \cdot 4.5$ GHz. As a result, the shift of the resonance frequencies due to the dipolar coupling is negligible, $\Delta\omega/2\pi \simeq 0.06$ GHz. The effect of dynamic dipolar interaction is more pronounced for the level of mode hybridization. For instance, at the frequency $\omega_A \approx \omega_a$, the ratio between the precession

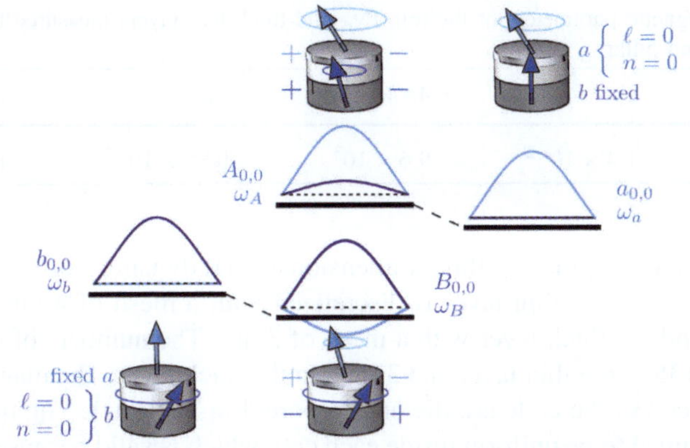

Fig. 1.3 Schematic representation of the coupled dynamics between two different magnetic disks. When the two disks are dynamically coupled through the dipolar interaction, the binding state B corresponds to the two layers oscillating in anti-phase at ω_B, with the precession occurring mostly in the thick layer, whereas the anti-binding state A corresponds to the layers oscillating in phase at ω_A, with the precession mostly in the thin layer

amplitudes in the two layers is given by

$$|c_b/c_a|_{\omega_A} = \Delta\omega/(\gamma h_{a,b}) \simeq \frac{\Omega}{\omega_a - \omega_b}. \tag{1.6}$$

For the experimental parameters, $\Omega/(\omega_a - \omega_b) \approx 0.1$, i.e., the precession amplitude in the disk b is about 10 % of that in the disk a. Thus, although the dipolar coupling induces a small spectral shift (second order in the coupling parameter (1.5)), its influence in the relative precession amplitude is significant (first order in the coupling parameter (1.6)).

1.2.3 *Micromagnetic Simulations vs. Mechanical-FMR Experiments*

Although the analytical formalism presented above allows estimating the spectrum, several approximations have been made. In particular, we have assumed total pinning at the disks boundary for the SW modes and no variation of the precession profile along the disks thicknesses (2D model), and we have neglected the dependence on the mode index ν of the dynamic dipolar coupling. Instead of developing a more complex analytical formalism, we have performed a complete calculation of the SW spectra inside our nano-pillar sample using the open source micromagnetic simulation package Nmag [34]. Nmag is a finite element solver based on the general purpose multi-physics library Nsim. It is developed by the group of H. Fangohr and T. Fischbacher in the School of Engineering Sciences at the University of Southampton.

Table 1.1 Magnetic parameters of the thin Py_a and thick Py_b layers measured by mechanical-FMR in the nano-pillar

$4\pi M_a$ (G)	α_a	$4\pi M_b$ (G)	α_b	γ (rad s^{-1} G^{-1})
8.0×10^3	1.4×10^{-2}	9.6×10^3	0.85×10^{-2}	1.87×10^7

In this simulation, the full three-dimensional (3D) dynamics of the bi-layer system is calculated. The thin layer is discretized with a mesh of 4 nm (equal to its thickness), and the thick layer with a mesh of 3 nm. The numbers of nodes are respectively 6135 in the thin layer and 37598 in the thick layer. The magnetic parameters introduced in the code are the ones reported in Table 1.1. The magnetization vector is assumed to be uniform inside each cell, which is valid because the cell size is smaller than the exchange length $\Lambda \simeq 5$ nm in Permalloy. We emphasize that the simulations incorporate the perturbing presence of the magnetic sphere attached on the MRFM cantilever. Moreover, the 10 nm thick Cu spacer is replaced by vacuum, so that the layers are only coupled trough the dipolar interaction (spin diffusion effects are absent).

The first step is to calculate the equilibrium configuration of the normally magnetized nano-pillar at $H_{\text{fix}} = 10$ kOe. The external magnetic field is applied exactly along $\hat{z}$ and the spherical probe with a magnetic moment $m = 2 \times 10^{-10}$ emu is placed on the axial symmetry axis at a distance $s = 1.3$ μm above the upper surface of the nano-pillar (see Fig. 1.1). The convergence criterion introduced in the code is $1/M_s\, d\boldsymbol{M}/dt < 1°$/ns. The result reveals that the equilibrium configuration is almost uniformly saturated along $\hat{z}$. Still, a small radial flaring (<5°) of the magnetization from $\hat{z}$ is observed at the periphery of the thick and thin layers.

In order to calculate the SW spectrum corresponding to a given excitation of this equilibrium state at $H_{\text{fix}} = 10$ kOe, we record the time decay response of the local magnetization to a small perturbation of the magnetic configuration, and we Fourier transform the ringdown of the transverse magnetization [35]. Two different initial conditions are simulated. For the conventional FMR spectrum, in which a uniform RF field is used to excite SW modes, we use the following perturbation vector field:

$$\boldsymbol{m} = \vartheta \sum_{n<6} J_0(k_{0n}\rho)\hat{\boldsymbol{x}}. \tag{1.7}$$

This form corresponds to an excitation which puts the same energy in the lowest 6 n-index modes with azimuthal index $\ell = 0$. For the spectrum excited by an RF current flowing through the nano-pillar, which creates an orthoradial RF Oersted field in the magnetic volume, we use:

$$\boldsymbol{m} = \vartheta \sum_{n<6} J_1(k_{1n}\rho)(-\sin\phi\hat{\boldsymbol{x}} + \cos\phi\hat{\boldsymbol{y}}). \tag{1.8}$$

At time $t = 0$, we add at every mesh the perturbation vector field defined in (1.7) or in (1.8) to the local unit vector along the equilibrium magnetization. The first one

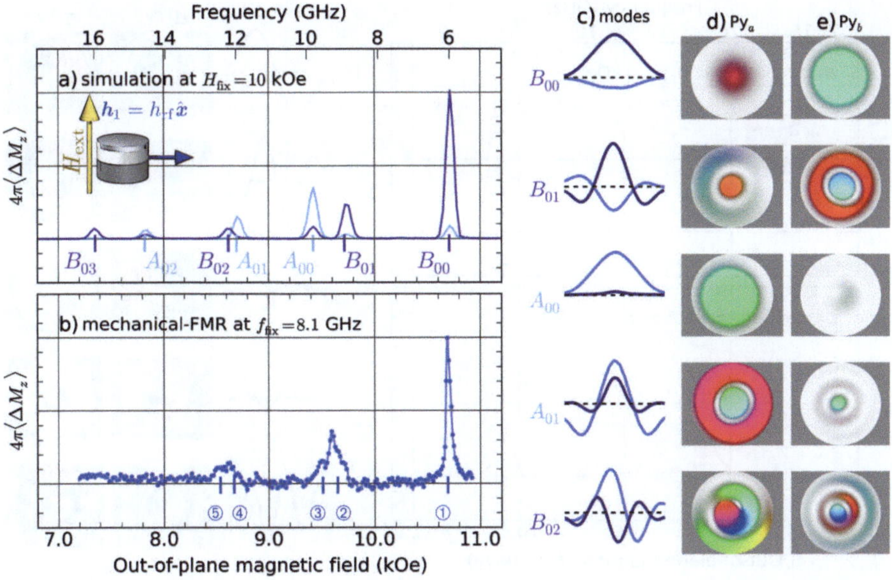

Fig. 1.4 Panel (**a**) is the numerically calculated spectral response to a uniform excitation field $\boldsymbol{h}_1 \propto \hat{\boldsymbol{x}}$, from a 3D micromagnetic simulation performed at $H_{\text{fix}} = 10$ kOe. The peaks are labeled according to their precession profiles shown in (**c**), (**d**), and (**e**). A *light* (*dark*) *color* is used to indicate the energy stored in the thin (thick) layer. Panel (**b**) shows the experimental spectrum measured by mechanical-FMR when exciting the nano-pillar by a homogeneous RF magnetic field at $f_{\text{fix}} = 8.1$ GHz

corresponds to a uniform tilt (i.e., in the same azimuthal direction) of the magnetic moments located in the center region of the two disks. The second one corresponds to an orthoradial tilt of the magnetization at the periphery of the nano-pillar. The perturbation angle $\vartheta = 0.01 \ll 1$ leads to a deviation from the local equilibrium axis of less than $1°$ degree, which is small enough to ensure that non-linear effects are weak in the simulated dynamics. We then compute the magnetization decay inside the whole sample. The ringdown of the transverse magnetization is recorded in a 10 ns time window with a sampling interval of 5 ps.

At every time step, a spatial average of the in-plane component of the magnetization is recorded. In the case of the initial condition of (1.7), we construct the complex reduced magnetization $\langle \boldsymbol{m} \cdot \hat{\boldsymbol{x}} \rangle$ inside each layer $j = a, b$. In the case of the initial condition of (1.8), a different spatial average is used, $\langle \boldsymbol{m} \cdot (\hat{\boldsymbol{x}} + i\,\hat{\boldsymbol{y}})e^{-i\phi} \rangle$, where ϕ is the local azimuthal coordinate of the mesh node. Then, the power density spectrum of the time decay of the full averaged reduced magnetization is calculated in each layer using a complex Fourier transform.

The comparison between the simulated and experimental spectra corresponding to a uniform RF field excitation and to an RF current flowing through the nano-pillar is presented in Figs. 1.4(a)–(b) and 1.5(a)–(b), respectively. The frequency scale of the simulation is in correspondence with the field-sweep scale of the experiments

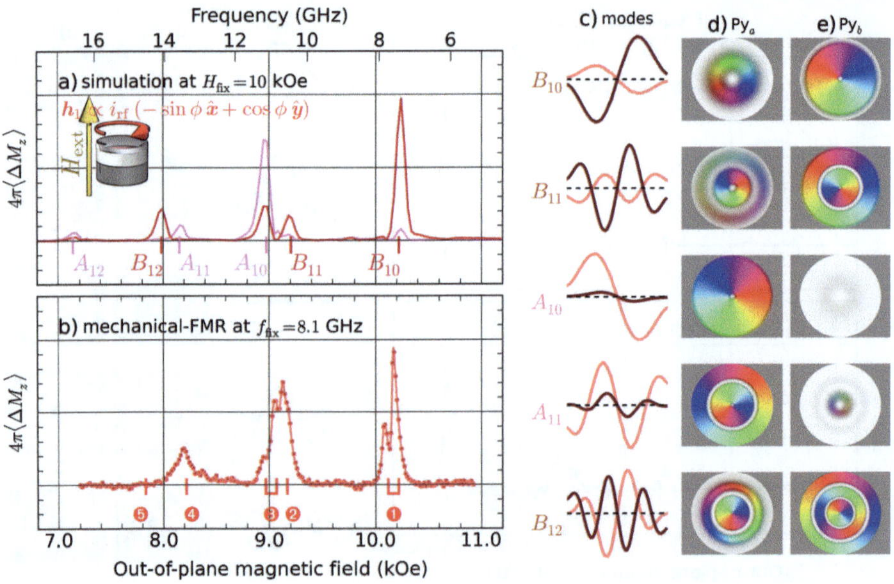

Fig. 1.5 Panel (**a**) is the simulated spectral response to an orthoradial excitation field $\boldsymbol{h}_1 \propto -\sin\phi\hat{\boldsymbol{x}} + \cos\phi\hat{\boldsymbol{y}}$. Panel (**b**) is the experimental spectrum measured by mechanical-FMR for an RF current excitation. Panels (**c**)–(**e**) show the simulated precession patterns of the eigen-vectors

performed at fixed RF frequency $f_{\text{fix}} = 8.1$ GHz through the affine transformation $H_{\text{ext}} - H_{\text{fix}} = 2\pi(f - f_{\text{fix}})/\gamma$, as seen from the frequency and field scales above Fig. 1.4(a) and below Fig. 1.4(b), respectively. Since in the simulations, we have access to the dynamics inside each layer, a light (dark) color is used to indicate the power spectrum of the thin (thick) layer. The resonance peaks are labeled according to the modes precession profiles, which are visualized by calculating the Fourier transform of the magnetization decay at every mesh node. The results are shown in the right side of Figs. 1.4 and 1.5, along with the precession profiles along the median direction. The 2D views of the spatial distribution of the transverse magnetization in the thin Py_a and thick Py_b layers are shown using the same color code as in Fig. 1.2.

Let us first focus on Fig. 1.4 which enables identifying the SW modes excited by a uniform RF magnetic field. The five peaks labeled on the simulated spectrum are in correspondence with the experimental resonance peaks. From the color code panels of Fig. 1.4(d)–(e), we see that these five modes have the same phase along the azimuthal direction, which is the character of the $\ell = 0$ index. The largest peak in the simulation occurs at the same field as the experimental peak at $H_{①}$. This lowest energy mode corresponds to the most uniform mode with the largest wave-vector and no node along the radial direction, thus it has the index $n = 0$. For this mode, the thick layer is mainly precessing, with the thin layer oscillating in anti-phase, as can be seen from the representations of its spatial profile in Fig. 1.4(c), so it bears the binding index B. Therefore, it should be labeled B_{00}, which confirms the character of the lowest energy collective mode in the perpendicular geometry discussed in Sect. 1.2.2. We also note that the relative amplitude of the precession in the two layers found in the simulation is roughly in agreement with the predicted value from the analytical model, see (1.6). The same analysis can be made for the second peak, labeled B_{01}, which occurs close to the experimental peak at $H_{②}$. It also corresponds to a resonance mainly in the thick layer, and its color representation shows that this is the first radial harmonic ($n = 1$), with one line of node in the radial direction. Again, the thin layer is oscillating in anti-phase, with the same radial index $n = 1$, as clearly shown by the mode profile along the median direction. The third peak is labeled A_{00} and is located close to the experimental peak at $H_{③}$. It corresponds this time to a uniform ($n = 0$) precession mainly located in the thin layer, in agreement with the experimental analysis presented in [17]. In this mode, the thick layer is also slightly vibrating in phase with the thin layer (anti-binding index A). The other two modes, A_{01} and B_{02}, correspond to higher radial harmonics of $\ell = 0$ modes in the nano-pillar, at $H_{④}$ and $H_{⑤}$. Due to the proximity between these two modes, the precession profile of the mode B_{02} appears somewhat hybridized with the precession profile of the mode A_{01}. Finally, one can also check from the simulations the independence of the precession profiles on the thickness (within the uncertainty of the calculated profile, which is about ± 2 %). This confirms the validity of the 2D approximation and explains the performance of the analytical model [17].

Let us now turn to Fig. 1.5(b), corresponding to the spectroscopic response to an RF current of same frequency 8.1 GHz flowing perpendicularly through the nano-pillar. The SW spectrum is acquired under the *exact same conditions* as for standard FMR, i.e., the spherical magnetic probe of the mechanical-FMR detection is kept at the same location above the sample. The striking result is that the position of the peaks in Figs. 1.4(b) and 1.5(b) do not coincide. More precisely, there seems to be a translational correspondence between the two spectra, which are shifted in field by about 0.5 kOe from each other. The comparison with Fig. 1.5(a) enables us to identify the SW modes excited by the orthoradial RF Oersted field produced by the RF current. Here again, the peaks of the simulated power spectrum in Fig. 1.5(a) are in good correspondence with the experimental resonance peaks in Fig. 1.5(b). Repeating the same analysis as above, the peaks follow the same sequence of A/B, n indexation and only differ by their ℓ-index. This explains the translational correspondence between the SW spectra of Figs. 1.4(b) and 1.5(b). Finally, we point out that the dipolar pinning at the boundaries of each disk in the modes displayed in Fig. 1.5(c) is not trivial [36]. The general trend observed here is that the thin layer is less pinned than the thick layer for symmetric modes, and vice versa.

Thus, these full 3D micromagnetic simulations allow the identification of the SW modes probed experimentally by both a uniform RF magnetic field and an RF current flowing through the nano-pillar, i.e., of their respective selection rules. They also give a deeper insight on the collective nature of the magnetization dynamics in the nano-pillar discussed in Sect. 1.2.2.

1.3 Conclusion

In summary, we used the MRFM technique [20] to study the SW eigen-modes in the prototype of a STNO – a normally magnetized nano-pillar composed of two magnetic layers in dipolar interaction. We were able to compare the SW spectra of this passive STNO excited by a uniform in-plane RF magnetic field and by an RF current flowing perpendicularly through the layers. We found that distinctly different SW modes (having azimuthal indices $\ell = 0$ and $\ell = +1$, respectively) are excited by the two above mentioned excitation methods. We also developed a simple analytic theory providing a comprehensive labeling of all the SW eigen-modes of a magnetic nano-pillar in the studied axially symmetric case. This labeling requires three independent indices: the usual azimuthal and radial indices ℓ and n used for the SW modes of a single magnetic disk, and an additional index referring to the binding or anti-binding (B or A) dipolar coupling between the two magnetic disks.

The obtained experimental and analytic results were also compared to the results of 3D micromagnetic simulations obtained with the Nmag package. We learned that the ℓ-index, related to the azimuthal symmetry of the SW modes, is the discriminating parameter for the selection rules of the SW mode excitation. We believe that our results are important for the optimization of the characteristics of nano-spintronic devices, and in particular STNOs, and for the experimental determination of the STNO parameters.

Acknowledgements We thank V. Cros, J. Grollier, H. Hurdequint, C. Ulysse, G. Faini, V.S. Tiberkevich, and A.N. Slavin for their help and support. This research was partially supported by the European Grant Master (NMP-FP7 212257) and by the French Grant Voice (ANR-09-NANO-006-01).

References

1. S. Wolf, D.D. Awschalom, R.A. Buhrman, J. Daughton, S. von Molnar, M.L. Roukes, A.Y. Chtchelkanova, D.M. Treger, Science **294**, 1488 (2001)
2. M.N. Baibich, J.M. Broto, A. Fert, F.N.V. Dau, F. Petroff, P. Etienne, G. Creuzet, A. Friederich, J. Chazelas, Phys. Rev. Lett. **61**(21), 2472 (1988)
3. G. Binasch, P. Grunberg, F. Saurenbach, W. Zinn, Phys. Rev. B **39**(7), 4828 (1989)
4. J. Slonczewski, J. Magn. Magn. Mater. **159**, L1 (1996). doi:10.1016/0304-8853(96)00062-5
5. L. Berger, Phys. Rev. B **54**(13), 9353 (1996)
6. V.E. Demidov, S.O. Demokritov, B. Hillebrands, M. Laufenberg, P.P. Freitas, Appl. Phys. Lett. **85**(14), 2866 (2004). doi:10.1063/1.1803621
7. G. Woltersdorf, O. Mosendz, B. Heinrich, C.H. Back, Phys. Rev. Lett. **99**(24), 246603 (2007). doi:10.1103/PhysRevLett.99.246603
8. G. de Loubens, V.V. Naletov, O. Klein, J. Ben Youssef, F. Boust, N. Vukadinovic, Phys. Rev. Lett. **98**(12), 127601 (2007)
9. G. de Loubens, V.V. Naletov, M. Viret, O. Klein, H. Hurdequint, J. Ben Youssef, F. Boust, N. Vukadinovic, J. Appl. Phys. **101**, 09F514 (2007)
10. G. Gubbiotti, M. Madami, S. Tacchi, G. Carlotti, H. Tanigawa, T. Ono, J. Phys. D, Appl. Phys. **41**(13), 134023 (2008). doi:10.1088/0022-3727/41/13/134023
11. P.S. Keatley, V.V. Kruglyak, A. Neudert, E.A. Galaktionov, R.J. Hicken, J.R. Childress, J.A. Katine, Phys. Rev. B **78**(21), 214412 (2008). doi:10.1103/PhysRevB.78.214412

12. A. Slavin, V. Tiberkevich, IEEE Trans. Magn. **45**(4), 1875 (2009). doi:10.1109/TMAG.2008.2009935
13. A.V. Nazarov, H.S. Cho, J. Nowak, S. Stokes, N. Tabat, Appl. Phys. Lett. **81**(24), 4559 (2002). doi:10.1063/1.1521578
14. N. Stutzke, S.L. Burkett, S.E. Russek, Appl. Phys. Lett. **82**, 91 (2003). doi:10.1063/1.1534386
15. J.C. Sankey, P.M. Braganca, A.G.F. Garcia, I.N. Krivorotov, R.A. Buhrman, D.C. Ralph, Phys. Rev. Lett. **96**(22), 227601 (2006)
16. W. Chen, J.M.L. Beaujour, G. de Loubens, A.D. Kent, J.Z. Sun, Appl. Phys. Lett. **92**(1), 012507 (2008). doi:10.1063/1.2827570
17. V.V. Naletov, G. de Loubens, G. Albuquerque, S. Borlenghi, V. Cros, G. Faini, J. Grollier, H. Hurdequint, N. Locatelli, B. Pigeau, A.N. Slavin, V.S. Tiberkevich, C. Ulysse, T. Valet, O. Klein, Phys. Rev. B **84**, 224423 (2011). doi:10.1103/PhysRevB.84.224423
18. Z. Zhang, P.C. Hammel, P.E. Wigen, Appl. Phys. Lett. **68**, 2005 (1996)
19. V. Charbois, V.V. Naletov, J. Ben Youssef, O. Klein, J. Appl. Phys. **91**, 7337 (2002)
20. O. Klein, G. de Loubens, V.V. Naletov, F. Boust, T. Guillet, H. Hurdequint, A. Leksikov, A.N. Slavin, V.S. Tiberkevich, N. Vukadinovic, Phys. Rev. B **78**(14), 144410 (2008). doi:10.1103/PhysRevB.78.144410
21. G. de Loubens, V.V. Naletov, O. Klein, Phys. Rev. B **71**(18), 180411 (2005)
22. V.V. Naletov, G. de Loubens, V. Charbois, O. Klein, V.S. Tiberkevich, A.N. Slavin, Phys. Rev. B **75**(14), 140405 (2007)
23. G. de Loubens, A. Riegler, B. Pigeau, F. Lochner, F. Boust, K.Y. Guslienko, H. Hurdequint, L.W. Molenkamp, G. Schmidt, A.N. Slavin, V.S. Tiberkevich, N. Vukadinovic, O. Klein, Phys. Rev. Lett. **102**(17), 177602 (2009). doi:10.1103/PhysRevLett.102.177602
24. R.E. Arias, D.L. Mills, Phys. Rev. B **79**, 144404 (2009)
25. A.G. Gurevich, G.A. Melkov, *Magnetization Oscillations and Waves* (CRC Press, Boca Raton, 1996)
26. K.Y. Guslienko, S.O. Demokritov, B. Hillebrands, A.N. Slavin, Phys. Rev. B **66**, 132402 (2002)
27. J.F. Dillon, J. Appl. Phys. **31**(9), 1605 (1960)
28. R.W. Damon, J.R. Eshbach, J. Phys. Chem. Solids **19**, 308 (1961)
29. M. Belmeguenai, T. Martin, G. Woltersdorf, M. Maier, G. Bayreuther, Phys. Rev. B **76**(10), 104414 (2007). doi:10.1103/PhysRevB.76.104414
30. G. Gubbiotti, M. Kostylev, N. Sergeeva, M. Conti, G. Carlotti, T. Ono, A.N. Slavin, A. Stashkevich, Phys. Rev. B **70**(22), 224422 (2004). doi:10.1103/PhysRevB.70.224422
31. J. Ben Youssef, A. Layadi, J. Appl. Phys. **108**(5), 053913 (2010). doi:10.1063/1.3476269
32. B. Kardasz, O. Mosendz, B. Heinrich, Z. Liu, M. Freeman, J. Appl. Phys. **103**(7), 07C509 (2008). doi:10.1063/1.2834399
33. O. Dmytriiev, T. Meitzler, E. Bankowski, A. Slavin, V. Tiberkevich, J. Phys. Condens. Matter **22**, 136001 (2010)
34. T. Fischbacher, M. Franchin, G. Bordignon, H. Fangohr, IEEE Trans. Magn. **43**(6), 2896 (2007)
35. R.D. McMichael, M.D. Stiles, J. Appl. Phys. **97**(10), 10J901 (2005)
36. M.P. Kostylev, A.A. Stashkevich, N.A. Sergeeva, Y. Roussigné, J. Magn. Magn. Mater. **278**, 397 (2004). doi:10.1016/j.jmmm.2003.11.400

Chapter 2
Bottom up Magnonics: Magnetization Dynamics of Individual Nanomagnets

P.S. Keatley, P. Gangmei, M. Dvornik, R.J. Hicken, J. Grollier, C. Ulysse, J.R. Childress, and J.A. Katine

Abstract A review is provided of recent time-resolved scanning Kerr microscopy experiments and micromagnetic simulations of magnetization dynamics in single nanomagnets, and in pairs of dipolar coupled nanomagnets. Two nanomagnet systems are presented. In a single 440 nm square, the effect of the pulsed field amplitude on the relative amplitudes of center- and edge-type confined spin wave modes is explored. In addition, the difference in the damping of dynamics observed in different magnetic ground states is discussed. In pairs of 300 nm disks, the relative effect of structural imperfections and static dipolar fields on the splitting of edge-modes is investigated as a function of disk separation.

2.1 Introduction

Planar nanomagnets are the constituent elements of magnonic crystals. Individual nanomagnets support a variety of confined spin wave eigenmodes that can be accessed by selecting the appropriate magnetic ground state with the application of a magnetic field [1]. In magnonic crystals, confined spin waves of individual nanomagnets interact via dipolar interactions forming collective excitations that

P.S. Keatley · P. Gangmei · M. Dvornik · R.J. Hicken (✉)
College of Engineering, Mathematics and Physical Sciences, School of Physics,
University of Exeter, Stocker Road, Exeter, EX4 4QL, UK
e-mail: R.J.Hicken@exeter.ac.uk

J. Grollier
Unité Mixte de Physique CNRS/Thales and Université Paris Sud 11, 1 av. A. Fresnel,
91767 Palaiseau, France

C. Ulysse
Laboratoire de Photonique et de Nanostructures (LPN), CNRS PHYNANO team,
Route de Nozay, 91460 Marcoussis, France

J.R. Childress · J.A. Katine
San Jose Research Center, Hitachi Global Storage Technologies, 3403 Yerba Buena Road,
San Jose, CA 95135, USA

S.O. Demokritov, A.N. Slavin (eds.), *Magnonics*, Topics in Applied Physics 125,
DOI 10.1007/978-3-642-30247-3_2, © Springer-Verlag Berlin Heidelberg 2013

are no longer described by the eigenmodes of the individual nanomagnets [2]. Recently researchers have demonstrated that arrays of interacting nanomagnets exhibit magnonic bands, bandgaps, and dispersion similar to other artificially periodic systems [3, 4].

The internal magnetic field of an in-plane magnetized nanomagnet is non-uniform and supports a quasi-uniform mode with large amplitude at the center of the nanomagnet (center mode) and localized modes at the edges of the nanomagnet perpendicular to the direction of the bias magnetic field (edge modes) [5, 6]. Since center and edge modes can coexist, the number of collective excitations is enhanced, while their frequencies and spatial character may be continuously tuned by applying a magnetic field [1]. Therefore, arrays of nanomagnets, or magnonic crystals, offer a highly agile class of metamaterial with potential for greater functionality than their non-magnetic counterparts in which, for example, the frequency of collective plasmonic modes are tailored by adjusting the array dimensions [7].

The dynamic interactions between neighboring nanomagnets, and therefore the collective excitations, are governed by the character of the confined spin waves of the individual nanomagnets. In this work, we explore the effects of excitation amplitude, magnetic parameters, magnetic ground state, structural imperfections, and static dipolar coupling on the spin wave spectrum of a nanomagnet. These effects cannot be measured within arrays due to the presence of collective modes, and because of inhomogeneous broadening and splitting due to structural and magnetic variation of nanomagnets throughout the array [1, 8, 9]. Instead, to isolate these effects it is necessary to measure individual nanomagnets [10, 11]. Here we use time-resolved scanning Kerr microscopy (TRSKM) to detect the individual response of a single 440 nm square, and each disk within pairs of 300 nm disks. We use micromagnetic simulations to confirm our understanding of the magnetization dynamics of the individual nanomagnets.

2.2 Experimental Methods and Sample Details

Picosecond magnetization dynamics of individual nanomagnets were measured using TRSKM with nanoscale spatial resolution[1] and enhanced mechanical stability.[2] The experimental set-up is shown schematically in Fig. 2.1. The dynamics were excited by in-plane pulsed field, or out-of-plane microwave (RF) field excitations $\mathbf{h}(t)$

[1]The spot size was determined by measuring the reflectivity as the laser spot was scanned across the edge of a thin film metallic structure fabricated on a Si substrate. The derivative of the change in reflectivity is Gaussian and the measured full-width at half maximum provides an estimate of the spatial resolution of the system. At a probe wavelength of 800 and 400 nm, the spot size is ~500 and ~300 nm, respectively.

[2]The mechanical drift was determined by measuring a scanning reflectivity image of an array of 640 nm squares with 240 nm edge-to-edge separation. The distortion of the image over the period of an hour due to mechanical drift was not more than 200 nm.

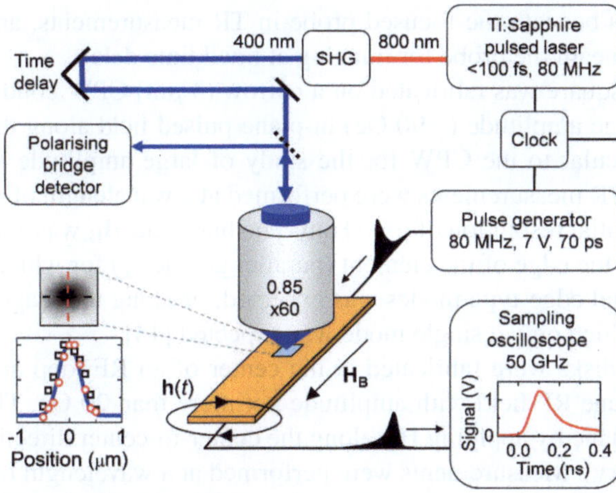

Fig. 2.1 A schematic of the TRSKM, and a TR image of the 440 nm square are shown. The $1 \times 1\ \mu m^2$ image was acquired at a precession antinode where *black* (*white*) corresponds to a Kerr rotation of ~1 mdeg (zero). The size of the square appears larger than 440 nm due to the convolution of the ~300 nm FWHM Gaussian laser spot with the square. The calculated convolution of a 440 nm wide 'top-hat' is shown below the image (*blue curve*) and overlaid with line scans across the image (symbol color corresponds to line scan color)

by fabricating the nanomagnets either on the center conductor of a coplanar waveguide (CPW), as shown in Fig. 2.1, or at the center of a microwave antenna loop (RF loop), not shown. The dynamics were detected stroboscopically by measuring the polar Kerr rotation of reflected optical probe pulses resulting from the change in the out-of-plane component of the dynamic magnetization. A 4 ns optical delay line was used to precisely set the phase between the excitation field and the magneto-optical probe pulse. The probe was expanded ×5 to reduce beam divergence before being focused to a diffraction limited spot using a high numerical aperture (0.85, ×60) microscope objective lens. Second harmonic generation outside the laser head allowed measurements to be performed at probe wavelengths of 800 nm or 400 nm.

Two nanomagnet systems were fabricated using dc magnetron sputtering, electron-beam lithography and ion milling. A single $440 \times 440\ nm^2$ square CoFe(1 nm)/NiFe(5 nm)/CoFe(1 nm) nanomagnet, and pairs of 300 nm diameter d Cu(6 nm)/$Ni_{80}Fe_{20}$(15 nm)/Cu(3 nm)/Al_2O_3(2 nm) disks with nominal edge-to-edge separation s corresponding to s/d of 2, 1, 0.6, and 0.3.[3] A single disk ($s/d \gg 2$) was also measured. The composition of the square is similar to that used in the free layer of a hard disk magnetoresistive sensor [12], while the composition of the disks is typical of the free layer of a spin transfer oscillator [13]. The samples were mounted on a piezoelectric scanning stage that was used to position individ-

[3]The 300 nm disks were fabricated by J. Grollier and C. Ulysse. The 440 nm square was fabricated by J.R. Childress and J.A. Katine.

ual nanomagnets beneath the focused probe in TR measurements, and to scan the nanomagnets beneath the probe for imaging at fixed time delay.

The 440 nm square was fabricated on a narrow (4 μm) CPW conductor in order to generate a large amplitude (~90 Oe) in-plane pulsed field along the edge of the square perpendicular to the CPW for the study of large amplitude magnetization dynamics [11]. TR measurements were performed at a wavelength of 400 nm corresponding to a spatial resolution of ~300 nm. The bias field $\mathbf{H}_B$ was applied in-plane and along either the edge of the element (parallel geometry) for which the observation of center- and edge-type modes was expected, or along the diagonal (diagonal geometry) for which only a single mode was expected [14].

The 300 nm disks were fabricated at the center of an RF loop and excited using an out-of-plane RF field with amplitude not more than 20 Oe. The disks were magnetized in-plane by applying $\mathbf{H}_B$ along the center-to-center direction of the pair (parallel geometry). Measurements were performed at a wavelength of 800 nm corresponding to a spatial resolution of ~500 nm. Measurements with ~300 nm spatial resolution were not possible due to significant attenuation of the optical probe by residual electron-beam resist at a wavelength of 400 nm. While the spatial resolution was inferior to that used for the 440 nm square, the 800 nm probe was found to be more sensitive to the smaller amplitude dynamics of the pairs of disks, and provided a sufficiently small spot size to detect magnetization dynamics of each disk independently. In addition to TR measurements, the RF excitation allowed phase-resolved ferromagnetic resonance (FMR) spectra to be acquired by sweeping the applied field through ± 1 kOe at a fixed phase of precession.

2.3 Results and Discussion

In Fig. 2.2(a), typical TR signals acquired from the 440 nm square (black and red traces) and the single 300 nm disk (blue trace) are shown. Fast Fourier transform spectra calculated from the TR signals are shown in Fig. 2.2(b). For the 440 nm square in the parallel geometry (black trace) only weak beating was observed in the TR signal indicating that either the center or edge mode dominates the dynamic response resulting in a single dominant peak in the FFT spectrum. In the diagonal geometry (red trace), the TR signals were more heavily damped resulting in a broader peak in the FFT spectrum. By matching the relaxation of simulated TR traces to that observed experimentally, the Gilbert damping parameter α was found to be 0.03 and 0.05 in the parallel and diagonal geometry, respectively [11].

The observed difference in the linewidth and relaxation is understood to be the result of different magnon–magnon scattering for the parallel and diagonal magnetic ground states. In experiments presented here, only modes that couple strongly to a uniform pulsed field are excited and detected. However, square nanomagnets can support a large number of asymmetric modes of similar frequency [1] that offer many channels for dissipation via magnon–magnon scattering. The efficiency of such scattering depends upon the ground state, the internal field, and the frequencies

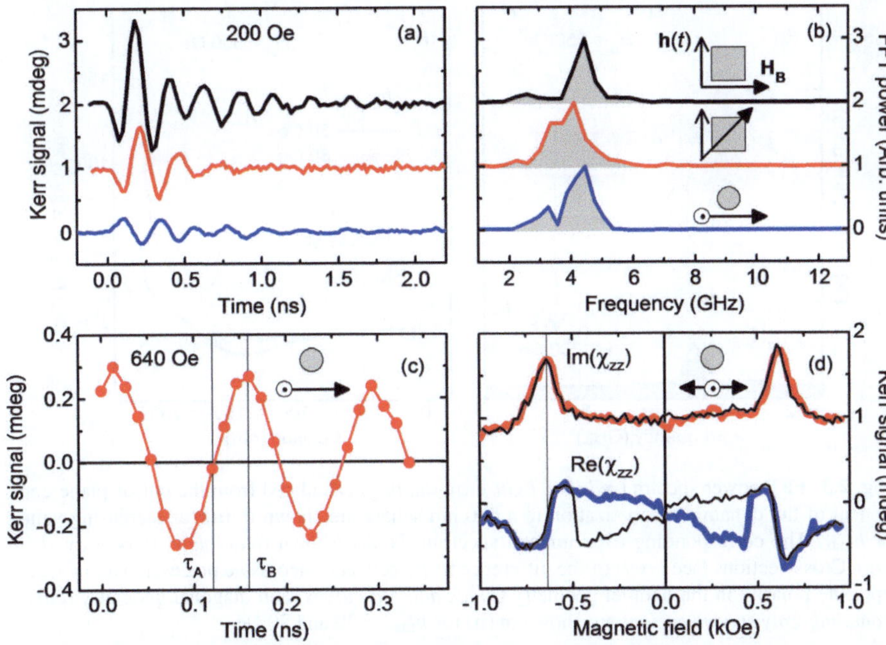

Fig. 2.2 Typical TR signals acquired from a single 440 nm square (*black and red traces*) and a 300 nm disk (*blue trace*) in response to a pulsed field are shown in (**a**) along with corresponding FFT spectra and field geometries in (**b**). In (**c**), a TR signal acquired from an individual 300 nm disk in response to an RF field is shown. By setting the time delay to a node (τ_A) or an antinode (τ_B), Re(χ_{zz}) and Im(χ_{zz}) FMR curves were are acquired by sweeping the in-plane magnetic field

of the modes [15], and therefore leads to differences in the relaxation for the parallel and diagonal geometries.

In Fig. 2.3(a), the normalized FFT power spectra calculated from micromagnetic simulations [16] of the 440 nm square are shown for increasing values of the maximum pulsed field amplitude h_{max}. Details of the simulations are described in [11]. In order to reproduce the frequency and linewidth observed in the experimental spectrum, the value of the saturation magnetization M_s was reduced from M_s at the center of the square to $0.8M_s$ at the edges with a smooth profile, while the Gilbert damping parameter was set to 0.03. As h_{max} increases from 20 to 80 Oe, the amplitude of the dominant lower frequency edge mode appears to be suppressed in favor of increasing amplitude of the higher frequency center mode. At 80 Oe, the simulated spectrum is very similar to that observed experimentally. Weaker coupling of the edge mode to the large amplitude field [17] and modification of the internal magnetic field H_i lead to the effective suppression of the edge mode.

Cross-sections of the component of H_i parallel to $\mathbf{H}_B$ were extracted from the simulations at $H_B = 300$ Oe for precession antinodes of opposite polarity. The change in H_i is shown in Fig. 2.3(b) for $h_{max} = 30$ (black curve) and 80 Oe (red curve). For $h_{max} = 80$ Oe, the large amplitude excitation of the center mode leads to a greater modification of H_i in the region of the edge mode than for $h_{max} = 30$ Oe.

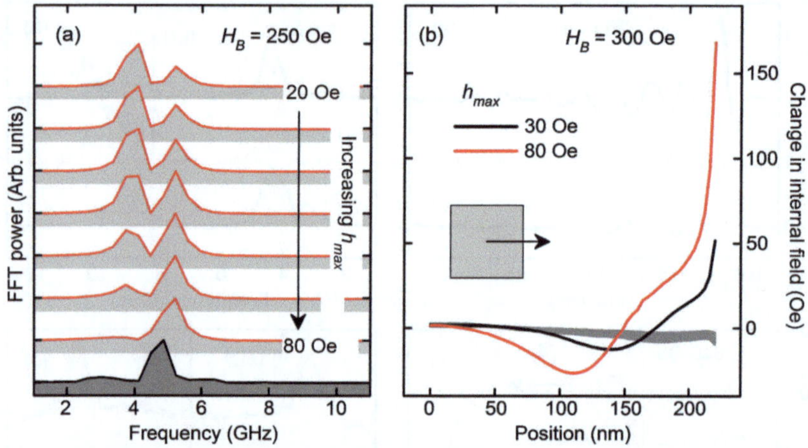

Fig. 2.3 FFT power spectra (*red line*, *light gray shading*) calculated from the out-of-plane component of the dynamic magnetization in a 440 nm square are shown in (**a**) for increasing values of $h_{\max}$. The corresponding experimental spectrum is also shown (*black line*, *dark gray shading*). Cross-sections (see *inset*) of the difference in H_i between successive precession antinodes of opposite polarity in the parallel geometry (*black and red curves*) and diagonal geometry (*curves bounding gray shaded region*) are shown in (**b**) for $h_{\max} = 30$ and 80 Oe

Therefore, the non-uniform profile of H_i that gives rise to edge mode localization is continuously modified due to the large amplitude precession associated with the center mode leading to reduced amplitude and enhanced linewidth of the edge mode. In contrast, line scans in the diagonal geometry show little variation of H_i at 30 Oe or at 80 Oe (gray shaded region) and therefore the dynamic response is relatively insensitive to $h_{\max}$ [11].

TR signals acquired from a single 300 nm disk (e.g., Fig. 2.2(a) (blue trace)) have different initial amplitude and phase to that of the square due to the different amplitude and orientation of $\mathbf{h}(t)$ in the RF loop and the CPW. The TR signal also shows evidence of beating, indicating the presence of at least two modes as observed in the FFT spectrum. However, a mismatch in the characteristic impedance of electrical connections to the RF loop resulted in reflections of the current pulse, coherent artifacts in the TR signals, and complicated splitting in the FFT spectra at different values of H_{B}. Therefore, an RF field was used to excite magnetization dynamics in the disks, for which any impedance mismatch will affect only the excitation amplitude.

In Fig. 2.2(c), a typical TR signal acquired from a 300 nm disk in response to a 7.2 GHz RF field at $H_{\mathrm{B}} = -640$ Oe is shown corresponding to the center mode resonance. By setting the time delay to that of a node τ_{A} or an antinode τ_{B} of the TR signal, ferromagnetic resonance (FMR) curves corresponding to the real and imaginary parts of the magnetic susceptibility tensor component χ_{zz} are obtained by sweeping the in-plane magnetic field ($\mathrm{Re}(\chi_{\mathrm{zz}})$ and $\mathrm{Im}(\chi_{\mathrm{zz}})$, Fig. 2.2(d)).

In Fig. 2.4(a), the spin wave spectrum of a single disk is shown. The greyscale plot is compiled of FMR curves corresponding to $\mathrm{Im}(\chi_{\mathrm{zz}})$ acquired at different fre-

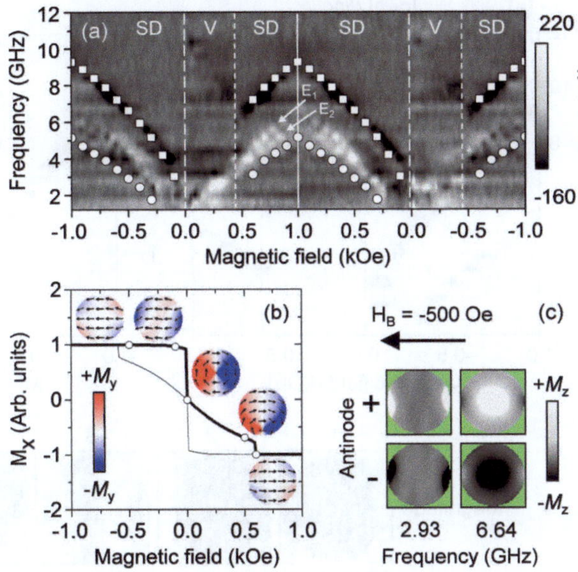

Fig. 2.4 The measured spin wave spectrum of a single 300 nm disk is shown in (**a**). The simulated frequencies of center and edge modes are overlaid as *squares and circles*, respectively. The spectrum is divided into regions corresponding to quasi-uniform single domain (SD) or vortex (V) ground states identified at different points (*open symbols*) from −1 to +1 kOe (*bold curve*) on the simulated hysteresis loop in (**b**). In (**c**), the simulated spatial character of the center and edge modes for an ideal 275 nm disk is shown

quencies of the RF field ranging from 1 to 12 GHz in steps of 0.4 GHz. The spectrum is unwrapped around +1 kOe (vertical solid white line) and divided into well-defined regions corresponding to quasi-uniform single domain (SD) or vortex (V) ground states identified from the simulated hysteresis loop in Fig. 2.4(b).

In the following discussion, we focus on the center and edge mode resonances of the quasi-uniform SD state. The center and edge modes are observed as a well-defined higher frequency branch (black) and a less well defined lower frequency branch (black/white), respectively. Micromagnetic simulations of an ideal 275 nm disk[4] confirm the character of the observed modes; the frequencies of the center and edge modes calculated from the simulations are overlaid on the experimental spectra, while the simulated spatial character is shown in Fig. 2.4(c). Details of the simulations will be described elsewhere [18].

In Fig. 2.4(a), the simulated frequency of the center mode (square) is in excellent agreement with the experimental spectra, while differences are observed in the number and frequencies of the edge modes. For example, in the single domain state for $H > 0$ two edge modes are observed experimentally (E1 and E2), while only a single edge mode is present in the simulations. The greater number of modes in the experimental spectrum is a result of structural distortions [8, 9].

In Fig. 2.5, the simulated spin wave spectrum of the ideal 275 nm disk (Fig. 2.5(a)) is compared to that of the real shaped disk (Fig. 2.5(b)). The shape was taken from the SEM image shown in Fig. 2.5(b). In Fig. 2.5(b), two lower

[4]The diameter was chosen to be slightly different to the nominal experimental value of 300 nm since some variation in diameter of this order is observed in the SEM images and because the vortex state forms more easily in the simulations for this smaller diameter.

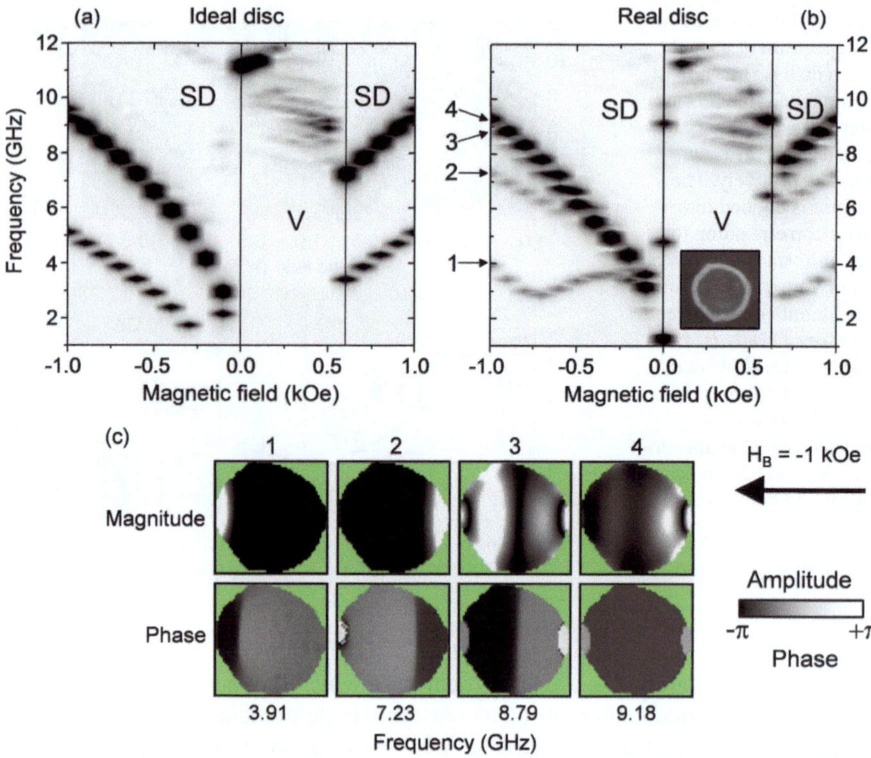

Fig. 2.5 The simulated spectra of an ideal 275 nm disk (**a**) and the real shaped disk taken from the *inset* SEM image (**b**) are shown. The spectra are divided into regions corresponding to quasi-uniform single domain (SD) or vortex (V) ground states. In (**c**), the simulated spatial character of the center and edge modes of the real shaped disk is shown in images of the magnitude and phase of the pixel dependent fast Fourier transform (FFT) of the out of plane component of the dynamic magnetization. For clarity, the images of FFT magnitude for modes 1 and 2, and modes 3 and 4 are normalized to 20 % and 50 % of the maximum value of the FFT magnitude, respectively

frequency modes (1, 2) appear below the center mode branch as observed in the experimental spectrum. The complicated behavior of the lower frequency edge mode (1) is a result of a large saturation field ($\sim$700 Oe) in its region of localization leading to the observed minimum in frequency at -700 Oe [19]. Furthermore, the center mode appears to be split into two modes (3, 4) and shows slightly different field dependence to that of the ideal disk due to the interaction with mode 2 at -500 Oe, and mode 1 at -100 Oe. These interactions are not clearly observed in the experiment; however, the experimental spectrum does exhibit complicated behavior of the center and edge modes at low values of the applied field. While the simulation shows the correct number of edge modes when the real shape of the disks is considered, the poor quantitative agreement may be a consequence of the ion milling fabrication process that is reported to modify the magnetic parameters at the edges of nanomagnets due to ion implantation [19].

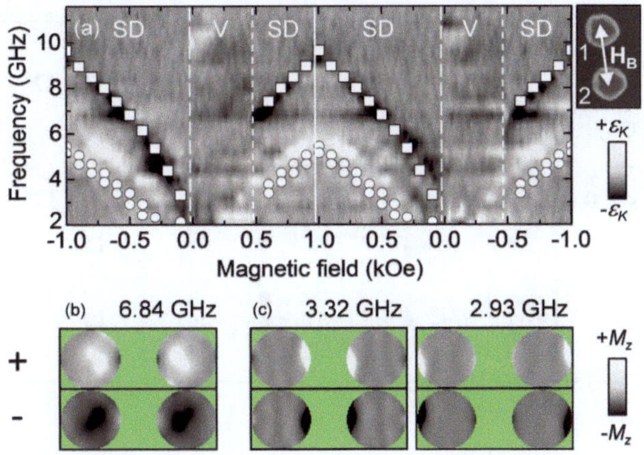

Fig. 2.6 The experimental spin wave spectrum acquired from disk 1 of a pair of 300 nm disks with $s/d = 0.6$ in the parallel geometry is shown in (**a**). The simulated frequencies of center and edge modes are overlaid as *squares and circles* respectively in the quasi-uniform single domain state (SD), while their spatial character is shown in (**b**) and (**c**), respectively

In addition to the shape of the disk, the edge modes E1 and E2 in Fig. 2.4(a) also seem to be sensitive to the field history. Edge mode E1 is observed as a white branch for both signs of the applied field, while E2 has opposite phase with respect to the excitation field when the external field is reversed and appears as a white branch for $H > 0$ and a black branch for $H < 0$. These differences in the edge mode spectra point to small differences in the ground state near to the edge of the disk that result from the non-perfect circular shape.

In Fig. 2.5(c), Fourier images of magnitude and phase show the spatial character of modes 1 to 4 at -1 kOe in the real shaped disk. The spatial plots show that modes 1 and 2 are edge modes. The large splitting is due to the very different shape at the edge of the disk, which significantly changes the magnetic ground state and internal field in the region of the edge mode localization. Due to the 'isolation' of the center mode from the edge of an ideal disk, it is usually expected that the shape has a reduced influence on the center mode. However, while the center mode branch is very similar to that of the ideal disk in Fig. 2.4(a), the spatial character is very different and complicated further by the spectral overlap of modes 3 and 4. The spatial plots show that mode 3 is an anti-symmetric mode with a nodal line passing through the disk perpendicular to $\mathbf{H}_\mathrm{B}$. The uniform excitation field is able to couple to the anti-symmetric mode because of the asymmetry of the internal field throughout the disk due to the real shape. For real shaped nanomagnets, the sensitivity of the center and edge modes to the precise shape of the disk present significant challenges for the controlled dynamic dipolar coupling of nanomagnets.

In Fig. 2.6(a), the spin wave spectrum acquired from one disk of a pair with $s/d = 0.6$ is shown for the parallel geometry. The spectrum is acquired from disk 1 shown in the adjacent SEM image. The response of nominally identical disks excited by a uniform microwave field is expected to be identical. However, the center mode is found to have larger amplitude in disk 1 than in disk 2 (not shown). The lower frequency edge mode branch (white) is broad and poorly defined in both disks. The simulated frequencies of the center (squares) and edge modes (circles) extracted from micromagnetic simulations of a pair of 275 nm disks are overlaid

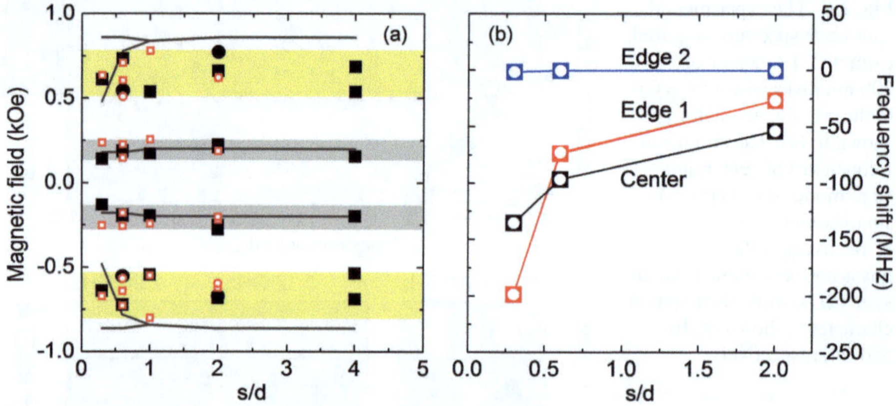

Fig. 2.7 The measured center and edge mode resonance fields of pairs of disks with different separation are shown in (**a**) for the parallel geometry at an RF frequency of 4.4 GHz. Data for a single disk are shown at $s/d = 4$, while data for disk 1 and disk 2 of a pair are shown as *filled black squares* and *open red squares*, respectively. Simulated resonance fields are shown as *gray curves*. A second pair was measured for separation values $s/d = 0.6$ and 2, shown as *circular symbols*. The range of resonance fields for the center and edge modes are shown as *gray and yellow bands*, respectively. In (**b**), the frequency shift of center and edge modes due to dynamic dipolar interactions is shown as a function of s/d for the parallel geometry

on the experimental spectra. In the parallel geometry, the static interaction causes splitting of the edge modes in the simulation. The small splitting is consistent with the edge mode branch of the experimental spectra that may consist of two modes that are not resolved experimentally.

In Fig. 2.6(b), the simulated spatial character of the center and edge modes is shown at $H_B = 500$ Oe. The non-uniform static dipolar field of one disk contributes to the internal field of the other. The asymmetry of the internal field modifies the spatial character of the center mode from the cylindrically symmetric profile of the single disk (Fig. 2.4(c)). The difference in the static field at opposite edges of the disks (~50 Oe for $s/d = 0.6$) splits the edge mode of the single disk into two modes; a mode with large amplitude at the inner edges of the disks (3.32 GHz), and a mode with large amplitude at the outer edges of the disks (2.93 GHz). The splitting of the modes is small and is therefore difficult to isolate from the effect of structural distortions that are present in the experiment.

In Fig. 2.7(a), the dependence of the resonance fields of the center and edge modes upon the separation (s/d) in the parallel geometry is shown. The excitation frequency of 4.4 GHz allowed both modes to be observed in the single domain state between 1 kOe and remanence. The resonance fields were extracted from the FMR curves by identifying the value of the magnetic field at the node (resp., peak) in the real (resp., imaginary) curves, for example at $H_B = -640$ Oe in Fig. 2.2(d). The most striking feature of Fig. 2.7(a) is that the range of resonance fields for center mode (gray band) is smaller than that of the edge mode (yellow band), which is understood to be due to the relative 'isolation' of the center mode from structural distortions at the edges and dipolar interactions.

Micromagnetic simulations show the expected resonance fields as a function of s/d (gray curves). However, the variation of the measured resonance fields due to the effects of shape, size, and ion milling demonstrates that it is not straightforward to observe the effect of static interactions on the resonance fields as a function of separation, particularly for the edge modes.

While the resonance field, splitting and spatial character of confined spin wave modes depend upon the shape, size, magnetic parameters, and static dipolar interaction, the collective excitations of arrays of nanomagnets require dynamic dipolar interactions to enforce a phase relationship between the modes of neighboring nanomagnets. Comparison of micromagnetic simulations of a pair of disks with the magnetization of one disk either fixed or free to precess revealed the effect of dynamic dipolar interactions on the center and edge mode resonances. The dynamic dipolar fields generated by center and edge modes of one disk are relatively weak in comparison with the static effects and generate frequency shifts of modes in the second disk, Fig. 2.7(b). Since the frequency shifts are smaller than the experimental linewidth, the effect of dynamic dipolar interactions on the resonance field of confined spin waves is difficult to observe experimentally. Instead, measurements of the relative phase of precession in the different elements are required to isolate and characterize the dynamic interactions and will be reported elsewhere [18].

2.4 Summary

In summary, we have reviewed recent TRSKM experiments and micromagnetic simulations of magnetization dynamics in single nanomagnets, and in pairs of dipolar coupled nanomagnets. We have shown that excitation amplitude, magnetic parameters, magnetic ground state, structural imperfections, and static dipolar coupling have a significant effect on the spin wave spectrum of individual nanomagnets. In contrast, the dynamic dipolar interaction was found to have little effect on the frequency of confined spin wave modes despite being responsible for collective excitations within arrays. Understanding the variety of contributions to the magnetization dynamics of individual nanomagnets is the key to controlling the coupled dynamics of multiple nanomagnets in order to tailor collective excitations for future application in magnonic metamaterials.

Acknowledgements The authors gratefully acknowledge the financial support from the EU grant MASTER NMP-FP7-212257, and the UK Engineering and Physical Sciences Research Council.

References

1. P.S. Keatley, V.V. Kruglyak, A. Neudert, E.A. Galaktionov, R.J. Hicken, J.R. Childress, J.A. Katine, Phys. Rev. B **78**, 214412 (2008)

2. V.V. Kruglyak, P.S. Keatley, A. Neudert, R.J. Hicken, J.R. Childress, J.A. Katine, Phys. Rev. Lett. **104**, 027201 (2010)
3. S. Tacchi, M. Madami, G. Gubbiotti, G. Carlotti, H. Tanigawa, T. Ono, M.P. Kostylev, Phys. Rev. B **82**, 024401 (2010)
4. R. Zivieri, F. Montoncello, L. Giovannini, F. Nizzoli, S. Tacchi, M. Madami, G. Gubbiotti, G. Carlotti, A.O. Adeyeye, Phys. Rev. B **83**, 054431 (2011)
5. V.V. Kruglyak, A. Barman, R.J. Hicken, J.R. Childress, J.A. Katine, Phys. Rev. B **71**, 220409(R) (2005)
6. G. Gubbiotti, M. Madami, S. Tacchi, G. Carlotti, A.O. Adeyeye, S. Goolaup, N. Singh, A.N. Slavin, J. Magn. Magn. Mater. **316**, 338 (2007)
7. C.P. Burrows, W.L. Barnes, Opt. Express **18**, 3187–3198 (2010)
8. J.M. Shaw, T.J. Silva, M.L. Schneider, R.D. McMichael, Phys. Rev. B **79**, 184404 (2009)
9. H.T. Nembach, J.M. Shaw, T.J. Silva, W.L. Johnson, S.A. Kim, R.D. McMichael, P. Kabos, Phys. Rev. B **83**, 094427 (2011)
10. A. Barman, S. Wang, J.D. Maas, A.R. Hawkins, S. Kwon, A. Liddle, J. Bokor, H. Schmidt, Nano Lett. **6**, 2939–2944 (2006)
11. P.S. Keatley, P. Gangmei, M. Dvornik, R.J. Hicken, J.R. Childress, J.A. Katine, Appl. Phys. Lett. **98**, 082506 (2011)
12. J.R. Childress, R.E. Fontana Jr., C. R. Phys. **6**, 997–1012 (2005)
13. A. Dussaux, B. Georges, J. Grollier, V. Cros, A.V. Khvalkovskiy, A. Fukushima, M. Konoto, H. Kubota, K. Yakushiji, S. Yuasa, K.A. Zvezdin, K. Ando, A. Fert, Nat. Commun. **1**, 8 (2010)
14. V.V. Kruglyak, P.S. Keatley, R.J. Hicken, J.R. Childress, J.A. Katine, Phys. Rev. B **75**, 024407 (2007)
15. H. Schultheiss, C.W. Sandweg, B. Obry, S. Hermsdorfer, S. Schafer, B. Leven, B. Hillebrands, J. Phys. D, Appl. Phys. **41**, 5 (2008)
16. M. Donahue, D.G. Porter, OOMMF user's guide, Version 1.0. NISTIR 6376 (National Institute of Standards and Technology, Gaithersburg, MD 1999), see http://math.nist.gov/oommf
17. V.E. Demidov, M. Buchmeier, K. Rott, P. Krzysteczko, J. Münchenberger, G. Reiss, S.O. Demokritov, Phys. Rev. Lett. **104**, 217203 (2010)
18. P.S. Keatley, P. Gangmei, M. Dvornik, R.J. Hicken, C. Ulysse, J. Grollier, Isolating the dynamic dipolar interaction between a pair of nanoscale ferromagnetic disks, submitted for publication
19. R.D. McMichael, B.B. Maranville, Phys. Rev. B **74**, 024424 (2006)

Chapter 3
Features of Chaotic Spin Waves in Magnetic Film Feedback Rings

Aaron M. Hagerstrom and Mingzhong Wu

Abstract The features of chaotic spin waves in magnetic film feedback rings were studied. Experiments made use of an yttrium iron garnet thin film-based active feedback ring. The excitation of chaotic spin waves relied on the three-wave interactions of short- and long-wavelength backward volume spin waves. The features of the chaotic spin waves were evaluated through the calculation of the ambiguity functions (AFs) of the measured signals. The calculation results indicate that one can tune the AF properties of the chaotic spin waves via the ring gain. In particular, an increase in the ring gain leads to the suppression of side lobes in the AF diagram, both in number and amplitude, but produces negligible effects on the main AF peak. The results on cross ambiguity functions are consistent with the chaotic nature of the chaotic spin waves. These results indicate the potential of magnetic film-based chaotic oscillators for chaotic radar applications.

3.1 Introduction

Studies of chaotic spin waves are of both fundamental and practical significance. From a fundamental point of view, the studies not only contribute to the understanding of chaos and nonlinear dynamics in general, but also promote the understanding of nonlinear magnetization dynamics in magnetic systems in particular. In practical terms, the studies yield new approaches for the development of microwave chaotic oscillators. Such oscillators are critically needed by high-resolution chaotic radars [1, 2], radars for complex target detection [3, 4], and microwave chaotic communications [5, 6].

Previous studies in the field of chaotic spin waves have been focused mainly on the use of yttrium iron garnet (YIG) thin film-based feedback rings to demonstrate the excitation of chaotic spin waves through different nonlinear processes. Specifically, it has been demonstrated that one can realize chaotic spin-wave excitations through three-wave interactions between surface and backward volume

A.M. Hagerstrom · M. Wu (✉)
Department of Physics, Colorado State University, Fort Collins, CO 80523, USA
e-mail: mwu@lamar.colostate.edu

S.O. Demokritov, A.N. Slavin (eds.), *Magnonics*, Topics in Applied Physics 125,
DOI 10.1007/978-3-642-30247-3_3, © Springer-Verlag Berlin Heidelberg 2013

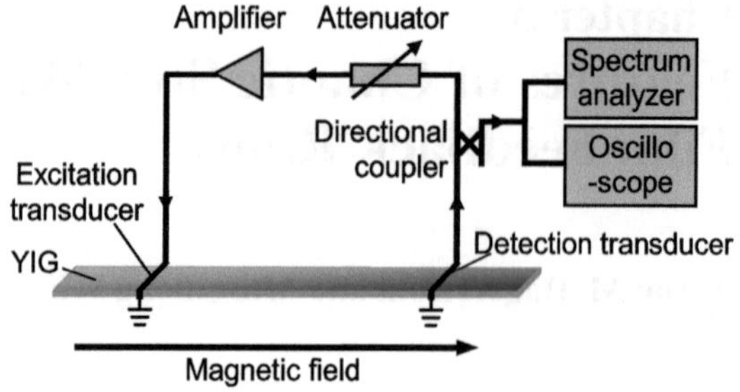

Fig. 3.1 Schematic of experimental setup

spin waves [7–9], three-wave interactions between short-wavelength and long-wavelength backward volume spin waves [10], and four-wave interactions of surface spin waves [11]. Very recently, experiments have also been carried out that demonstrated the tuning of the fractal dimension of spin-wave chaos in feedback rings via the ring gain coefficient [12]. In spite of those studies, however, no work has been done to explore the features of chaotic spin waves in terms of radar and communication applications.

This chapter reports on the properties of chaotic spin waves in the context of radar applications where one uses chaotic microwaves as carrier waves. Experiments made use of a YIG film feedback ring in a backward volume spin wave configuration [13, 14]. The excitation of chaotic spin waves relied on the three-wave interactions of short-wavelength and long-wavelength backward volume spin waves [10, 15]. The features of the chaotic spin waves were evaluated through the calculation of the ambiguity functions of the measured chaotic signals. It should be noted that the ambiguity function (AF) is a two-dimensional (2D) function of delay time and Doppler frequency, and a 2D AF diagram of a certain signal provides useful information on the potential of the signal as a radar carrier wave [16, 17]. The main purpose of this work was to study the AF properties of the chaotic spin waves. The emphasis was placed on the tuning of the AF properties via the ring gain, the effects of the signal duration on the AF properties, and the cross ambiguity function of different signals.

3.2 Experiments

The experimental arrangement is the same as that described in [10]. Figure 3.1 shows a schematic diagram of the YIG film feedback ring. This ring consists of a YIG thin film strip and two microstrip line transducers placed over the YIG strip to excite and detect spin waves. The output signal from the detection transducer is fed back to the excitation transducer through a tunable microwave attenuator and a broadband microwave amplifier. The YIG strip is magnetized to saturation by a static magnetic field which is in the plane of the YIG film strip and parallel to the length of the strip. This film-field configuration supports the propagation of backward volume spin waves along the YIG strip [13, 14]. The magnetic field is set to be relatively low so that the three-wave nonlinear interactions between the short-wavelength and long-wavelength backward volume spin waves are allowed [18]. It

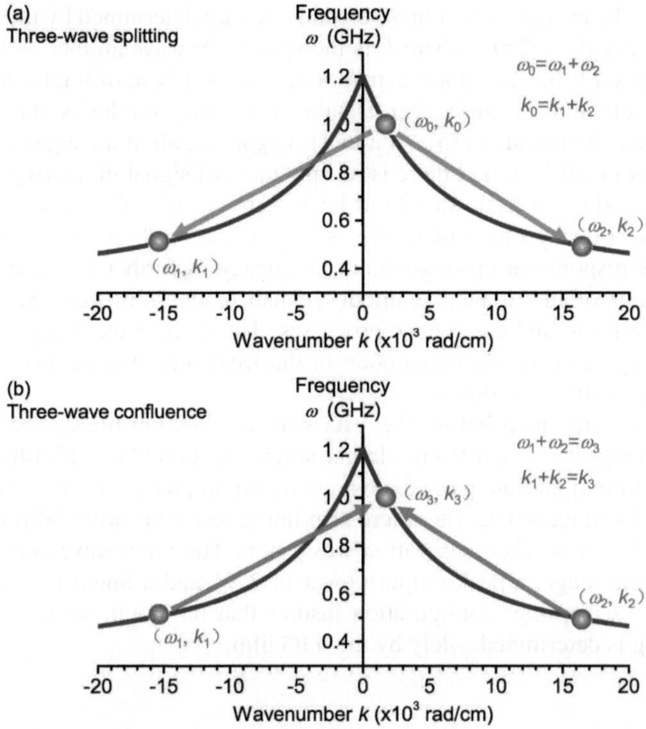

Fig. 3.2 Three-wave interactions of backward volume spin waves. (**a**) Three-wave splitting process. (**b**) Three-wave confluence process. The dispersion curves are calculated for a 5-μm-thick YIG film and a static magnetic field of 100 Oe

is these three-wave interactions that drive the excitation of chaos in the ring. The ring signal is sampled through a directional coupler, with feeds to a spectrum analyzer for frequency-domain analysis and a broadband fast oscilloscope for temporal signal measurements.

The three-wave processes mentioned above include three-wave splitting and three-wave confluence. Such processes are illustrated schematically in Fig. 3.2. The splitting process involves the annihilation of a long-wavelength spin wave and the formation of two new short-wavelength spin waves, with the frequencies of the new waves being about half of the frequency of the initial wave. The confluence process is opposite to the splitting process. It involves the annihilation of two short-wavelength spin waves and the formation of a new long-wavelength spin wave, with the frequency of the new wave about twice the frequencies of the initial waves. The net effect of the interplay of the splitting and confluence processes is a broad spectrum in the frequency domain and a corresponding chaotic response in the time domain.

Turn now to the chaotic excitation through these three-wave processes in a feedback ring system shown in Fig. 3.1. A feedback ring has a number of resonance eigenmodes. The frequencies of these eigenmodes are determined by the phase condition $k(\omega) \cdot l + \phi_0 = 2\pi n$, where k is the spin-wave wavenumber, ω is the spin-wave frequency, l is the transducer separation, ϕ_0 is the phase shift introduced by the feedback circuit, and n is an integer. Detailed discussion on the excitation of these eigenmodes can be found in [15]. At a low ring gain G, all of the eigenmodes experience an overall net loss, and there is no spontaneous signal in the ring. If the ring gain is increased to a certain threshold level, here taken as $G = 0$, the eigenmode with the lowest decay rate will start to self-generate, and one will obtain a continuous wave response at this eigenmode frequency. A further increase in G leads to the excitation of additional eigenmodes, enhancement of modes, and subsequent three-wave splitting and confluence processes. The three-wave processes result in the broadening of each ring eigenmode in the frequency domain and the chaotic ring responses in the time domain.

For the data presented below, the YIG strip was 5.0 μm thick, 1.88 mm wide, and 60 mm long. It was cut from a larger single-crystal (111) YIG film grown on a (111) gadolinium gallium garnet substrate by liquid phase epitaxy. The YIG film had unpinned surface spins. The microstrip line transducers were 50-μm-wide and 2-mm-long elements. Their separation was 5 mm. The microwave amplifier had a 30 dB dynamic range, a peak output power of 2 W, and a linear response from 2 to 8 GHz. This amplifier configuration insures that the nonlinear response of the feedback ring is determined solely by the YIG film.

3.3 Frequency- and Time-Domain Signals

Figure 3.3 shows representative ring signals measured at two different ring gain levels, as indicated. The left graphs show the power-frequency spectra of the signals, and the right graphs give the corresponding time-domain waveforms. As shown in graph (a), the spectrum for the $G = 1.5$ dB signal shows four dominant peaks. Each of these peaks does not represent a single frequency but consists of a broad spectrum centered on a certain ring eigenfrequency. Although this aspect is not clearly shown in graph (a), it becomes very clear when one zooms into a much smaller frequency range [10]. The corresponding time-domain signal clearly shows chaotic behavior. The fractal dimension of this signal is about 4.8. The details on the computation of this dimension were reported in [10]. As shown in graph (b), an increase in G to 3.0 dB leads to enrichment in the chaotic dynamics. This enrichment manifests itself in the broadening of the initial peaks and the formation of more broad peaks in the spectrum and a higher fractal dimension value of about 12.5 for the time-domain signal [10].

3.4 Ambiguity Functions

As mentioned above, the ambiguity function (AF) is a 2D function of delay time τ and Doppler frequency f. In the context of radar signal processing, τ and f are

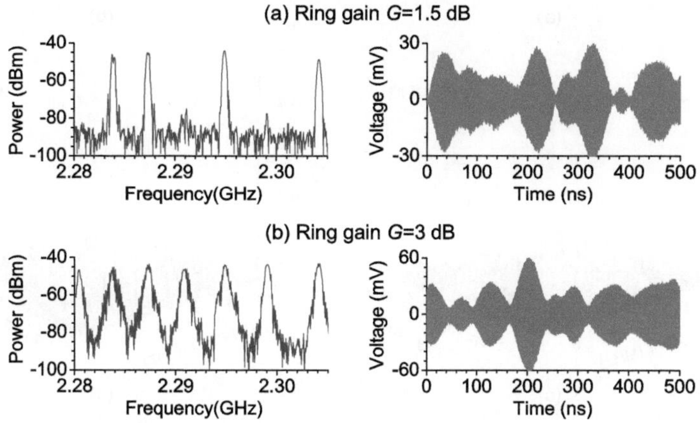

Fig. 3.3 Ring signals obtained at two different ring gain levels, as indicated

the delay time and Doppler frequency shift, respectively, of the received signal in comparison with the transmitted signal. For the 2D AF diagrams shown below, the ambiguity function $\chi(\tau, f)$ was calculated as [16]

$$\chi(\tau, f) = \frac{\sum_{i=1}^{N} u^*(t_i)u(t_i+\tau)e^{-j2\pi f t_i}}{\sum_{i=1}^{N} u^*(t_i)u(t_i)} \tag{3.1}$$

where $u(t_i) = V(t_i) + j\hat{V}(t_i)$, $V(t_i)$ is the measured time-domain data, and $\hat{V}(t_i)$ is the Hilbert transform of $V(t_i)$. The symbol $*$ in (3.1) denotes complex conjugate. For a chaotic signal, the AF diagram has a strong peak centered at the origin and many weak side lobes. In terms of radar applications, the width of the main peak along the delay time axis determines the radar range resolution, while the width along the Doppler frequency axis determines the speed resolution. In general, the narrower the peak is, the higher the resolution is. In addition, fewer and weaker side lobes are desirable.

3.5 Tuning of Ambiguity Function Properties via Ring Gain

Figure 3.4 shows the AF diagrams obtained for four different ring signals. These signals were measured at different ring gain levels, as indicated in Fig. 3.4. The durations ($T = t_N - t_1$) of the signals used in the AF calculations were 1.6 μs. In each diagram, the horizontal axes show the delay time and the Doppler frequency, and the vertical axis shows the square of the AF amplitude. It is evident from the diagrams in Fig. 3.4 that an increase in G results in the suppression of both the number and amplitude of the AF side lobes but produces negligible effects on the main peak. This qualitative conclusion is further supported by more quantitative results shown in Fig. 3.5. In Fig. 3.5, graphs (a) and (b) show the "full width at half maximum"

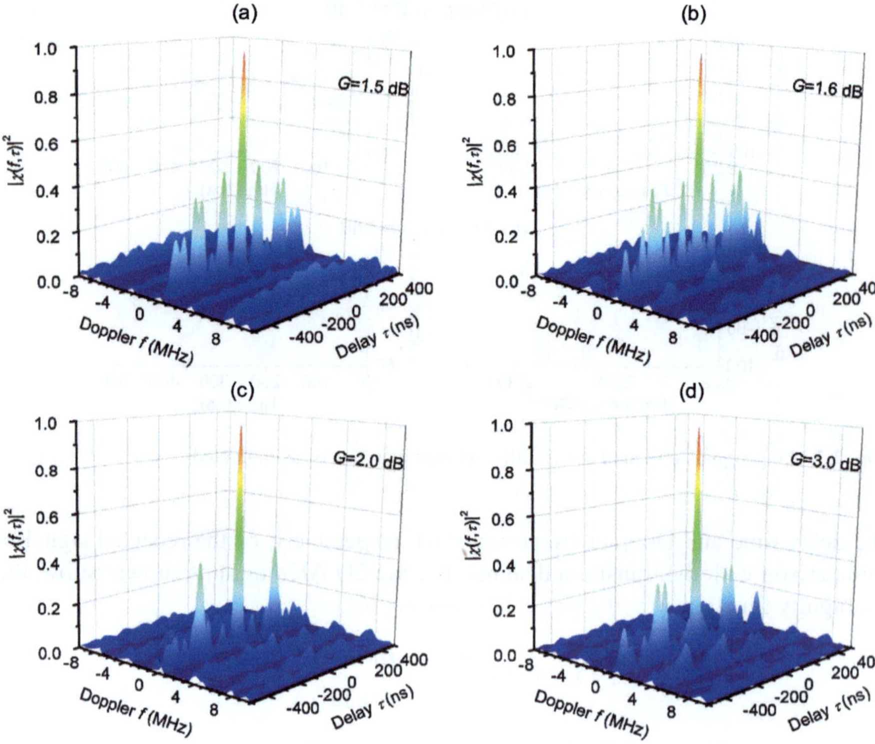

Fig. 3.4 Ambiguity function diagrams for ring signals obtained at different ring gain levels, as indicated

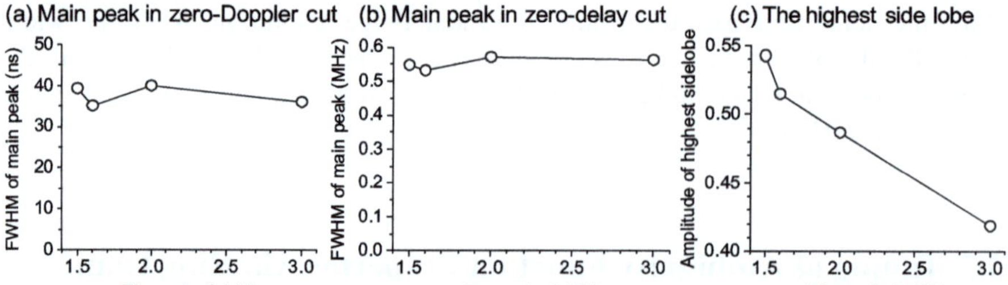

Fig. 3.5 Graphs (**a**) and (**b**) show the "full width at half maximum" of the main AF peaks in the zero-Doppler cut and the zero-delay cut, respectively, as a function of ring gain. Graph (**c**) shows the amplitude of the highest AF side lobe as a function of ring gain

(FWHM) of the main AF peaks in the zero-Doppler cut and the zero-delay cut, respectively, in the diagrams shown in Fig. 3.4. These FWHM values are shown as a function of the ring gain G. Graph (c) shows the amplitude of the highest side lobe in the AF diagram as a function of G. These data show two things: (i) A change in G produces insignificant effects on the FWHM values of both the main AF peak in the zero-Doppler cut and that in the zero-delay cut. (ii) The amplitude of the strongest side lobe decreases significantly with G. These results support the results shown in Fig. 3.4.

Two points should be mentioned: (i) The results from Figs. 3.4 and 3.5 clearly demonstrate that one can tune the AF properties of chaotic spin waves simply through a change in the ring gain. Note that previous work has demonstrated the tuning of the fractal dimension of chaotic spin waves via the ring gain [12]. These results together show the easy tuning of the complexity and features of chaotic spin waves in feedback rings through the ring gain. Future work is of great interest that makes use of other control parameters, such as the magnetic field and the transducer separation, to tune the properties of chaotic spin waves. (ii) In the context of chaotic radar, the results from Figs. 3.4 and 3.5 indicate that, by increasing the ring gain, one can effectively suppress the side lobes while maintaining both the range and speed resolutions.

3.6 Effects of Signal Duration on Ambiguity Function

For the AF data shown in Figs. 3.4 and 3.5, the calculation used a fixed signal duration (T). Figures 3.6 and 3.7 show the AF data calculated for the same signal ($G = 3.0$ dB) but different signal durations. Graphs (a), (b), and (c) in Fig. 3.6 show 2D AF diagrams obtained for different durations, as indicated. Graphs (a) and (b) in Fig. 3.7 show the zero-Doppler cuts and the zero-delay cuts, respectively. These cuts were obtained for different signal durations, as indicated. In graph (c), the circles show the FWHM values for the main peaks in the zero-delay cuts, and the dashed line serves as a guide to eye.

Three results are evident from the data shown in Figs. 3.6 and 3.7: (i) A change in the signal duration produces negligible effects on the main peak in the zero-Doppler cut. (ii) One can significantly reduce the width of the main peak in the zero-delay cut by increasing the signal duration. (iii) The FWHM vs. $1/T$ response for the main peak in the zero-delay cut shows linear behavior. In the context of chaotic radar, these results indicate that, by increasing the signal duration, one can linearly improve the radar speed resolution with no degradation of the range resolution. Note that the linear response shown in Fig. 3.7(c) was also observed for chaotic optical signals [17].

3.7 Cross Ambiguity Function

Figure 3.8 shows representative cross ambiguity function (CAF) diagrams for chaotic spin-wave signals. The CAF data were calculated as

$$\chi(\tau, f) = \frac{\sum_{i=1}^{N} u_1^*(t_i) u_2(t_i + \tau) e^{-j2\pi f t_i}}{\sum_{i=1}^{N} u_1^*(t_i) u_2(t_i)} \tag{3.2}$$

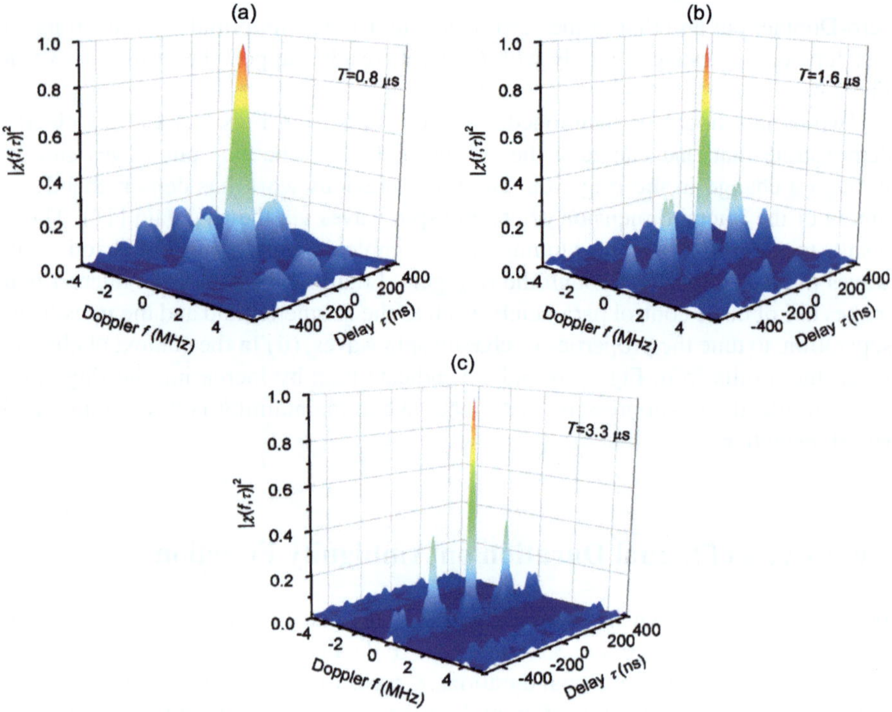

Fig. 3.6 Ambiguity function diagrams calculated for the same ring signal ($G = 3.0$ dB) but different signal durations (T), as indicated

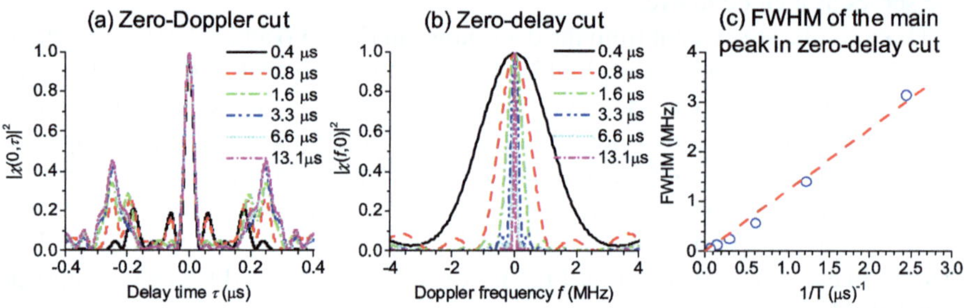

Fig. 3.7 Ambiguity function data calculated for the same ring signal ($G = 3.0$ dB) but different signal durations (T). Graphs (**a**) and (**b**) show the zero-Doppler cuts and the zero-delay cuts, respectively, for different T values, as indicated. Graph (**c**) shows the "full width at half maximum" (FWHM) of the main peak in the zero-delay cut as a function of $1/T$

where $u_1(t_i) = V_1(t_i) + j\hat{V}_1(t_i)$ and $u_2(t_i) = V_2(t_i) + j\hat{V}_2(t_i)$. $V_1(t_i)$ and $V_2(t_i)$ are the signals measured at different time windows. For the CAF data shown in graph (a), $V_1(t_i)$ and $V_2(t_i)$ were recorded at time windows of 0.0–0.4 μs and 0.8–1.2 μs, respectively. For those shown in graph (b), $V_1(t_i)$ and $V_2(t_i)$ were recorded

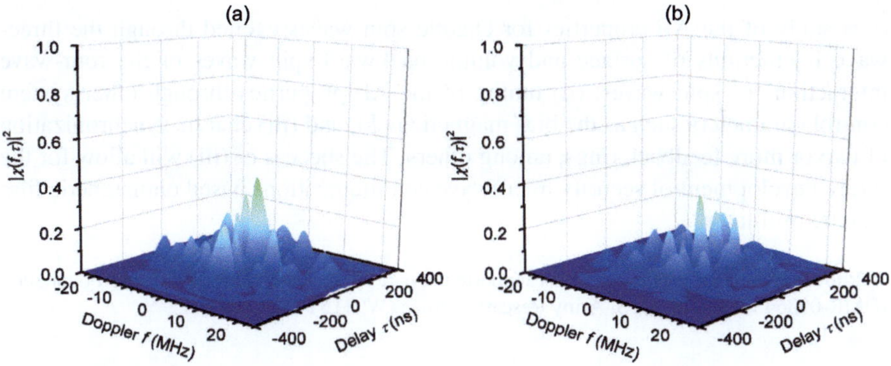

Fig. 3.8 Cross ambiguity functions of two different signals. For the data in graph (**a**), the two signals were measured at time windows of 0.0–0.4 μs and 0.8–1.2 μs, respectively. For the data in graph (**b**), the two signals were measured at time windows of 0.0–0.4 μs and 8.2–8.6 μs, respectively. All the signals were measured at $G = 3.0$ dB

at time windows of 0.0–0.4 μs and 8.0–8.6 μs, respectively. All these signals were measured at $G = 3.0$ dB.

The diagrams in Fig. 3.8 show three results: (i) Both diagrams consist of a number of weak peaks. This indicates that there exists a certain level of correlation in the measured chaotic signal. Note that, in the case of a pure random noise signal, one expects no notable peaks in the CAF diagram; in the case of a periodic signal, the strong correlation results in very strong peaks in the CAF diagram. (ii) The peaks in both the diagrams are randomly distributed. (iii) A significant increase in the separation between the measurement time windows does not produce a significant change in both the number and amplitude of the CAF peaks. These results together are consistent with the chaotic nature of the measured ring signal.

3.8 Conclusion

This work studied the ambiguity function (AF) properties of chaotic spin waves self-generated in a magnetic film feedback ring. The main results are as follows: (i) One can tune the AF properties of the chaotic spin waves simply through a change in the ring gain. In particular, an increase in the ring gain leads to the suppression of side lobes in the AF diagram, both in number and amplitude, but produces negligible effects on the main AF peak. (ii) An increase in the signal duration T results in a significant narrowing of the main peak in the zero-delay cut, but leads to no changes in the main peak in the zero-Doppler cut. Moreover, the width vs. $1/T$ response for the main peak in the zero-delay cut shows linear behavior. (iii) The results on the cross ambiguity functions are consistent with the chaotic nature of the chaotic spin waves. These results demonstrate the potential of magnetic film-based chaotic oscillators for chaotic radar applications. Future work that is of great interest includes

(i) a study of the AF properties for chaotic spin waves excited through the three-wave interactions of surface and volume backward spin waves or the four-wave interactions of spin waves, (ii) tuning of the AF properties through other system control parameters such as the bias magnetic field, and (iii) chaotic synchronization of two or more feedback rings, among others. The success of (iii) will allow for the future development of security microwave communications based on magnetic film chaotic oscillators.

Acknowledgements This work was supported in part by the U. S. National Science Foundation (DMR-0906489) and the U. S. Army Research Office (W911NF-11-C-0075).

References

1. V. Venkatasubramanian, H. Leung, IEEE Signal Process. Lett. **12**, 528 (2005)
2. Z. Liu, X. Zhu, W. Hu, F. Jiang, Int. J. Bifurc. Chaos **17**, 1735 (2007)
3. T.L. Carroll, Chaos **17**, 033103 (2007)
4. T.L. Carroll, IET Radar Sonar Navig. **2**, 256 (2008)
5. A.S. Dmitriev, B.Y. Kyarginsky, A.I. Panas, S.O. Starkov, Int. J. Bifurc. Chaos Appl. Sci. Eng. **13**, 1495 (2003)
6. A.S. Dmitriev, B.Y. Kyarginskii, A.I. Panas, D.Y. Puzikov, S.O. Starkov, Tech. Phys. Lett. **29**, 72 (2003)
7. V.E. Demidov, N.G. Kovshikov, Tech. Phys. Lett. **24**(4), 274 (1998)
8. V.E. Demidov, N.G. Kovshikov, Tech. Phys. Lett. **24**, 647 (1998)
9. M. Wu, B.A. Kalinikos, C.E. Patton, Phys. Rev. Lett. **95**, 237202 (2005)
10. A.M. Hagerstrom, W. Tong, M. Wu, B.A. Kalinikos, R. Eykholt, Phys. Rev. Lett. **102**, 207202 (2009)
11. M. Wu, A.M. Hagerstrom, R. Eykholt, A. Kondrashov, B.A. Kalinikos, Phys. Rev. Lett. **102**, 237203 (2009)
12. A.M. Hagerstrom, M. Wu, R. Eykholt, B.A. Kalinikos, Phys. Rev. B **83**, 104402 (2011)
13. P. Kabos, V.S. Stalmachov, *Magnetostatic Waves and Their Applications* (Chapman & Hall, London, 1994)
14. D.D. Stancil, A. Prabhakar, *Spin Waves – Theory and Applications* (Springer, New York, 2009)
15. M. Wu, Nonlinear spin waves in magnetic film feedback rings, in *Solid State Physics*, vol. 62, ed. by R. Camley, R. Stamps (Academic Press, Burlington, 2011), pp. 163–224
16. C.E. Cook, M. Bernfeld, *Radar Signals: An Introduction to Theory and Application* (Artech House, Norwood, 1993)
17. F. Lin, J. Liu, IEEE J. Quantum Electron. **40**, 1732 (2004)
18. A.G. Gurevich, G.A. Melkov, *Magnetization Oscillations and Waves* (CRC Press, Boca Raton, 1996)

Chapter 4
Magnon Coherent States and Condensates

Sergio M. Rezende

Abstract It is known that spin waves that have quanta called magnons can be excited in magnetic materials by means of several processes. Most spin-wave phenomena involve a large number of magnons and can be well described by the semi-classical Landau–Lifshitz equation. Perhaps this is one reason for the coherent magnon states, defined in analogy to the coherent photon states that are widely used in quantum optics, to be not well known by the magnetism community. In this paper, we review the concept of coherent magnon states and show that they are the quantum states generated in a linear microwave driving process. We also address the recently observed Bose–Einstein condensation of magnons in magnetic films under strong microwave driving. Using a microscopic theoretical model that relies on the cooperative mechanisms made possible by magnon interactions, we explain quantitatively the spontaneous generation of quantum coherence and magnetic dynamic order observed above a critical microwave power level.

4.1 Introduction

In recent years, very exciting experimental observations of spin dynamics in magnetic films, multilayers, and nanostructures have produced a revival of interest in phenomena involving spin waves whose quanta are called magnons. Special attention has been given to the excitation of spin dynamics at microwave frequencies by means of a dc current in magnetic metallic nanostructures [1–8], spin pumping and inverse spin Hall effect in ferromagnetic/normal metal bilayers [9–13], room-temperature Bose–Einstein condensation of a magnon gas driven by microwave pumping in the insulating ferromagnetic yttrium-iron-garnet (YIG) [14–19], generation and amplification of spin waves in YIG films through the spin Hall and spin Seebeck effects [20–23]. The wealth of phenomena involving spin waves has given rise to a new subfield of research in magnetism, which has been called magnonics [24, 25], in analogy to the well known field of photonics.

S.M. Rezende (✉)
Departamento de Física, Universidade Federal de Pernambuco, Recife, PE 50670-901, Brazil
e-mail: rezende@df.ufpe.br

S.O. Demokritov, A.N. Slavin (eds.), *Magnonics*, Topics in Applied Physics 125,
DOI 10.1007/978-3-642-30247-3_4, © Springer-Verlag Berlin Heidelberg 2013

The concept of spin waves as elementary excitations in magnetically ordered materials was introduced in the early 1930s by Bloch [26], who showed that they are the low-lying excitations of the spin system. In 1934, Heller and Kramers [27] developed the semi-classical picture of a spin wave consisting of spins precessing around the equilibrium direction with a phase angle varying along the propagation direction. Soon after, Landau and Lifshitz published their famous paper introducing the torque equation of motion for the magnetization and predicting the magnetic resonance phenomenon [28]. In 1940, Holstein and Primakoff [29] introduced a boson formulation that became the most widely used quantum mechanical formalism for studying spin waves, naturally leading to the concept of magnons. By the 1950s, spin waves were thought to be interesting objects, but not quite exciting because they seemed to be mainly a mathematical entity with a small role in the low temperature behavior of the magnetization.

With the development of microwave spectroscopy and ferromagnetic resonance (FMR) techniques in the early 1950s, ferrite materials started to attract much attention due to their novel properties and potential for technological applications in microwave devices [30, 31]. Two intriguing effects observed at high microwave powers by Bloembergen, Damon and Wang [32, 33], the premature saturation of the FMR and the appearance of a subsidiary resonance, were explained in the late 1950s by Suhl [34] based on the nonlinear excitation of spin waves. The evidences of spin waves as real physical entities stimulated a variety of microwave experiments in the early 1960s, and new schemes for the nonlinear driving of spin waves were discovered, such as the parallel pumping process [35, 36]. Single crystal YIG grown in bulk or film shape became the prototype material to study spin-wave phenomena due to its very low magnetic losses. Several schemes were devised to generate spin waves by pulsed microwave radiation [37, 38] and to study their properties [39, 40]. In the 1970s and 1980s, spin-wave phenomena in YIG films were extensively studied, motivated by their interesting physical properties and potential applications in microwave signal processing devices [41–46]. With the recent discoveries of new spin dynamic phenomena mentioned earlier, magnonics has become one of the hottest fields of magnetism.

Many spin dynamic phenomena involve a large number of magnons so that they can be described by the semi-classical Landau–Lifshitz equation (LLE) for the magnetization $\vec{M}(\vec{r},t)$, $d\vec{M}/dt = \gamma\vec{M} \times \vec{H}_t$, where γ is the gyromagnetic ratio and $\vec{H}_t$ represents the total field acting on $\vec{M}$, containing the static external magnetic field $\vec{H}$ and effective fields due to the spin interactions, such as exchange, dipolar, and anisotropy [44–46]. The LLE can be solved analytically for simple situations or can be treated numerically with micromagnetic modeling. If we write $\vec{M}(\vec{r},t) = \hat{z}M_z + \hat{x}m_x + \hat{y}m_y$, where z denotes the equilibrium direction and m_x and m_y are the small-signal transverse magnetization components, the LLE with exchange and dipolar fields in a uniform field gives for $m^+ = m_x + im_y$ the solution $m^+ = m_0 \exp(i\vec{k}\cdot\vec{r} - i\omega_k t)$. This represents a traveling spin wave consisting of spins precessing around the equilibrium direction with frequency ω_k and a phase angle varying along the direction of the wave vector $\vec{k}$. The form of the function $\omega_k(\vec{k})$, called dispersion relation, depends on the sample shape and the values of the field and magnetic parameters. Spin phenomena that are intrinsically quantum in nature cannot be described by classical equations and require other approaches, such as the one discussed in the next section.

4.2 Coherent Magnon States

The most used quantum treatment of the collective spin excitations employs transformations aimed at expressing the spin operators in terms of magnon operators [29, 42–46]. Initially, the components of the local magnetization at site i are written in terms of boson operators for creation and annihilation of spin deviation, a_i^+ and a_i, such as $m^+ = \gamma\hbar(N/V)(2S)^{1/2}(a_i - a_i^+ a_i a_i)$, where N is the number of spins S in the volume V. Then ones uses Fourier and Bogoliubov transformations to express the spin deviation operators into magnon creation and annihilation operators c_k^+ and c_k, where u_k and v_k are the well known coefficients of the Holstein–Primakoff transformations that diagonalize the Hamiltonian of the system and satisfy the relation $u_k^2 - v_k^2 = 1$. If the nonlinear interactions are neglected, the spin-wave excitations form a system of independent harmonic oscillators, governed by the unperturbed Hamiltonian $H_0 = \hbar\sum_k \omega_k c_k^+ c_k$. The eigenstates $|n_k\rangle$ of this Hamiltonian, which are also eigenstates of the number operator $n_k = c_k^+ c_k$, can be obtained by applying integral powers of the creation operator to the vacuum,

$$|n_k\rangle = \left[(c_k^+)^{n_k}/(n_k!)^{1/2}\right]|0\rangle, \tag{4.1}$$

where the vacuum state is defined by the condition $c_k|0\rangle = 0$. These stationary states have precisely defined number of magnons n_k but have uncertain phase. They form a complete orthonormal set which can be used as a basis for the expansion of any state of spin excitation. They are used in nearly all quantum treatments of thermodynamic properties, relaxation mechanisms, and other phenomena involving magnons. However, since $\langle n_k|n_k \pm 1\rangle = 0$, the eigenstates $|n_k\rangle$ have zero expectation value for the magnetization operators m_x and m_y and thus do not have a macroscopic wavefunction. In order to establish a correspondence between classical and quantum spin waves, one should use the concept of coherent magnon states [47, 48], defined in analogy to the coherent photon states introduced by Glauber [49]. A coherent magnon state is the eigenket of the circularly polarized magnetization operator $m^+ = m_x + im_y$. It can be written as the direct product of single-mode coherent states, defined as the eigenstates of the annihilation operator, $c_k|\alpha_k\rangle = \alpha_k|\alpha_k\rangle$, where the eigenvalue α_k is a complex number. Although the coherent states are not eigenstates of the unperturbed Hamiltonian and as such do not have a well defined number of magnons, they have nonzero expectation values for the magnetization m^+ with a well defined phase. Here we review a few important properties of the coherent states. First, we recall that they can be expanded in terms of the eigenstates of the unperturbed Hamiltonian [47–49],

$$|\alpha_k\rangle = e^{-|\alpha_k|^2/2}\sum_{n_k}(\alpha_k)^{n_k}/(n_k!)^{1/2}|n_k\rangle. \tag{4.2}$$

The probability of finding n_k magnons in the coherent state $|\alpha_k\rangle$ obtained directly from (4.2) is given by

$$\rho_{\text{coh}}(n_k) = \left|\langle n_k|\alpha_k\rangle\right|^2 = \left(|\alpha_k|^{2n_k}/n_k!\right)e^{-|\alpha_k|^2}. \tag{4.3}$$

This function is a Poisson distribution that exhibits a peak at the expectation value of the occupation number operator $\langle n_k\rangle = |\alpha_k|^2$ in the coherent state. It can be shown that coherent states are not orthogonal to one another, but they form a complete set, so that they constitute a basis for the expansion of an arbitrary state [49]. The distribution (4.3) is very different from the one prevailing in systems in thermal equilibrium, which cannot be described by pure quantum states. They are described by a mixture, with average number of magnons with energy $\hbar\omega_k$ in thermal equilibrium at a temperature T given by the Bose–Einstein (BE) distribution $\bar{n}_k = 1/[\exp(\hbar\omega_k/k_B T) - 1]$, where k_B is the Boltzmann constant. The probability of finding n_k magnons with energy $\hbar\omega_k$ in the system in thermal equilibrium is

$$\rho_{th}(n_k) = (\bar{n}_k)^{n_k}/(1+\bar{n}_k)^{n_k+1}. \tag{4.4}$$

Note that for large n_k (4.4) approaches the exponential function $\exp(-\bar{n}_k)$ and is very different from the peaked shape Poisson distribution (4.3). Another important property of a coherent state is that it can be generated by the application of a displacement operator to the vacuum [47–49],

$$|\alpha_k\rangle = D(\alpha_k)|0\rangle, \quad \text{where } D(\alpha_k) = \exp\left(\alpha_k c_k^+ - \alpha_k^* c_k\right). \tag{4.5}$$

In order to study the coherence properties of a magnon system, it is convenient to use the density matrix operator ρ and its representation as a statistical mixture of coherent states,

$$\rho = \int P(\alpha_k)|\alpha_k\rangle\langle\alpha_k|\, d^2\alpha_k, \tag{4.6}$$

where $P(\alpha_k)$ is a probability density, called P representation, satisfying the normalization condition $\int P(\alpha_k)\, d^2\alpha_k = 1$ and $d^2\alpha_k = d(\text{Re}\,\alpha_k)d(\text{Im}\,\alpha_k)$. It can be shown [49] that if ρ corresponds to a coherent state, $P(\alpha_k)$ is a Dirac δ-function, whereas if ρ represents a BE thermal distribution, $P(\alpha_k)$ is a Gaussian function.

To conclude this section, we calculate the expectation values of the components of the magnetization operators for a single-mode coherent state with eigenvalue $\alpha_k = |\alpha_k|\exp(i\phi_k)$. Using the relations between the magnetization components and the magnon operators, it is straightforward to show that

$$\left\langle m_x(\vec{r},t)\right\rangle = M(NS/2)^{-1/2}|\alpha_k|(u_k + v_k)\cos(\vec{k}.\vec{r} - \omega_k t + \phi_k), \tag{4.7a}$$

$$\left\langle m_y(\vec{r},t)\right\rangle = M(NS/2)^{-1/2}|\alpha_k|(u_k - v_k)\sin(\vec{k}.\vec{r} - \omega_k t + \phi_k). \tag{4.7b}$$

Equations (4.7a), (4.7b) show clearly that the coherent magnon states are the quantum states that correspond to the classical magnetization. The transverse components in (4.7a), (4.7b) together with $\hat{z}M_z$ correspond to the classical view of a spin wave, namely, the magnetization precesses with an ellipticity $m_x^{max}/m_y^{max} = (u_k + v_k)/(u_k - v_k)$ around the equilibrium direction and with a phase that varies along the direction of propagation. Note that the elliptical precession of the transverse magnetization with frequency ω_k results in an oscillating z-component with frequency $2\omega_k$. As is well known, it is this fact that makes it possible to excite spin-waves with a microwave field parallel to the static field [35, 36].

4.3 Linear Excitation of Magnons by a Microwave Field

Spin waves can be generated in magnetic samples by means of a microwave magnetic field applied perpendicularly to the static magnetic field. In the semi-classical view, the rf field exerts an oscillating torque on the spins that set them to precess around the equilibrium direction. In a YIG film, this is conveniently done by placing the sample on a microstrip patterned on a circuit board and fed by a signal from a microwave generator [42–44]. For the quantum formulation, we consider a driving field $\hat{x}h(\vec{r})\cos(\omega t)$ perpendicular to the static field $\hat{z}H$ and write the Hamiltonian as $H(t) = H_0 + H_1(t)$, where the perturbation arises from the Zeeman interaction with the driving field and can be shown [48] to be

$$H_1(t) = -\gamma\hbar(S/2)^{1/2}\sum_k\left[(u_k + v_k)h(\vec{k})e^{i\omega t}c_k + H.c.\right], \tag{4.8}$$

where

$$h(\vec{k}) = (N)^{-1/2}\sum_i e^{i\vec{k}.\vec{r}}h(\vec{r}_i) \tag{4.9}$$

is the space Fourier transform of the driving field. We assume that at $t \to -\infty$ the system is in the ground state $|0\rangle$ and that there is no damping. One can show [48] that at time t the state is $|t\rangle = U(t, -\infty)|0\rangle$, where $U(t)$ is the evolution operator in the Schrödinger picture,

$$U(t, -\infty) = \exp(-iH_0t/\hbar)\exp\left[\sum_k\left(\gamma_k c_k^{+} - \gamma_k^{*}c_k\right)\right], \tag{4.10}$$

where $\gamma_k(t)$ is a c-number function given by

$$\gamma_k(t) = i\gamma(S/2)^{1/2}\left(u_k + v_k^{*}\right)\int_{-\infty}^{t} h(\vec{k})^{*}e^{i(\omega_k-\omega)t}. \tag{4.11}$$

Performing the integration in (4.11) and replacing $\gamma_k(t)$ in (4.10), we find that if initially the system is in the vacuum state, at time t the state is

$$|t\rangle = U(t, -\infty)|0\rangle = \prod_k\left|\gamma_k(t)\exp(i\omega_k t)\right\rangle = \prod_k|\alpha_k\rangle, \tag{4.12a}$$

where the eigenvalue is

$$\alpha_k = \gamma(S/8)^{1/2}(u_k + v_k)h(\vec{k})/(\omega_k - \omega). \tag{4.12b}$$

Equation (4.12a) represents a product of the displacement operators defined in (4.5). This result shows that the transverse microwave driving field excites a superposition of coherent magnon states with eigenvalues given by (4.12b). Thus the driven states have wave vector $\vec{k}$ determined by the Fourier transform of the driving field and frequency $\omega_k = \omega$. The eigenvalue in (4.12b) is proportional to the amplitude of the driving field, characterizing a linear excitation process. For sufficiently high driving fields, nonlinear processes come into play [34, 46, 50].

Note that by using (4.12a), (4.12b) in (4.7a), (4.7b) one can find the transverse components of magnetization and obtain the same expressions calculated with the Landau–Lifshitz equation. This result represents another confirmation of the correspondence between the coherent states and the classical magnetization.

4.4 Microwave Excitation of Parametric Magnons in Thin Films

Spin waves can be nonlinearly excited in a magnetic material by means of several techniques employing microwave radiation, with the driving field applied either perpendicular or parallel to the static field [46, 50]. The pumped system is conveniently treated with a quantum approach in which the magnetic Hamiltonian contains Zeeman, exchange, and magnetic dipolar contributions that can be written as

$$H = H_0 + H_{\text{int}} + H'(t), \tag{4.13}$$

where H_0 is the unperturbed Hamiltonian, H_{int} represents the nonlinear interactions, and $H'(t)$ the external microwave driving. In the parallel pumping process, the driving Hamiltonian in (4.13) follows from the Zeeman interaction of the microwave pumping field $\hat{z}h\cos(\omega_p t)$ with the magnetic system, which can be expressed in terms of the magnon operators. Keeping only terms that conserve energy, one can show that for a YIG film magnetized in the plane [19]

$$H'(t) = (\hbar/2)\sum_k h\rho_k e^{-i\omega_p t} c_k^+ c_{-k}^+ + h.c., \tag{4.14}$$

where $\rho_k = \gamma\omega_M[(1-F_k)\sin^2\theta_k - F_k]/4\omega_k$ represents the coupling of the pumping field h (frequency ω_p) with the $\vec{k}$, $-\vec{k}$ magnon pair with frequency ω_k equal or close to half the pumping frequency $\omega_p/2$, $\omega_M = \gamma 4\pi M$, $F_k = (1 - e^{-kd})/kd$ is a form factor, d is the film thickness, and θ_k is the angle between the wave vector $\vec{k}$ in the plane and the field. For a thick film or bulk sample, the coupling is maximum for $\theta_k = 90°$ and vanishes for $\theta_k = 0$. However, in films with kd on the order of 1 or less, F_k is finite so that the parallel pumping field can drive spin waves with any value of θ_k as in the experiments of [14–18].

The Heisenberg equations of motion for the operators c_k and c_k^+ with the Hamiltonian $H = H_0 + H'(t)$ can be easily solved for a pumping field h applied at $t = 0$. The solution shows that if the field exceeds a critical value h_c the number of parametric magnons increases exponentially in time until saturation is reached due to nonlinear four-magnon interactions that can be represented by [19, 50–52]

$$H^{(4)} = \hbar \sum_{k,k'} \left(\frac{1}{2} S_{kk'} c_k^+ c_{-k}^+ c_{k'} c_{-k'} + T_{kk'} c_k^+ c_{k'}^+ c_k c_{k'} \right), \tag{4.15}$$

where the interaction coefficients are determined mainly by the dipolar and exchange energies. For the k-values relevant to the experiments the exchange contribution is negligible so that the coefficients in (4.15) are given approximately by $S_{kk'} = 2T_{kk'} = 2\omega_M / NS$. Using the Hamiltonian (4.13) with (4.14) and (4.15) as the driving and interaction terms, one can write the equations for the operators c_k and c_k^+ from which several quantities of interest can be obtained. One of them is the correlation function $\sigma_k = \langle c_k c_{-k} \rangle = n_k e^{i\varphi_k} e^{-i2\omega_k t}$ [51, 52], where n_k is the magnon number operator and φ_k the phase between the states of the pair. Considering a situation in which a single $\vec{k}$, $-\vec{k}$ parametric magnon pair is driven, it can be shown that for $h > h_c$, the steady-state population of the pair is [19, 50–52]

$$\langle n_k \rangle_{ss} = \left\{ \left[(h\rho_k)^2 - \eta_k^2 \right]^{1/2} - |\Delta\omega_k| \right\} / 2V_{(4)}, \tag{4.16}$$

where $V_{(4)} = S_{kk} + 2T_{kk} = 4\omega_M / NS$, $\Delta\omega_k = \omega_k - \omega_p/2$ is the detuning from the frequency of maximum coupling, and η_k is the magnon relaxation rate that was introduced phenomenologically in the equations of motion. It can also be shown that the phase φ_k varies from $-\pi/2$ to π as h increases from h_c to infinity. In the range of powers used in [14–18], $\varphi_k \sim -\pi/2$.

Equation (4.16) shows that magnon pairs with frequencies in a narrow range around $\omega_p/2$ are pumped by the microwave field when $h > h_c = (\eta_k^2 + \Delta\omega_k^2)^{1/2} / \rho_k$. Note that for $\omega_k = \omega_p/2$ the critical field is $h_c = \eta_k / \rho_k$. In the experiments of [14–18], the minimum h_c corresponds to a critical power p_c in the range of 100 μW to 1 mW determined by the experimental geometry and the spin-lattice relaxation rate in YIG, $\eta_{SL} \sim 2 \times 10^6\ \mathrm{s}^{-1}$. However, when very short microwave pulses are used, much higher power levels are required to reduce the rise time and to build up large magnon populations. In this case, the relaxation rate that prevails in the dynamics is dominated by magnon–magnon scattering $\eta_m \sim 25\eta_{SL} = 5 \times 10^7\ \mathrm{s}^{-1}$ [19], so one can define a critical field $h_{c1} = \eta_m / \rho_k = h_c \eta_m / \eta_{SL}$ for driving magnons out of equilibrium from the magnon heat bath. Since the driving power p is proportional to h^2, one can write from (4.16) an expression for the steady-state total number of parametric magnons with $\omega_k \approx \omega_p/2$,

$$N_p = r_p n_H \left[(p - p_{c1}) / p_{c1} \right]^{1/2}, \tag{4.17}$$

where $n_H \equiv \eta_m / 2V_{(4)} = \eta_m NS / 8\omega_M$ and r_p is a factor that represents the number of pumped modes weighted by a factor relative to the number of magnons of the mode with maximum coupling.

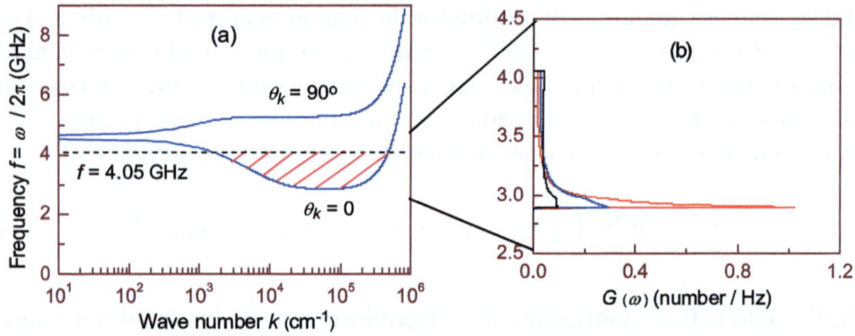

Fig. 4.1 Characteristics of spin waves propagating in a 5 μm thick YIG film with an in-plane field $H = 1.0$ kOe. (**a**) Dispersion relation for waves propagating at an angle θ_k with the field. (**b**) Spectral density $G(\omega)$ calculated numerically as in [19] for several values of the chemical potential μ in frequency units ($f_{\text{min}} = 2.898$ GHz and h is Planck's constant): (*black line*) $\mu = 0$; (*blue line*) $\mu/h = 2.718$ GHz; (*red line*) $\mu/h = 2.868$ GHz

4.5 Bose–Einstein Condensation of a Microwave Driven Interacting Magnon Gas

In this section, we present a brief account of a theoretical model [19] for the formation of Bose–Einstein condensation (BEC) of magnons at room-temperature in films of yttrium-iron garnet (YIG) driven by microwave radiation and its application to interpret the experiments of Demokritov and co-workers [14–18]. BEC occurs when a macroscopic number of bosons occupies the lowest available energy quantum states and has been observed in a few physical systems at very low temperatures [53, 54]. The room temperature experiments of [14–18] have materialized earlier proposals for producing BEC of magnons [55] and demonstrated powerful techniques for observing its properties.

The experiments were done at room temperature in epitaxial crystalline YIG films magnetized in the plane. In these films, the combined effects of the exchange and magnetic dipolar interactions among the spins produce a magnon dispersion relation $\omega(\vec{k})$ that has a minimum ω_{k0} at $k_0 \sim 10^5$ cm^{-1} as shown in Fig. 4.1(a). As will be shown, the energy minimum away from $k = 0$ produces a peak in the density of states that is essential for the formation of the magnon condensate.

4.5.1 Dynamics of the Microwave Driven Magnon Gas in k Space

In the experiments of [14–18], magnon pairs are parametrically driven in a YIG film at room temperature by parallel-pumping with short microwave pulses with frequency f_p and studied by Brillouin light scattering (BLS). If the microwave power exceeds a threshold value, there is a large increase in the population of the magnons with frequencies in a narrow range around $f_p/2$. Then the energy of these primary magnons quickly redistributes through modes with lower frequencies down to the minimum frequency $f_{\min} = \omega_{k0}/2\pi$ as a result of magnon scattering events. This produces a hot magnon gas in the shaded region in Fig. 4.1(a) that remains decoupled from the lattice for several hundred nanoseconds due to the long spin-lattice relaxation time. However, as observed with time and wave vector resolved BLS techniques [17], if the microwave power exceeds a second critical value, much larger than the one for parallel pumping, the non-equilibrium magnon population evolves in time to condense in a narrow region in phase space around the minimum frequency and develops quantum coherence. The theoretical model presented here provides rigorous support for the formation of the BEC of magnons in YIG films [19]. The theory treats the dynamics of the magnon gas in wave vector space and is based on the microscopic interactions. We show that the cooperative mechanisms made possible by the nonlinear magnetic interactions explain the spontaneous generation of quantum coherence and magnetic dynamic order in a macroscopic scale when the driving microwave power exceeds a critical value. The results fit very well the experimental BLS data.

We assume that the magnon gas resulting from the redistribution of the primary parametric magnons is in quasi-equilibrium. If we initially neglect the interaction between magnons, we can write the occupation number of the state with energy $\hbar\omega$ at temperature T according to the BE distribution

$$n_{\mathrm{BE}}(\omega, \mu, T) = 1/\{\exp[(\hbar\omega - \mu)/k_B T] - 1\}, \tag{4.18}$$

where μ is the chemical potential. Considering that the number of magnons N_p pumped into the system by the microwave driving in the frequency range from ω_{k0} to $\omega_p/2$ is much larger than the thermal number, we can write $N_p = \int D(\omega) n_{\mathrm{BE}}(\omega, \mu, T)\, d\omega$, where $D(\omega)$ is the magnon density of states and the integral is carried out over the range $[\omega_{k0}, \omega_p/2]$. As the microwave power is raised, the total number of magnons increases so that the chemical potential rises. At a high enough power, μ reaches the energy corresponding to ω_{k0}, resulting in an overpopulation of magnons with that frequency so that the gas is spontaneously divided in two parts, one with the magnons distributed according to (4.18) and the other with the magnons accumulated in states near the minimum energy.

While the thermodynamic interpretation of the experiments of [14–18] is satisfactory and explains qualitatively several observed features, it fails in providing quantitative results to compare with data and does not explain the observed emergence of quantum coherence in the BEC. Here we show that the cooperative action of the magnon gas through the four-magnon interaction provides the mechanism for the onset of quantum coherence in the BEC. The theory relies in part on some assumptions based on the experimental observations and on some approximations to allow an analytical treatment of the problem. They are ultimately justified by the good agreement between theory and experimental data.

We assume that after the hot magnon reservoir is formed by the redistribution of the primary magnons, the correlation between the phases of the magnon pairs lasts for a time that can be as large as $4/\eta_m$, which is about 100 ns in the experiments. This is sufficient time for the four-magnon interaction to come into play for establishing a cooperative phenomenon to drive a specific k mode. The effective driving

Hamiltonian for this process is obtained from (4.15) by taking averages of pairs of destruction operators of reservoir magnons with wave vectors $\vec{k}_R$ to form correlation functions so that one can write

$$H'(t) = \hbar \sum_{k_R} \frac{1}{2} S_{kk_R} n_{k_R} e^{i\varphi_{k_R}} e^{-i2\omega_{k_R} t} c_k^+ c_{-k}^+ + h.c. \tag{4.19}$$

Equation (4.19) has a form similar to the Hamiltonian (4.14) for parallel pumping, revealing that under appropriate conditions magnon pairs can be pumped out of equilibrium in the gas. Consider that the population of the primary magnons is distributed among the N_R modes k_R in the magnon reservoir, so that with (4.17) we can write an expression for the average population of the modes as a function of microwave power p,

$$n_R = r n_H \left[(p - p_{c1})/p_{c1}\right]^{1/2}, \tag{4.20}$$

where $r = r_p/N_R$. The number of magnons in each state k_R can be written approximately as $n_{k_R} = f_{\mathrm{BE}}(\omega_{kR}) n_R$, where $f_{\mathrm{BE}}(\omega) = n_{\mathrm{BE}}(\omega)\Delta\omega_R / \int n_{\mathrm{BE}}\, d\omega$ and $\Delta\omega_R = \omega_p/2 - \omega_{k0}$. So the relevant quantity for determining the frequency dependence of the coefficient in (4.19) is the spectral density $G(\omega) = D(\omega) f_{\mathrm{BE}}(\omega)$. This has a peak at ω_{k0} which becomes sharper as the chemical potential rises and approaches the minimum energy, as illustrated in Fig. 4.1(b). Thus as the microwave pumping power increases and $(\hbar\omega_{k0} - \mu)/k_B T$ becomes very small, the peak in $G(\omega)$ dominates the coefficient in (4.19). This establishes a cooperative action of the modes with ω_{k_R} close to ω_{k0} that can drive magnon pairs. Considering that the pumping is effective for frequencies ω_{k_R} in the range $\omega_{k0} \pm \eta_m$, one can write from (4.19) an effective Hamiltonian for driving k_0, $-k_0$ magnon pairs as,

$$H'_{\mathrm{eff}}(t) \cong \hbar (h\rho)_{\mathrm{eff}} e^{-i2\omega_{k0} t} c_{k_0}^+ c_{-k_0}^+ + H.c., \tag{4.21}$$

where $(h\rho)_{\mathrm{eff}} = -iG(\omega_{k0})\eta_m V_{(4)} n_R/2$ represents an effective driving field and the factor $-i$ arises from the phase between pairs that is approximately $-\pi/2$ in the range of power of [14–18]. From the analysis in Sect. 4.4, one can see that there is a critical number of reservoir modes above which they act cooperatively to drive the k_0, $-k_0$ magnons. Since (4.21) has the same form as (4.14), the population of the k_0 mode driven by the effective field and saturated by the four-magnon interaction is calculated in the same manner as for the direct parallel-pumping process. From (4.16) with $\Delta\omega_k = 0$, we have

$$n_{k0} = \left[\left|(h\rho)_{\mathrm{eff}}\right|^2 - \eta_m^2\right]^{1/2} / 2V_{(4)}. \tag{4.22}$$

With (4.17) and (4.22) one can write the population of the k_0 mode in terms of the microwave power p,

$$n_{k_0} = n_H \left[(p - p_{c2})/(p_{c2} - p_{c1})\right]^{1/2}, \tag{4.23}$$

where $p_{c2} = p_{c1}\{1 + 16/[r\eta_m G(\omega_{k0})]^2\}$ is the critical power for the formation of the BEC and is much larger than p_{c1}. Equation (4.23) reveals that for $p \geq p_{c2}$ the k_0 magnons are pumped-up out of equilibrium as a result of a spontaneous cooperative action of the reservoir modes and, as will be shown next, they are in coherent magnon states. Note that with (4.22) and (4.23) one can write the effective magnon-pair driving field as

$$(h\rho)_{\text{eff}} = -i\eta_m\big[(p - p_{c2})/(p_{c2} - p_{c1}\big]^{1/2}, \qquad (4.24)$$

which is a very convenient equation to express the driving field in terms of the microwave pumping power p, the critical power p_{c2} for the formation of the BEC, the critical power p_{c1} for parallel pumping, and the intermagnon relaxation rate η_m.

4.5.2 Coherence of the Magnon Condensate

In order to study the coherence properties of the k_0, $-k_0$ pair-mode pumped above threshold p_{c2}, we use methods of statistical mechanics appropriate for a boson system governed by the Hamiltonian (4.13), with the driving term of (4.21) and interacting with a heat-bath. From the equations of motion for the magnon operators in the representation of coherent states with eigenvalue $\alpha_k = a_k \exp(i\phi_k)$ one can show that the probability density $P(\alpha_k)$ in (4.6) obeys a Fokker–Plank equation that has a stationary solution [19, 56–59],

$$P(x) = C\exp\left(\frac{1}{2}Ax^2 - \frac{1}{6}x^6\right), \qquad (4.25)$$

where C is a normalization constant that makes the integral of the probability density $P(x)$ equal to unity, $x = (2/n_H^2\bar{n}_{k0})^{1/6}a_k$ represents a normalized magnon amplitude, and A is a dimensionless parameter proportional to the microwave power,

$$A = (2n_H/\bar{n}_{k0})^{2/3}(p - p_{c2})/(p_{c2} - p_{c1}). \qquad (4.26)$$

From (4.26) we see that for $p < p_{c2}$, the parameter A is negative. In this case, $P(x)$ given by (4.25) behaves as a Gaussian distribution, typical of systems in thermal equilibrium with incoherent magnon states, as shown in Fig. 4.2. On the other hand, for $p > p_{c2}$, $A > 0$ and $P(x)$ consists of two components, a coherent one convoluted with a much smaller fluctuation with Gaussian distribution. Since $P(x)$ has a variance proportional to A^{-1}, for $A \gg 1$ it becomes a delta-like function, as illustrated in Fig. 4.2, characteristic of coherent magnon states [19, 56–59] with average number of magnons given by $x_0^2 = A^{1/2}$. From (4.26) we see that this corresponds to a magnon number a_0^2 which is precisely the value n_{k_0} given by (4.23). This means that the magnons ω_{k0} driven cooperatively by the reservoir modes are in quantum coherent states, that have, from (4.7a), (4.7b) a small-signal magnetization

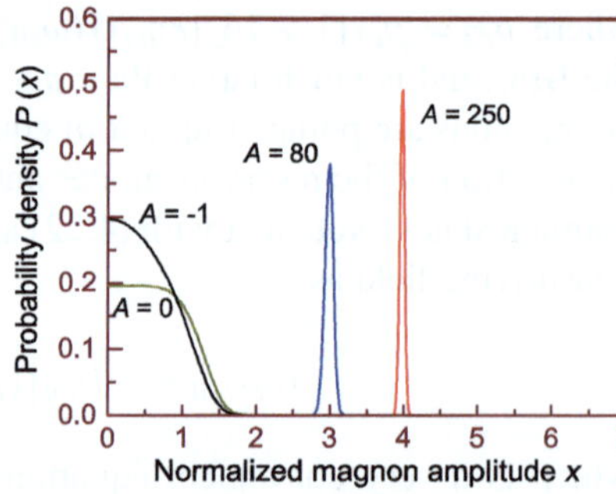

Fig. 4.2 Probability density characteristic of a microwave driven interacting magnon gas for several values of the parameter A. Negative values correspond to $p < p_{c2}$; $A = 0$ corresponds to $p = p_{c2}$; $A = 80$ and 250 correspond to $p/p_{c2} = 1.015$ and 1.047, respectively

$m^+ \propto (p - p_{c2})^{1/4}$. This important result reveals that m^+ is the order parameter of the BEC of the magnon gas.

The calculations presented so far are valid for magnon pairs with frequencies and wave vectors in the vicinity of ω_{k0} and k_0, $-k_0$. The dynamics for several modes can be treated approximately assuming that the condensate consists of p_{k0} modes all governed by the single-mode equations. A formal definition of this number will be given in Sect. 4.5.3. As will be shown later the experimental data are well fit by this model with $p_{k0} = 4.4 \times 10^3$, which is a very small number compared to the number of reservoir states $N_R \sim 10^9$ and thus characterizes a condensation in a few quantum states. With (4.23) we can express the number of magnons in the condensate in terms of the microwave power as

$$N_0 = p_{k0} n_H \big[(p - p_{c2})/(p_{c2} - p_{c1})\big]^{1/2}, \tag{4.27}$$

valid for $p \geq p_{c2}$. It can be shown [19] that as the microwave power increases above the critical value, the number of condensate magnons approaches the total number of particles while the number of uncondensed magnons vanishes. This is also a typical feature of a BEC.

A crucial test for the model theory is a quantitative comparison between data and the results of the model. In the experiments of [16], the coherence properties of the excited magnons states emerge clearly in the behavior of the intensity of the BLS peak at $f_{\min}$. As argued in [16], for incoherent scatterers the BLS intensity is proportional to their number, whereas for coherent scatterers it is proportional to the number squared. To compare theory with data we express the BLS intensity in terms of the microwave power p in two regimes, $p < p_{c2}$ and $p \geq p_{c2}$. Figure 4.3 shows a fit to data using $I^{\text{inc}} = c_1(p - p_{c1})^{1/2}$ and $I^{\text{coh}} = c_2(p - p_{c2})$, with $c_1 = 6.7$, $c_2 = 370.0$, and $p_{c2} = 2.8$ W. As shown in [19], the parameters are consistent with the values $r_p = 5 \times 10^2$ and $p_{k0} = 4.4 \times 10^3$ used to fit other experimental data. The good fit of theory to data demonstrates unequivocally that the spontaneous coherence develops above a critical microwave power level p_{c2}. To obtain a pictorial view of the condensate in k space, we assume that the small-signal transverse magnetization m^+ is described by a Gaussian function in frequency with a peak at $f_{\min}$ and linewidth 0.07 MHz [19]. Using the dispersion relation, we can express the frequency in terms of the wave vector components k_x and k_z to obtain the distribution of the BEC magnetization in k space shown in the inset of Fig. 4.3.

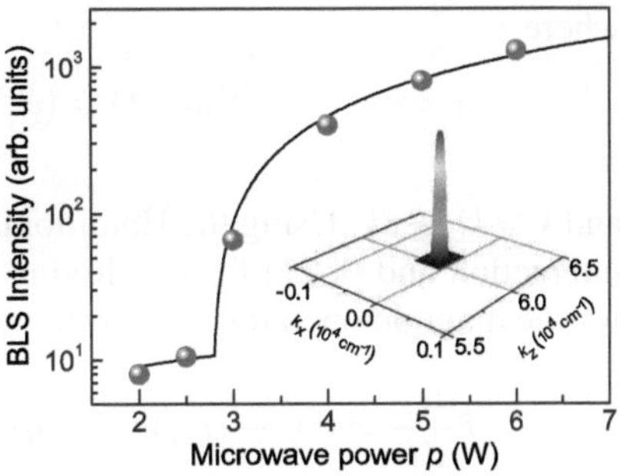

Fig. 4.3 Fit of the theoretical result (*line*) to the experimental data (*symbols*) of [16] for the BLS peak intensity at f_{min} as a function of microwave power. The inset shows a pictorial view of the condensate by the distribution in k space of the transverse magnetization

4.5.3 Wavefunction of the Magnon Condensate

Another important test for the theoretical model presented here for the dynamics of the magnon system in k space is to check its ability to provide an equation for the BEC wavefunction in configuration space that can be used to interpret measurements of the spatial distribution of the magnon condensate driven by a nonuniform microwave field [18]. In BEC systems, the condensate wavefunction is proportional to the order parameter [53, 54]. So we define the eigenfunction for the BEC of magnons as

$$\psi(\vec{r},t) = (1/V)^{1/2} \sum_k e^{i\vec{k}.\vec{r}} \alpha_k, \tag{4.28}$$

where α_k are the coherent state eigenvalues for k-states around k_0 and V is the volume of the sample. It is important to note that the eigenfunction (4.28) is a function of time and space and since $|\alpha_k| = \langle n_k \rangle^{1/2}$ it implies that

$$\int d^3 r \psi^* \psi = \sum_k n_k = N_0, \tag{4.29}$$

where N_0 is the number of magnons in the condensate given by (4.27). Consider that the wavefunction (4.28) corresponds to a narrow packet in wave vector space with a central wave number k_0 and with frequency around the minimum ω_{k0}, so that the frequency of the states involved can be written as

$$\omega_k \cong \omega_{k0} + \lambda_x k_x^2 + \lambda_z k_z^2, \tag{4.30}$$

where $\lambda_x = [\partial^2 \omega_k / \partial k_x^2]_{k0}/2$ and $\lambda_z = [\partial^2 \omega_k / \partial k_z^2]_{k0}/2$, x and z being the coordinates in the film plane. Since the central mode corresponds to a pair $\vec{k}_0$, $-\vec{k}_0$, one can introduce a slowly varying envelope $\psi_0(\vec{r},t)$ in a frame rotating with frequency ω_{k0} such that

$$\psi(\vec{r},t) = 2\cos(\vec{k}_0.\vec{r})\psi_0(\vec{r},t)e^{-i\omega_{k0}t}, \tag{4.31}$$

where

$$\psi_0(\vec{r},t) = \left(e^{i\omega_{k0}t}/V^{1/2}\right)\sum_{\delta k} e^{i\delta\vec{k}.\vec{r}}\alpha_k \tag{4.32}$$

and $\vec{k} = \vec{k}_0 + \delta\vec{k}$. Using the Hamiltonian for the system with a general four-magnon interaction and (4.21) for the driving term, one can write the Heisenberg equation for the magnon operator c_k as [60]

$$\begin{aligned} i\frac{dc_k}{dt} &= (\omega_k - i\eta_m)c_k - |h\rho|_{\text{eff}}e^{-i2\omega_{k0}t}c_{-k}^{+} \\ &\quad + V_{(4)}\sum_{2,3,4}\left(c_2^{+}c_3c_4 + c_1^{+}c_3c_4 + c_4^{+}c_1c_2 + c_3^{+}c_1c_2\right)\Delta(\vec{k}), \end{aligned} \tag{4.33}$$

where $V_{(4)} = 4T_{1234} = 4\omega_M/NS$, the summation runs over wave vectors $\vec{k}$, $-\vec{k}$, $\Delta(\vec{k})$ is the appropriate Kronecker delta for momentum conservation, and the relaxation rate η_m, assumed to be the same for the wave vectors involved, was introduced phenomenologically. If ones assumes that only one pair mode $\vec{k}_0$, $-\vec{k}_0$ is driven by the collective action, (4.33) and the corresponding one for the operator c_{-k0}^{+} can be solved in steady-state to give the population of the k_0-mode driven by the effective field and saturated by the four-magnon interaction. Considering that the four-magnon interaction process must conserve energy, one can do calculations similar to those for spin-wave solitons [61–64] using (4.30)–(4.32) and show [60] that (4.33) leads to

$$\begin{aligned} i\partial\psi_0/\partial t &= -i\eta_m\psi_0 - \lambda_x\partial^2\psi_0/\partial x^2 - \lambda_z\partial^2\psi_0/\partial z^2 \\ &\quad + 2V_{(4)}V|\psi_0|^2\psi_0 - |h\rho|_{\text{eff}}\psi_0^{*}. \end{aligned} \tag{4.34}$$

This equation has the form of the Gross–Pitaevskii equation (GPE) used to describe the wavefunction of other BEC systems [53, 54]. Without the last term, it is also the nonlinear Schrödinger equation used to study solitons [61–64]. Note that the usual GP equation has a term describing the non uniform external potential, such as the trapping potential in atomic gas systems [53, 54]. This is not present in (4.34) which has instead a non-uniform driving term $|h\rho|_{\text{eff}}\psi_0^{*}$ due to the spatial variation of the pumping microwave field. In order to check the consistency of the current approach with the results obtained from the analysis of the dynamics in wave vector space, let us obtain the solution of (4.34) in steady-state assuming a uniform driving field. With $d/dt = d/dx = d/dz = 0$ in (4.34) we obtain

$$(2V_{(4)})^2V|\psi_0|^2|\psi_0|^2 = |h\rho|_{\text{eff}}^2 - \eta_m^2, \tag{4.35}$$

which, of course, is valid only for $|h\rho|_{\text{eff}}^2 - \eta_m^2 \geq 0$, or $p \geq p_{c2}$. Integrating (4.35) in the volume V and using (4.24) and (4.29), one can see that the number of magnons in the condensate is the same as in (4.27) as long as the factor p_{k0} obeys the relation

$$\int d^3r|\psi_0|^2|\psi_0|^2 = N_0^2/\left(p_{k0}^2V\right). \tag{4.36}$$

Note that (4.36) provides a formal definition of the factor p_{k0} which was introduced earlier in an ad-hoc manner. We now assume that the pumping is uniform in the x direction to reduce the problem to one dimensional. This is the situation in the experiments of [18] where a thin wire or tape placed on top and close to the YIG film is used to generate the driving magnetic field. One can write the wavefunction in the same form as in other BEC systems [53, 54]

$$\psi_0 = (N_0)^{1/2}\chi(z), \tag{4.37}$$

where $\chi(z)$ is a normalized wavefunction independent of the number of magnons in the condensate. To find $\chi(z)$, we use the GPE (4.34) without the driving and relaxation terms and consider that their effect is expressed in the number of magnons in the condensate given by (4.27). With $d/dt = d/dx = 0$ in (4.34), we obtain

$$\lambda_z d^2\chi/dz^2 - b|\chi|^2\chi = 0, \tag{4.38}$$

where the coefficient b obtained with (4.27) considering $p_{c2} \gg p_{c1}$ becomes $b = \eta_m(p/p_{c2} - 1)^{1/2}$. Notice that there is a characteristic length associated with (4.38) given by $\xi = (\lambda_z/b)^{1/2}$, which is identified as the healing length [53, 54]. With parameter values appropriate for the YIG film used in the experiments of [14–18], $H = 1$ kOe, $4\pi M = 1.76$ kG, exchange stiffness $D = 4.5 \times 10^{-9}$ Oe cm^2, we obtain from the dispersion relation $\lambda_z = 0.29\ \text{s}^{-1}\,\text{cm}^2$. Assuming $p = 2p_{c2}$ and using $\eta_m = 5 \times 10^7\ \text{s}^{-1}$, we find for the healing length $\xi = 0.75$ μm. This value is very small compared to the typical length scale of the spatial variation of the driving field. Thus we can use (4.38) to calculate the spatial distribution of the condensate for a non-uniform pumping power $p(z)$. This has been done for a parallel pumping field produced by a microwave current in a thin wire close to a YIG film and the result is in good agreement with the experimental data [18].

In closing, we point out that the magnon condensation in a microwave driven YIG film provides a unique situation in which the condensate wavefunction is governed by a Gross–Pitaevskii equation that was derived from a microscopic treatment of the boson dynamics in wave vector space. This is not common to other BEC systems [53, 54]. We also note that other treatments of the Bose–Einstein condensation of magnons rely on phenomenological models adapted from other BEC systems which require the introduction of ad-hoc terms or parameters to connect them to the experimental situation [65].

4.6 Summary

We have reviewed the concept of coherent magnon states and presented a theoretical model for a microwave-driven interacting magnon gas in YIG films at room temperature that provides rigorous support for the formation of Bose–Einstein condensation of magnons in the experiments of Demokritov and co-workers [14–18]. The model describes the dynamics of the magnon gas in wave vector space and relies on the cooperative action of magnons with frequencies close to the minimum of the dispersion relation through the nonlinear four-magnon interactions. The theory provides the basic requirements for the characterization of a BEC, namely: (a) The onset of the BEC is characterized by a phase transition that takes place as the microwave power p is increased and exceeds a critical value p_{c2}; (b) The magnons in the condensate are in coherent states and as such they have non-zero small-signal transverse magnetization $m^+ \propto (p - p_{c2})^{1/4}$ that is the order parameter of the BEC. This result agrees with the arguments of [66] that coherence is a condition for the formation of a BEC; (c) For $p > p_{c2}$ the magnon system separates in two parts, one in thermal equilibrium with the reservoir and the other with $N_0 \propto (p - p_{c2})^{1/2}$ coherent magnons having frequencies and wave vectors in a very narrow region of phase space. As the microwave power increases above p_{c2}, N_0 approaches the total number of magnons pumped into the system characterizing unequivocally a Bose–Einstein condensation. The results of the model fit quite well the experimental Brillouin light scattering data; (d) The equations of motion for the magnon operators in k space lead to an equation for the spatial wavefunction of the magnon condensate which has the form of a Gross–Pitaevskii equation that describes the wavefunction in other BEC systems.

Acknowledgements The author thanks the Brazilian agencies CNPq, FINEP, CAPES, and FACEPE for supporting this work.

References

1. J.C. Slonczewski, J. Magn. Magn. Mater. **159**, L1 (1996)
2. J.C. Slonczewski, J. Magn. Magn. Mater. **195**, L261 (1999)
3. L. Berger, Phys. Rev. B **54**, 9353 (1996)
4. S.I. Kiselev, J.C. Sankey, I.N. Krivorotov, N.C. Emley, R.J. Schoelkopf, R.A. Buhrman, D.C. Ralph, Nature (London) **425**, 308 (2003)
5. S. Kaka, M.R. Pufall, W.H. Rippard, T.J. Silva, S.E. Russek, J.A. Katine, Nature (London) **437**, 389 (2005)
6. S.M. Rezende, F.M. de Aguiar, A. Azevedo, Phys. Rev. Lett. **94**, 037202 (2005)
7. S.M. Rezende, F.M. de Aguiar, A. Azevedo, Phys. Rev. B **73**, 094402 (2006)
8. A. Slavin, P. Kabos, IEEE Trans. Magn. **41**, 1264 (2005)
9. Y. Tserkovnyak, A. Brataas, G.E. Bauer, Phys. Rev. Lett. **88**, 117601 (2002)
10. Y. Tserkovnyak, A. Brataas, G.E. Bauer, Phys. Rev. B **66**, 224403 (2002)
11. A. Brataas, Y. Tserkovnyak, G.E.W. Bauer, B.I. Halperin, Phys. Rev. B **66**, 060404(R) (2002)
12. A. Azevedo, L.H. Vilela Leão, R.L. Rodriguez-Suarez, A.B. Oliveira, S.M. Rezende, J. Appl. Phys. **97**, 10C715 (2005)
13. E. Saitoh, M. Ueda, H. Miyajima, G. Tatara, Appl. Phys. Lett. **88**, 182509 (2006)
14. S.O. Demokritov, V.E. Demidov, O. Dzyapko, G.A. Melkov, A.A. Serga, B. Hillebrands, A.N. Slavin, Nature **443**, 430 (2006)
15. V.E. Demidov, O. Dzyapko, S.O. Demokritov, G.A. Melkov, A.N. Slavin, Phys. Rev. Lett. **99**, 037205 (2007)
16. V.E. Demidov, O. Dzyapko, S.O. Demokritov, G.A. Melkov, A.N. Slavin, Phys. Rev. Lett. **100**, 047205 (2008)
17. V.E. Demidov, O. Dzyapko, M. Buchmeier, T. Stockhoff, G. Schmitz, G.A. Melkov, S.O. Demokritov, Phys. Rev. Lett. **101**, 257201 (2008)

18. O. Dzyapko, V.E. Demidov, M. Buchmeier, T. Stockhoff, G. Schmitz, G.A. Melkov, S.O. Demokritov, Phys. Rev. B **80**, 060401(R) (2009)
19. S.M. Rezende, Phys. Rev. B **79**, 174411 (2009)
20. Y. Kajiwara, K. Harii, S. Takahashi, J. Ohe, K. Uchida, M. Mizuguchi, H. Umezawa, K. Kawai, K. Ando, K. Takanashi, S. Maekawa, E. Saitoh, Nature (London) **464**, 262 (2010)
21. Z. Wang, Y. Sun, M. Wu, V. Tiberkevich, A. Slavin, Phys. Rev. Lett. **107**, 146602 (2011)
22. E. Padrón-Hernández, A. Azevedo, S.M. Rezende, Phys. Rev. Lett. **107**, 197203 (2011)
23. E. Padrón-Hernández, A. Azevedo, S.M. Rezende, Appl. Phys. Lett. **99**, 192511 (2011)
24. V.V. Kruglyak, S.O. Demokritov, D. Grundler, J. Phys. D, Appl. Phys. **43**, 264001 (2010)
25. A.A. Serga, A.V. Chumak, B. Hillebrands, J. Phys. D, Appl. Phys. **43**, 264002 (2010)
26. F. Bloch, Z. Phys. **61**, 206 (1930)
27. G. Heller, H.A. Kramers, Proc. R. Acad. Sci. (Amsterdam) **37**, 378 (1934)
28. L. Landau, E. Lifshitz, Phys. Z. Sowjetunion **8**, 153 (1935)
29. T. Holstein, H. Primakoff, Phys. Rev. **58**, 1098 (1940)
30. J. Smit, H.P.J. Wijn, *Ferrites* (Wiley, New York, 1959)
31. B. Lax, K.J. Button, *Microwave Ferrites and Ferrimagnetics* (McGraw-Hill Books, New York, 1962)
32. N. Bloembergen, R.W. Damon, Phys. Rev. **85**, 699 (1952)
33. N. Bloembergen, S. Wang, Phys. Rev. **93**, 72 (1954)
34. H. Suhl, J. Phys. Chem. Solids **1**, 209 (1957)
35. F.R. Morgenthaler, J. Appl. Phys. **31**, 1647 (1960)
36. E. Schloemann, J.J. Green, U. Milano, J. Appl. Phys. **31**, 386S (1960)
37. J.R. Eshbach, J. Appl. Phys. **34**, 1298 (1963)
38. W. Strauss, J. Appl. Phys. **36**, 118 (1965)
39. S.M. Rezende, F.R. Morgenthaler, Appl. Phys. Lett. **10**, 184 (1967)
40. S.M. Rezende, F.R. Morgenthaler, J. Appl. Phys. **40**, 524 (1969)
41. M.G. Cottam (ed.), *Linear and Nonlinear Spin Waves in Magnetic Films and Superlattices* (World Scientific, Singapore, 1994)
42. P. Kabos, V.S. Stalmachov, *Magnetostatic Waves and Their Applications* (Chapman and Hall, London, 1994)
43. A.G. Gurevich, G.A. Melkov, *Magnetization Oscillations and Waves* (CRC Press, Boca Raton, 1996)
44. D.D. Stancil, A. Prabhakar, *Spin Waves – Theory and Applications* (Springer, New York, 2009)
45. A.I. Akhiezer, V.G. Bar'yakhtar, S.V. Peletminskii, *Spin Waves* (North-Holland, Amsterdam, 1968)
46. R.M. White, *Quantum Theory of Magnetism*, 3rd edn. (Springer, Berlin, 2007)
47. S.M. Rezende, N. Zagury, Phys. Lett. A **29**, 47 (1969)
48. N. Zagury, S.M. Rezende, Phys. Rev. B **4**, 201 (1971)
49. R.J. Glauber, Phys. Rev. **131**, 2766 (1963)
50. S.M. Rezende, F.M. de Aguiar, Proc. IEEE **78**, 893 (1990)
51. V.E. Zakharov, V.S. L'vov, S.S. Starobinets, Usp. Fiz. Nauk **114**, 609 (1974)
52. V.E. Zakharov, V.S. L'vov, S.S. Starobinets, Sov. Phys. Usp. **17**, 896 (1975)
53. L. Pitaevskii, S. Stringari, *Bose-Einstein Condensation* (Clarendon, Oxford, 2003)
54. A.J. Leggett, *Quantum Liquids: Bose Condensation and Cooper Pairing in Condensed-Matter Systems* (Oxford University Press, Oxford, 2006)
55. G.A. Melkov, V.L. Safonov, A.Y. Taranenko, S.V. Sholom, J. Magn. Magn. Mater. **132**, 180 (1994)
56. C.B. de Araújo, Phys. Rev. B **10**, 3961 (1974)
57. C.B. de Araújo, S.M. Rezende, Phys. Rev. B **9**, 3074 (1974)
58. H. Haken, Rev. Mod. Phys. **47**, 67 (1975)
59. P. Meystre, M. Sargent III, *Elements of Quantum Optics* (Springer, Berlin, 1992)
60. S.M. Rezende, Phys. Rev. B **81**, 020414(R) (2010)

61. A.N. Slavin, B.A. Kalinikos, N.G. Kovshikov, in *Nonlinear Phenomena and Chaos in Magnetic Materials*, ed. by P.E. Wigen (World Scientific, Singapore, 1994). Chap. 9
62. A.K. Zvezdin, A.F. Popkov, Zh. Èksp. Teor. Fiz. **84**, 606 (1983)
63. A.K. Zvezdin, A.F. Popkov, Sov. Phys. JETP **57**, 350 (1983)
64. V.I. Karpman, *Nonlinear Waves in Dispersive Media* (Pergamon, New York, 1973)
65. B.A. Malomed, O. Dzyapko, V.E. Demidov, S.O. Demokritov, Phys. Rev. B **81**, 024418 (2010)
66. D. Snoke, Nature (London) **443**, 403 (2006)

Part II
Probing and Manipulation of Magnons with Femtosecond Light and Polarized Electrons: Experiment and Simulations

Chapter 5
The Role of Angular Momentum in Ultrafast Magnetization Dynamics

Andrei Kirilyuk, Alexey V. Kimel, and Theo Rasing

Abstract Many aspects of magnetization dynamics are related to the fact that a certain amount of angular momentum is associated with magnetic moments. Here the dynamics of angular momentum is considered in ferrimagnetic rare-earth–transition-metal alloys that consist of two antiferromagnetically coupled sublattices, where both magnetization and angular momenta are temperature dependent. For certain compositions, such ferrimagnets can exhibit a magnetization compensation temperature T_M where the magnetizations of the sublattices cancel each other, and similarly, an angular momentum compensation temperature T_A where the net angular momentum vanishes. At the latter point, the frequency of the homogeneous spin precession diverges. As a consequence, ultrafast heating of a ferrimagnet across its compensation points may result in subpicosecond magnetization reversal. Moreover, the magnetization can be manipulated and even reversed by a single 40 femtosecond laser pulse, without any applied magnetic field. This optically induced ultrafast magnetization reversal is the combined result of laser heating of the magnetic system and the exchange interaction between the sublattices. This novel reversal pathway is shown to crucially depend on the net angular momentum, reflecting the balance of the two oppositely magnetized sublattices.

5.1 Introduction

The role of the angular momentum in magnetism has intrigued the magnetic community since 1915, when Einstein and de Haas [1] and Barnett [2] published their experimental observations of the interconnection between the magnetization of a

A. Kirilyuk (✉) · A.V. Kimel · T. Rasing
Institute for Molecules and Materials, Radboud University Nijmegen, Heyendaalseweg 135, 6525 AJ Nijmegen, The Netherlands
e-mail: A.Kirilyuk@science.ru.nl

A.V. Kimel
e-mail: A.Kimel@science.ru.nl

T. Rasing
e-mail: Th.Rasing@science.ru.nl

S.O. Demokritov, A.N. Slavin (eds.), *Magnonics*, Topics in Applied Physics 125,
DOI 10.1007/978-3-642-30247-3_5,

macroscopic body and its rotation. Therefore, as derived by Landau and Lifshitz in 1935 [3], the dynamics of the magnetic moments is described by a simple mechanical relation between the change of angular momentum and the torque applied to the system. The law of conservation of the angular momentum is therefore an important issue to be considered when sudden changes of the magnetization are treated. In particular, the laser-induced ultrafast demagnetization of ferromagnets, demonstrated to happen at a femtosecond time scale already in 1996 [4], is still a subject of hot debates [5]. Whether the angular momentum is dissipated into the lattice via phonons and defects [6], or whether it is carried away by hot electrons [7] or the photons [8] – such questions are still on the forefront of ultrafast magnetism.

The dynamics of magnetic moments becomes even more intriguing in multisublattice magnets where an important role is played by the exchange of angular momentum between the sublattices [9]. In such systems, the time scale of the magnetization dynamics becomes dependent on the balance of the angular momentum between the sublattices [10], particularly when these are antiferromagnetically coupled. In pure antiferromagnets, moreover, the dynamic behavior changes its character completely, demonstrating Newton-like dynamics with acceleration and inertia [11].

The goal of this chapter is to discuss the recent progress in our understanding of the magnetization dynamics in multisublattice magnets. In particular, we consider a ferrimagnet rare-earth–transition-metal alloy GdFeCo, where the Gd and FeCo sublattices are antiferromagnetically coupled. We show that the compensation of the angular momentum between the sublattices leads to a considerable acceleration of the dynamics, suggesting an ultrafast route to magnetic switching. Such a switching, indeed, can be demonstrated to happen within a picosecond, which is far shorter than in any of the ferromagnetic samples. The understanding of the exact mechanism of such switching, however, is very challenging. We have shown that the actual switching is preceded by an intermediate, highly-nonequilibrium ferromagnetic state, which is driven by a balance of the angular momentum in the system at the subpicosecond time scale.

5.2 Precession in Ferrimagnetic Materials

Systems where both magnetic moment and angular momentum can be continuously tuned, for example, by changing temperature, form a unique playground to study the role of the angular momentum in laser-induced coherent spin precession and optical control of magnetism.

One set of examples of such a system, rare-earth–$3d$-transition-metal (RE-TM) ferrimagnetic compounds, were once widely used materials for magneto-optical (MO) recording. Depending on their composition, RE-TM ferrimagnets can exhibit a magnetization compensation temperature T_M, where the magnetizations of the RE and TM sublattices cancel each other, and similarly, an angular momentum compensation temperature T_A, where the net angular momentum of the sublattices vanishes. The theory of ferrimagnetic resonance [12] predicts a strong temperature dependence of the dynamic behavior in such systems. In particular, the frequency of the homogeneous spin precession and the Gilbert damping parameter α are expected to diverge at the temperature T_A [13]. Such a combination of a high frequency and large damping of the spin precession would provide a means for ultrafast and ringing-free magnetization reversal via precessional motion [14].

In a ferrimagnetic system, the Landau–Lifshitz–Gilbert equation should be written for each ith sublattice ($i = \mathrm{RE}, \mathrm{TM}$). These equations are coupled by the presence of the exchange field $\mathbf{H}^{\mathrm{ex}}_{\mathrm{RE,TM}} = -\lambda_{\mathrm{ex}}\mathbf{M}_{\mathrm{TM,RE}}$ and give rise to two modes. Here λ_{ex} is a parameter that characterizes the strength of the exchange field. The ferromagnetic mode with $\omega_{\mathrm{FMR}} = \gamma_{\mathrm{eff}}\mathbf{H}^{\mathrm{eff}}$ can actually be described by a single LLG equation, but now employing an effective gyromagnetic ratio γ_{eff} [12, 15]:

$$\gamma_{\mathrm{eff}}(T) = \frac{M_{\mathrm{RE}}(T) - M_{\mathrm{TM}}(T)}{\frac{M_{\mathrm{RE}}(T)}{|\gamma_{\mathrm{RE}}|} - \frac{M_{\mathrm{TM}}(T)}{|\gamma_{\mathrm{TM}}|}} = \frac{M(T)}{A(T)} \tag{5.1}$$

and an effective Gilbert damping parameter α_{eff} [16]

$$\alpha_{\mathrm{eff}}(T) = \frac{\frac{\lambda_{\mathrm{RE}}}{|\gamma_{\mathrm{RE}}|^2} + \frac{\lambda_{\mathrm{TM}}}{|\gamma_{\mathrm{TM}}|^2}}{\frac{M_{\mathrm{RE}}(T)}{|\gamma_{\mathrm{RE}}|} - \frac{M_{\mathrm{TM}}(T)}{|\gamma_{\mathrm{TM}}|}} = \frac{A_0}{A(T)}, \tag{5.2}$$

where $M(T)$ and $A(T)$ are the net magnetic moment and net angular momentum, respectively; λ_{TM} and λ_{RE} are the Landau–Lifshitz damping parameters for the rare-earth and transition metal sublattices, respectively; A_0 is a constant under the assumption that the Landau–Lifshitz damping parameters are independent of temperature [13]. The validity of this assumption was confirmed over a wide temperature interval by FMR measurements in 3d-TM [17].

In addition to the ferromagnetic mode ω_{FMR}, spins in a ferrimagnetic system may oscillate with the exchange resonance frequency [18]

$$\omega_{\mathrm{ex}} = \lambda_{\mathrm{ex}}\big(|\gamma_{\mathrm{TM}}|M_{\mathrm{RE}} - |\gamma_{\mathrm{RE}}|M_{\mathrm{TM}}\big) = \lambda_{\mathrm{ex}}|\gamma_{\mathrm{RE}}||\gamma_{\mathrm{TM}}|A(T). \tag{5.3}$$

Equations (5.1) and (5.2) indicate a divergence of both the precession frequency and the Gilbert damping parameter of the FMR mode at the temperature T_A. Moreover, from (5.1), one notices that at the temperature T_M, the FMR frequency becomes zero. In contrast, (5.3) indicates that the exchange resonance branch softens at the angular momentum compensation temperature T_A [19], where the FMR mode diverges.

Laser-induced spin precession in the ferrimagnetic GdFeCo alloy was investigated using an all-optical pump-and-probe method [10]. The sample used in these experiments was a 20 nm thick GdFeCo layer, in a form of an amorphous alloy, having perpendicular magnetic anisotropy.

Figure 5.1 shows the temperature dependence of the magnetization precession frequency (a) and the Gilbert damping parameter (b). At $T = 220$ K, a significant increase is observed in both the precession frequency and the damping parameter. As expected from (5.1) and (5.2), the fact that both α_{eff} and ω_{FMR} peak at the same temperature clearly indicates the existence of angular momentum compensation

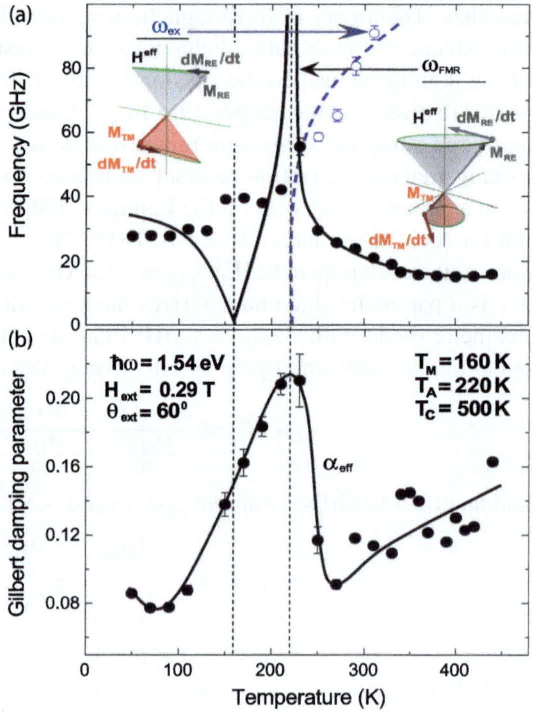

Fig. 5.1 Temperature dependence of: (**a**) the magnetization precession frequencies ω_{FMR} and ω_{ex} and (**b**) the Gilbert damping parameter α_{eff}. As temperature decreases from 310 K towards T_A, the exchange resonance mode ω_{ex} (*open circles*) softens and mixes with the ordinary FMR resonance ω_{FMR} (*closed circles*). The *solid lines* are a qualitative representation of the expected trend of the two resonance branches as indicated by (5.1) and (5.3). *Lines* are guides to the eye [10]

near this temperature of 220 K. The strong temperature dependence of γ_{eff} demonstrates the nonequivalent character of the gyromagnetic ratios of the two magnetic sublattices in GdFeCo. This non-equivalence also leads to the difference between the point of magnetization compensation T_M and angular momentum compensation T_A. In addition to the peak near T_A, an enhancement of α_{eff} has also been observed as the temperature is increased towards T_C. Again (5.2) predicts this enhancement under the assumption of temperature independent Landau–Lifshitz damping parameters λ_{RE} and λ_{TM}. This enhancement is consistent with earlier data [20, 21]. The observed strong increase of the precession frequency and the Gilbert damping when the temperature approaches T_A, is ideal for ultrafast ringing-free precessional switching in magnetic and magneto-optical recording.

Interestingly, in the temperature region just above T_A, Fourier analysis of the measured time dependencies reveals two frequencies, one decreasing and the other increasing with temperature. While the former one can be attributed to the FMR mode, such temperature behavior allows identifying the latter one as the exchange mode (see (5.3)), the frequency of which can be low around T_A but is usually very high everywhere else. Later, the coexistence of these two modes was studied in detail in a $Gd_{1-x}Co_x$ ferrimagnetic thin film on both sides of its angular momentum compensation composition [22]. It was shown that these two modes could be selectively excited by adjusting the pump-laser fluence, that is, the temperature of the spin reservoir in the pumped area of the sample.

5.3 Laser-Induced Magnetization Reversal

Magnetic switching, or full magnetization reversal, at a picosecond time scale is a challenge sought in many experiments. Technologically, laser-induced heating of a magnetic medium in the presence of an external field was used in thermomagnetic writing [23] to record a bit of information via magnetization reversal. At present, a modification of such a scheme is planned to be used in the heat-assisted magnetic recording (HAMR) [24]. Therefore, revealing the ultimate speed at which such recording event can be realized is a fundamentally interesting issue with possible consequences for the future development of magnetic recording and information processing. The acceleration of the precessional dynamics in GdFeCo alloys [10] discussed above, indicates a challenging possibility to use an all-optical scheme to achieve magnetic switching on a very short time scale [25].

Indeed, if in a ferrimagnet close to the compensation point the frequency of the magnetization precession diverges, one may expect that in such a ferrimagnetic material an application of a magnetic field would instantaneously flip the magnetization. In other words, at T_A, magnetization can be regarded as a mechanical system with no inertia which can be moved by the slightest torque. However, to verify this ultrafast switching, instantaneous application of a magnetic field is required, which in real experiments is not feasible. Instead, a DC magnetic field can be applied to the ferrimagnet parallel to the original magnetization direction, at a temperature $T < T_M$. When the temperature of the ferrimagnet increases above both T_M and T_A, its magnetization will reverse [26, 27]. Thus a femtosecond (fs) laser pulse heating of the sample might effectively act as an instantaneously applied magnetic field, allowing the investigation of the magnetization reversal speed near T_A.

The question arises: How fast is this switching process? To answer this question, hysteresis loops were measured at different time delays after the pump pulses heated the sample [25]. The results are shown in Fig. 5.2. One can clearly observe that the measured Faraday signal changes sign after about 700 fs. The sign change reflects the change of the FeCo sublattice direction towards the applied magnetic field, as the spin temperature of the ferrimagnetic system increases over T_M in the probed area. This observation unambiguously demonstrates that the magnetization reversal takes place on a sub-picosecond time scale. Note that this reversal time is considerably faster than that found in GdFeCo at temperatures above the compensation points [28]. On the other hand, the growth of the reversed domain to its full 100 % is determined by the cooling rate and takes place at a much longer time scale (~ns).

What are the implications of the observed sub-picosecond magnetization reversal? Although it is well known that ultrafast laser excitation of itinerant ferromagnets as Co, Ni, or Fe leads to a demagnetization on the femtosecond time scale, little is known about how fast the magnetic moments in metals as Gd can be excited with photon energies in the visible range. This is because in Gd, the optically excitable electrons of the $5d$–$6s$ bands carry only ≈9 % of the total moment while the localized $4f$ electrons dominate the magnetic spin moment. The localization of the magnetic moment together with the weak spin–lattice coupling that can be deduced from the small value of the magnetocrystalline anisotropy characteristic for Gd suggests that the transfer of the photon energy to the localized states should be a slow

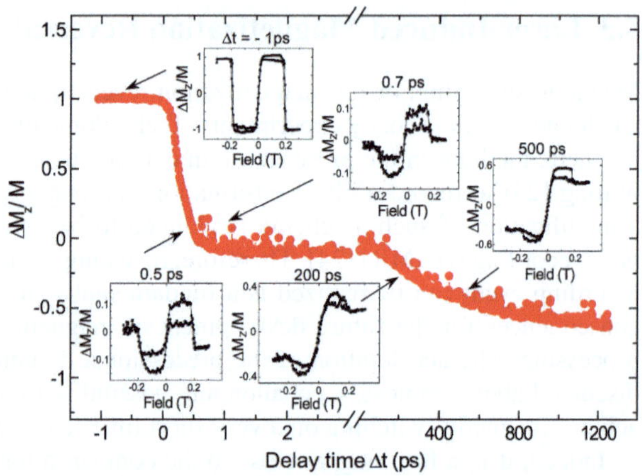

Fig. 5.2 Transient magnetization reversal dynamics measured for a pump fluence of 6.29 mJ/cm^2. *Insets* show hysteresis loops measured at distinct pump–probe delays. The loops demonstrate the magnetization reversal after about 700 fs [25]

process of the order of 100 ps [29]. In this context, it was recently claimed that laser excitation of a CoGd sample resulted in an independent excitation of the Co sublattice only [30]. In contrast to this, the sub-picosecond magnetization reversal over the compensation points demonstrated here, implies that both the Gd and FeCo sublattice magnetizations are considerably reduced on a sub-picosecond time scale. For Gd this represents a reduction of the magnetic moment larger than that contributed by the itinerant $5d6s$ electrons. A sub-picosecond access to the localized $4f$ spin moments is therefore required, thus revealing the important role played by the Gd $4f$ electrons in this fast reversal process. Such excitation time scale is indeed allowed in Gd by the strong exchange coupling between $5d$–$4f$, responsible for its ferromagnetic order [31]. The strength of this intra-atomic exchange is $\approx$100 meV, corresponding to $\approx$50 fs. From this it seems that indeed it should be possible to excite optically the localized $4f$ spin magnetic moment in Gd on a time scale comparable with that observed for the itinerant ferromagnets. The question is, however, whether the Gd sublattice really follows the FeCo one, or maybe a certain delay exists?

5.4 Transient Ferromagnetic State

As shown in the previous section, the whole process of demagnetization and subsequent switching of a two-sublattice ferrimagnet happens on a sub-picosecond time scale, i.e., *before* the system approaches thermodynamic equilibrium. To study this transient regime of spin dynamics where novel coupling phenomena can emerge, generally one faces two challenges: (i) probing the response of one magnetic sublattice relative to another; and (ii) bringing the moments out of equilibrium on a timescale corresponding to the exchange interaction. As follows from the consideration above, ferrimagnetic materials are ideal candidates to address both these problems. First, in contrast to elementary ferromagnets, such as Fe, Ni, or Co, where all the spins are equivalent, ferrimagnets consist of two (or more) non-equivalent and antiferromagnetically coupled spin sublattices. This non-equivalence of the sublattices can be combined with an element-specific detection technique like X-ray magnetic circular dichroism (XMCD) [32, 33]. This will allow us to "color" spins in the magnet and to probe the response of one sublattice relative to another. Second, as shown above, ultrafast heating of a ferrimagnet over its compensation point in an external magnetic field allows the initiation of the fastest spin reversal reported so far.

Because the heating of the ferrimagnetic GdFeCo alloy with femtosecond laser pulses triggers the reversal of the FeCo sublattice within 700 fs, as seen in a magneto-optical scheme, looking at the Gd sublattice during ultrafast spin reversal of its neighbor Fe would thus be a promising approach to generate and study a state with possible decoupled behavior and different dynamics of the two sublattices. For simplicity, in the following we focus on the behavior of the Gd and Fe moments and ignore the small percentage of Co in the transition metal sublattice. In order to probe the magnetizations of the two sublattices independently, we used the element-specific XMCD technique. The XMCD measurements were performed in transmission for a fixed X-ray light helicity and opposite orientations of the external magnetic field of 0.5 T. The photon energy of the X-ray light was tuned at the Fe L_3 and at the Gd M_5 absorption edges in order to measure their magnetic response separately. The results of static XMCD measurements as a function of magnetic field measured below and above T_M reveal the respective sign change of the Fe and Gd hysteresis loops on heating the alloy above T_M. These data thus confirm that XMCD serves as an element-specific probe of spins in this rare earth-transition metal alloy.

In order to trigger ultrafast spin dynamics in this alloy, the reversal of the magnetizations of the two sublattices is initiated by ultrafast heating of the sample using a linearly polarized 60-fs laser pulse with photon energy of 1.55 eV. The dynamics of the Fe and Gd magnetic moments were independently probed using 100-fs soft X-ray pulses [34]. Figure 5.3 shows the results of time-resolved measurements of the dynamics of the Fe and Gd sublattice magnetizations. The measurements were performed with an incident laser fluence of 4.4 mJ/cm^2 and a sample temperature of 83 K, which is well below $T_M = 250$ K.

First, the net magnetizations of both sublattices rapidly decrease. However, whereas the net magnetization of Fe has collapsed within 300 fs, the demagnetization of Gd takes as long as 1.5 ps. Remarkably, in spite of the exchange coupling between the rare earth and transition metal sublattices, they apparently lose their net magnetizations independently, as would be the case for two decoupled transition metal and rare earth ferromagnets [6]. To retrieve the time constants of the involved processes, we have used a double exponential fit function convoluted with the time resolution of the experiment of 100 fs, depicted by the Gaussian curve in Fig. 5.3. We find a time constant $t_{Fe} = 100 \pm 25$ fs for the Fe and $t_{Gd} = 430 \pm 100$ fs for the Gd magnetic moment characterizing the fast initial drop in the transient dichroic signal.

Second, the magnetizations of both sublattices switch their directions by crossing the zero signal level and rebuilding their net magnetic moments. Up to 10 ps,

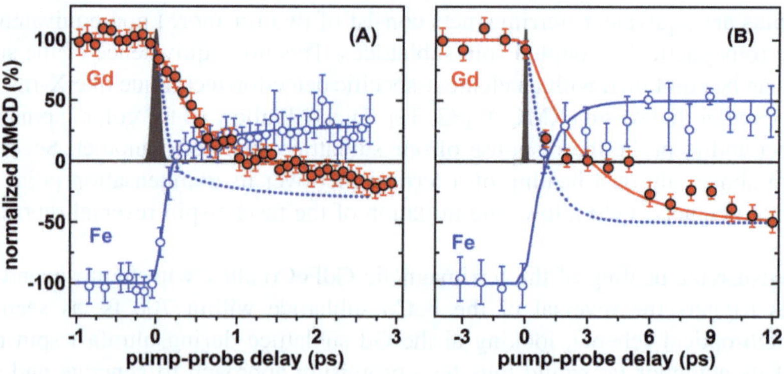

Fig. 5.3 Divergent dynamics of the Fe and Gd magnetic moments measured by time-resolved XMCD with fs time-resolution. (**a**) Transient dynamics of the Fe (*open circles*) and Gd (*solid circles*) magnetic moments measured within the first 3 ps. (**b**) The same as in (**a**) but on a 12 ps time scale. *Error bars* of the experimental data represent the statistical standard error. The measurements have been performed at a sample temperature of 83 K for an incident laser fluence of 4.4 mJ/cm^2. Experimental time resolution of 100 fs is depicted by the solid Gaussian profile. The *solid and dashed lines* are guides to the eye. The *dashed line* in both panels depicts the magnetization of Fe-sublattice taken with the opposite sign [9]

the Gd- and Fe-sublattices show distinctly different switching dynamics. Even more surprisingly, within the timescale between the zero crossings of the Fe and Gd moments (that is, between 300 fs and 1.5 ps), the net Fe and Gd moments are parallel aligned along the z axis despite antiferromagnetic coupling of the spins in the ground state. Note that the net Fe and Gd moments in the transient ferromagnet-like state are reaching rather large values, up to 25 % of the equilibrium magnetization. This, together with the substantial laser-induced increase in temperature, indicates a rather strong transient parallel alignment of the Fe and Gd moments. Thus we have entered an unexplored regime of magnetization dynamics, where two exchange-coupled magnetic sublattices are not in equilibrium with each other. One of the consequences, for example, is that the concept of a magnon (a quantized collective excitation of spins) in such a non-equilibrium ferrimagnetic state should be re-examined, as this state precedes the establishment of the collective Gd–Fe excitations and magnetic resonances in this material.

What is the reason of such behavior? Note that also for other fluences, different switching times of the two sublattices, and thus the ferromagnetic intermediate state, are observed. For example, slightly decreasing the fluence from 4.4 to 4 mJ/cm^2 leads to longer but still distinct switching times for Fe and Gd. This results in a longer lived transient ferromagnetic-like state, which spans the time range from 700 fs to about 4 ps for this lower fluence.

To obtain a better understanding of the origin of such strongly nonequilibrium spin dynamics, we have developed a model of a ferrimagnet comprising 10^6 localized atomistic exchange-coupled spins and performed numerical simulations of the dynamics after ultrashort laser excitation. The model takes into account realistic magnetic moments for Gd and Fe as well as the exchange constants extracted from

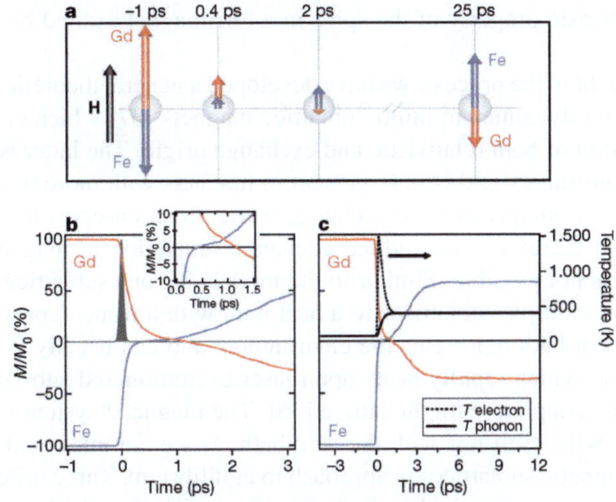

Fig. 5.4 Computed time-resolved dynamics of the Fe and Gd magnetic moments from localized atomistic spin model. (**a**) Cartoon-like illustration of the non-equilibrium dynamics of the Fe and Gd magnetizations with respect to an external magnetic field H. The lengths of the arrows are scaled to the magnitude of the transient XMCD signals shown in Fig. 5.3. Simulated dynamics for the first 3 ps (**b**) and the first 12 ps (**c**) after laser excitation. The calculations were performed for a peak electronic temperature of 1500 K with the corresponding transient electronic and phononic temperatures shown in (**c**). The transient magnetization changes are normalized to magnetization values at negative delays, that is, to equilibrium values [9]

experimentally measured temperature dependencies of the sublattice magnetizations. In particular, we used an effective Fe–Gd exchange constant of 4.77×10^{-21} J per spin (∼140 fs). Following [35], we couple the spin system to the electron temperature, which is calculated using the so-called two-temperature model [36].

Figure 5.4 shows the results of the simulations when applying a heat pulse with a maximum electron temperature of about 1500 K. First, the time evolution of the sublattice magnetization qualitatively reproduces the timescales for the demagnetization of each sublattice. The time required for the disappearance of the net magnetizations of the sublattices is found to be proportional to the ratio m/λ, where m and λ are respectively the magnetic moment of the ion and the damping constant of the sublattice. Second, the ferromagnetic-like state is also reproduced, in a time window close to that observed experimentally. In addition, the reversal of the sublattice magnetizations is found to occur via the mechanism of linear reversal [n13], that is, no transverse moment is observed. Note that even making the Gd–Gd exchange interaction in the simulations as strong as the Fe–Fe interaction does not lead to qualitative changes of the dynamics of the Fe and Gd sublattices. Moreover, for the sake of argument, the model ignores any differences in the electron temperature and the damping constants for Gd and Fe ($\lambda_{Fe} = \lambda_{Gd} = 0.05$). However, even after such simplifications, a good agreement between the experiment and the simulations is observed. This is a strong indication that the novel magnetization dynamics reported here are an intrinsic property of the spins in a ferrimagnet excited by an ultrashort stimulus.

To gain insight in the process, we have developed a general theoretical framework for ultrafast spin dynamics in multi-sublattice magnets [37] which contains longitudinal relaxation of both relativistic and exchange origin. The latter is the key new ingredient of our theory and is only present in magnets with more than one sublattice. The reason is simply that the exchange interaction conserves the total angular momentum and therefore longitudinal exchange relaxation in magnets with only one sublattice is not possible. Similar to the models for one sublattice [35, 38, 39], we coupled the multiple sublattices to a heat bath with a (time dependent) temperature T. The heat bath represents the environment and can usually be taken as the electron system, which rapidly heats upon laser excitation and subsequently cools down due to the coupling with the lattice [36]. The magnetic system is thus neither in equilibrium with itself nor with the heat bath, as we are interested in the relaxation of the magnetic sublattices in approach to equilibrium. Three different regimes were found, depending on whether the temperature of the heat bath is above, below, or in the vicinity of the critical temperature T_C of the multisublattice magnet. Therefore, immediately after the excitation, in the temperature dominated regime $T > T_C$, the sublattices behave independently, relaxing their angular momentum into the lattice. In contrast, at longer time scales, when $T < T_C$, the thermal effects could be neglected, the exchange coupling dominates the dynamics, and the total angular momentum of the two sublattices is conserved.

Both the experiments and the simulations show that the ferromagnetic-like state emerges owing to the substantially different demagnetization rates of the transition metal and rare earth sublattices – in particular, the fact that the transition metal reaches the state with zero net magnetization much faster than the rare earth component. Following this, the dynamics of the Fe spins occur on the background of a reducing magnetization of Gd. The exchange interaction between the Gd and Fe can flip an Fe spin when a Gd spin is reversed. As the exchange is antiferromagnetic, such a process will promote growth of the net Fe magnetization in the direction parallel to the remaining magnetization of Gd. The simulations indeed confirm that the growth of the Fe magnetization in this time domain occurs with a speed similar (only slightly lower) to that at which the Gd magnetization decreases (Fig. 5.4(b), inset).

5.5 Conclusions

To conclude, here we have discussed several observations in the laser-induced dynamics of ferrimagnetically ordered materials, related to the angular momentum associated with the magnetic moments. Thus, the reduction of the integrated angular momentum leads to a considerable acceleration of the magnetic precession, which can also be used for a sub-picosecond switching of the magnetic order. Moreover, our findings unexpectedly show that two magnetic sublattices may have totally different spin dynamics even on a timescale much longer than the characteristic time of the exchange interaction between them. In the newly discovered transient ferromagnetic-like state, the magnetic system stores a large amount of energy in the intra- and inter-sublattice exchange interactions. Thus it is reasonable to hypothesize that the much weaker interaction of the spins with an external magnetic field hardly influences the spin dynamics in this ultrafast regime, leading to novel switching scenarios in the absence of any magnetic field. The discovery of these dynamics is important for both the understanding of the physics of ultrafast magnetic phenomena on the timescale of the exchange interaction, and for the establishment of the fundamental limits on the speed of magnetic recording and information processing.

Acknowledgements This research has received funding from Nederlandse Organisatie voor Wetenschappelijk Onderzoek (NWO), Stichting voor Fundamenteel Onderzoek der Materie (FOM), the Dutch Nanotechnology initiative NanoNed, and EC FP7 contributions under grants NMP3-SL-2008-214469 (UltraMagnetron) and N 214810 (FANTOMAS).

References

1. A. Einstein, W. de Haas, Verh. Dtsch. Phys. Ges. **17**, 152 (1915)
2. S.J. Barnett, Phys. Rev. **6**, 239 (1915)
3. L. Landau, E. Lifshitz, Phys. Z. Union **8**, 153 (1935)
4. E. Beaurepaire, J.C. Merle, A. Daunois, J.Y. Bigot, Phys. Rev. Lett. **76**, 4250 (1996)
5. A. Kirilyuk, A.V. Kimel, T. Rasing, Rev. Mod. Phys. **82**, 2731 (2010)
6. B. Koopmans, G. Malinowski, F.D. Longa, D. Steiauf, M. Fähnle, T. Roth, M. Cinchetti, M. Aeschlimann, Nat. Mater. **9**, 259 (2010)
7. M. Battiato, K. Carva, P.M. Oppeneer, Phys. Rev. Lett. **105**, 027203 (2010)
8. G.P. Zhang, T.F. George, Phys. Rev. B **78**, 052407 (2008)
9. I. Radu, K. Vahaplar, C. Stamm, T. Kachel, N. Pontius, H.A. Dürr, T.A. Ostler, J. Barker, R.F.L. Evans, R.W. Chantrell, A. Tsukamoto, A. Itoh, A. Kirilyuk, T. Rasing, A.V. Kimel, Nature **472**, 205 (2011)
10. C.D. Stanciu, A.V. Kimel, F. Hansteen, A. Tsukamoto, A. Itoh, A. Kirilyuk, T. Rasing, Phys. Rev. B **73**, 220402(R) (2006)
11. A.V. Kimel, B.A. Ivanov, R.V. Pisarev, P.A. Usachev, A. Kirilyuk, T. Rasing, Nature Physics (2009)
12. R.K. Wangsness, Phys. Rev. **91**, 1085 (1953)
13. T. Kobayashi, H. Hayashi, Y. Fujiwara, S. Shiomi, IEEE Trans. Magn. **41**, 2848 (2005)
14. H.W. Schumacher, C. Chappert, P. Crozat, R.C. Sousa, P.P. Freitas, J. Miltat, J. Fassbender, B. Hillebrands, Phys. Rev. Lett. **88**, 227201 (2002)
15. C. Kittel, Phys. Rev. **76**, 743 (1949)
16. M. Mansuripur, *The Physical Principles of Magneto-Optical Recording* (Cambridge University Press, Cambridge, 1995)
17. S. Bhagat, P. Lubitz, Phys. Rev. B **10**, 179 (1974)
18. J. Kaplan, C. Kittel, J. Chem. Phys. **21**, 760 (1953)
19. R.K. Wangsness, Phys. Rev. **97**, 831 (1955)
20. T. Silva, J. Nibarger, B. Rippard, B. Stamps, APS Meet. **Mar.**, B23011 (2004)
21. Y. Li, K. Baberschke, Phys. Rev. Lett. **68**, 1208 (1992)
22. A. Mekonnen, M. Cormier, A.V. Kimel, A. Kirilyuk, A. Hrabec, L. Ranno, T. Rasing, Phys. Rev. Lett. **107**, 117202 (2011)
23. A.E. Berkowitz, W.H. Meiklejohn, IEEE Trans. Magn. **11**, 996 (1975)

24. M.H. Kryder, E.C. Gage, T.W. McDaniel, W.A. Challener, R.E. Rottmayer, G. Ju, Y.T. Hsia, M.F. Erden, IEEE Proc. **96**, 1810 (2008)
25. C. Stanciu, A. Tsukamoto, A.V. Kimel, F. Hansteen, A. Kirilyuk, A. Itoh, T. Rasing, Phys. Rev. Lett. **99**, 217204 (2007)
26. M. Aeschlimann, A. Vaterlaus, M. Lutz, M. Stampanoni, F. Meier, H.C. Siegmann, S. Klahn, P. Hansen, Appl. Phys. Lett. **59**, 2189 (1991)
27. M. Aeschlimann, A. Vaterlaus, M. Lutz, M. Stampanoni, F. Meier, J. Appl. Phys. **67**, 4438 (1990)
28. J. Hohlfeld, T. Gerrits, M. Bilderbeek, T. Rasing, H. Awano, N. Ohta, Phys. Rev. B **65**, 012413 (2001)
29. A. Vaterlaus, T. Beutler, F. Meier, Phys. Rev. Lett. **67**, 3314 (1991)
30. M. Binder, A. Weber, O. Mosendz, G. Woltersdorf, M. Izquierdo, I. Neudecker, J.R. Dahn, T.D. Hatchard, J.U. Thiele, C.H. Back, M.R. Scheinfein, Phys. Rev. B **74**, 134404 (2006)
31. U. Bovensiepen, Appl. Phys. A **82**, 395 (2006)
32. G. Schütz, W. Wagner, W. Wilhelm, P. Kienle, R. Zeller, R. Frahm, G. Materlik, Phys. Rev. Lett. **58**, 737 (1987)
33. J. Stöhr, Y. Wu, B.D. Hermsmeir, M.G. Samant, G.R. Harp, S. Koranda, D. Dunham, B.P. Tonner, Science **259**, 658 (1993)
34. C. Stamm, T. Kachel, N. Pontius, R. Mitzner, T. Quast, K. Holldack, S. Khan, C. Lupulescu, E.F. Aziz, M. Wietstruk, H.A. Dürr, W. Eberhardt, Nat. Mater. **6**, 740 (2007)
35. N. Kazantseva, U. Nowak, R.W. Chantrell, J. Hohlfeld, A. Rebei, Europhys. Lett. **81**, 27004 (2008)
36. S.I. Anisimov, B.L. Kapeliovich, T.L. Perelman, Sov. Phys. JETP **39**, 375 (1974)
37. J. Mentink, J. Hellsvik, D.V. Afanasiev, B.A. Ivanov, A. Kirilyuk, A.V. Kimel, O. Eriksson, M.I. Katsnelson, Th. Rasing, Phys. Rev. Lett. **107** (2011). doi:10.1103/PhysRevLett.108.057202
38. B. Koopmans, J.J. Ruigrok, F.D. Longa, W.J. de Jonge, Phys. Rev. Lett. **95**, 267207 (2005)
39. U. Atxitia, O. Chubykalo-Fesenko, N. Kazantseva, D. Hinzke, U. Nowak, R.W. Chantrell, Appl. Phys. Lett. **91**, 232507 (2007)

Chapter 6
Photo-Magnonics

Benjamin Lenk, Fabian Garbs, Henning Ulrichs, Nils Abeling, and Markus Münzenberg

Abstract In the framework of magnonics, all-optical femtosecond laser experiments are used to study spin waves and their relaxation paths. Magnonic crystal structures based on antidots allow the control over the spin-wave modes. In these two-dimensional magnetic metamaterials with periodicities in the wave-length range of dipolar spin waves, the spin-wave bands and dispersion are modified. Hence, a specific selection of spin-wave modes excited by laser pulses is possible. Different to photonics, the modes depend strongly on the strength of the magnetostatic potential at around each antidot site – the dipolar field. While this may lead to a mode localization, also for filling fractions around or below 10 %, Bloch states are found in low damping ferromagnetic metals. In this chapter, an overview of these mechanisms is given and the connection to spin-wave band spectra calculated from an analytical model is established. Namely, the plane-wave method yields flattened bands as well as band gaps at the antidot lattice Brillouin zone boundary.

6.1 Introduction

Being a quickly evolving research field, magnonics and magnonic materials have been investigated with different approaches [1, 2]. These can be in the time or frequency domain, as well as with very high spatial resolution down to nanometers or averaging over wider areas, then giving more general information on the system in question. For example, the spin-wave dispersion $\omega(k)$ can precisely be measured using a vector network analyzer (VNA) [3, 4] or Brillouin light scattering (BLS) setup [5, 6]. In such experiments, the selection of the excitation frequency or wave vector k, respectively, enables a detailed study of the magnetic mode spectrum in Fourier space.

In contrast, all-optical pump-probe techniques are neither k- nor frequency-selective. Instead, a broad continuum of spin-wave modes is populated on ultra-

B. Lenk · F. Garbs · H. Ulrichs · N. Abeling · M. Münzenberg (✉)
Institute of Physics, Georg-August-University of Göttingen, Göttingen, Germany
e-mail: mmuenze@uni-goettingen.de

B. Lenk
e-mail: blenk@gwdg.de

S.O. Demokritov, A.N. Slavin (eds.), *Magnonics*, Topics in Applied Physics 125,
DOI 10.1007/978-3-642-30247-3_6, © Springer-Verlag Berlin Heidelberg 2013

fast time scales in the femto- to picosecond range by absorption of an intense laser pulse [7]. The heat-induced disorder can be modeled by high-k spin-wave modes that subsequently relax into energetically lower-lying states. A population of respective spin-wave modes leads to a spatial distribution of energy by spin-wave propagation away from the spot of (optical) excitation [8]. The underlying processes for excitation, relaxation, and propagation span a large range of interaction energies (i.e., time scales). Connected transient dynamics have an effect on the dielectric tensor and thus can be optically investigated. Herein lies the large potential of (laser-) pulsed experiments: femto-, pico-, and nanosecond characteristics can be resolved in a fast and non-destructive manner.

Concerning optically excited spin waves, these condensate-like modes are intrinsically dependent on the matrix material properties. However, the reverse process is also possible: In a magnonic structure, i.e., a spin-wave metamaterial, the material properties can be tailored to produce spin-wave modes whose characteristic properties are decoupled from the ferromagnetic matrix. In the present chapter of this book, we shall restrict the discussion to two-dimensional systems. First, in Sect. 6.1.1, the continuous thin film case is briefly described, while Sect. 6.2 introduces all-optical experiments. The final Sects. 6.3–6.5 are dedicated to structured media for which the manipulation of the spin-wave spectrum will be exemplified. Finally, an outlook on possible devices based on spin-wave computing is presented.

6.1.1 Spin-Wave Modes in a Thin Ferromagnetic Film

For the case of a continuous film of thickness t, the Landau–Lifshitz–Gilbert equation of motion can be solved analytically [9, 10]. On sufficiently large length scales, the exchange interaction may be neglected. The resulting spin-wave modes are of dipolar character and have theoretically been studied by Damon and Eshbach (DE) in the 1960s. In the geometry of propagation perpendicular to the applied magnetic field (wave vector $k_{\mathrm{DE}} \perp H_{\mathrm{ext}}$), the dispersion takes the form [9]

$$\left(\frac{\omega_{\mathrm{DE}}}{\gamma\mu_0}\right)^2 = H_x\left(H_x + M_{\mathrm{S}} - \frac{2K_z}{\mu_0 M_{\mathrm{S}}}\right) + \frac{M_{\mathrm{S}}^2}{4}\left(1 - e^{-2|k_{\mathrm{DE}}|t}\right), \tag{6.1}$$

where ω_{DE} is the spin-wave frequency, M_{S} is the material's saturation magnetization, H_x is the in-plane component of the external field, and K_z accounts for an effective out-of-plane anisotropy. Respective wavelengths $\lambda_{\mathrm{DE}} = 2\pi/k_{\mathrm{DE}}$ are in the micron range. On much smaller length scales, the exchange interaction has to be considered, while the dipolar interaction can then be neglected. The dispersion for exchange-dominated spin waves reads

$$\left(\frac{\omega_{\perp}}{\gamma\mu_0}\right)^2 = \left(H_x + \frac{2A}{M_{\mathrm{S}}}k_{\perp}^2\right)\left(H_x + M_{\mathrm{S}} - \frac{2K_z}{\mu_0 M_{\mathrm{S}}} + \frac{2A}{M_{\mathrm{S}}}k_{\perp}^2\right), \tag{6.2}$$

where A is the exchange constant. In (6.2), it has already been considered that respective spin-wave lengths are only relevant in the direction perpendicular to the film plane. Standing spin waves of the order n can be excited where the wave vector is then quantized according to $k_\perp = n\pi/t$ – hence the naming perpendicular standing spin waves (PSSW). The geometric confinement in thin films with thickness in the nanometer range implies the possibility to independently study exchange and dipolar spin waves. In the lateral directions, the exchange interaction does not play a significant role, whereas in the direction perpendicular to the film plane it determines the spin waves potentially excited. From (6.1) and (6.2), another precessional mode can be deduced: An in-phase precession of all spins, i.e., the uniform mode of ferromagnetic resonance (Kittel mode [11]), will have the dispersion $\omega_K = \omega_{DE}|_{k=0} = \omega_\perp|_{k=0}$. Equations (6.1) and (6.2), together with the Kittel mode, constitute the basic possibilities of magnetic excitations in thin ferromagnetic films. In the following Sect. 6.2, a brief example of a continuous film investigated all-optically is given.

6.2 Samples and Experiments

Ferromagnetic thin films made of nickel or cobalt-iron-boron (CoFeB) are produced by electron-beam evaporation or magnetron sputtering [12]. The magnonic metamaterials' properties cannot be simply derived from the independent properties of continuous film and (periodically arranged) constituents. Instead, effects emerge that are intrinsic to the newly created material, i.e., the observed collective effects themselves define the respective system as being a metamaterial. Consequently, different periodic structures on a continuous film with otherwise identical properties yield different magnonic materials.

In photo-magnonics, all-optical experiments are performed using ultrashort laser pulses to both excite and detect spin-waves. A schematic drawing of the experiment is given in Fig. 6.1(a). It includes the intense optical pump pulses with a central wavelength of 800 nm and a temporal duration of 40 fs. Probe pulses with 5 % of the intensity at a variable time delay $\Delta\tau$ yield the electron or magnetization dynamics [13]. The excitation by the pump pulses takes place within picoseconds, including the ultrafast demagnetization ($\approx$100 fs) [14] and the relaxation of the highly damped high-k spin-wave modes to longer-lived ones ($\approx$20 ps) [7]. The respective processes can be experimentally accessed making use of the time-resolved magneto-optical Kerr effect (TRMOKE). Shown in Fig. 6.1(b) are the precessional modes observed with probe pulses that are time-delayed by $\Delta\tau$ with respect to the pump pulses. The observed modes have frequencies in the GHz range, thus oscillate on timescales up to nanoseconds. In consequence, the non-selectivity in terms of wave vector or frequency of the optical pump can be compared to a delta-like excitation of a classical oscillator which then will swing with its intrinsic eigen-frequency.

6.2.1 Thin-Film Magnetization Dynamics

The reference data set in Fig. 6.1(b) was recorded on a continuous CoFeB film of thickness $t = 50$ nm for $\mu_0 H_{ext} = 150$ mT. It shows the fast demagnetization [15]

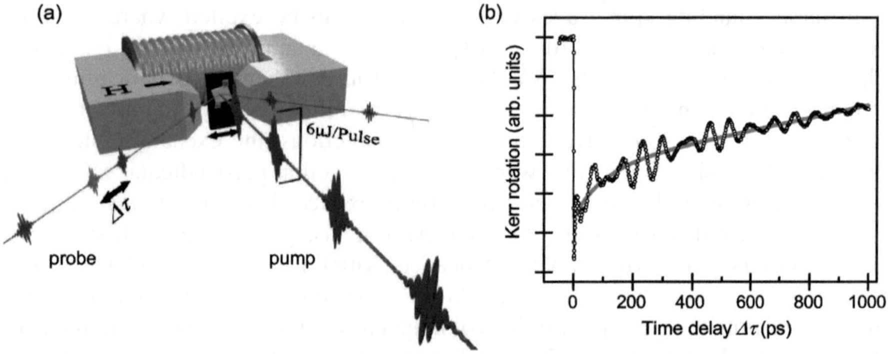

Fig. 6.1 A schematic of the time-resolved all-optical experiments is given in (**a**). Pump pulses with a high intensity excite the magnetization dynamics and probe pulses that are time-delayed by $\Delta\tau$ measure the pump-induced change in the Kerr rotation. A reference data set recorded on a 50 nm CoFeB film at $\mu_0 H_{\text{ext}} = 150$ mT is shown in (**b**). The *solid gray line* represents an exponential fit to the background which is subtracted prior to further analysis

upon absorption of the pump pulse (time delay $\Delta\tau = 0$) and the subsequent magnetization dynamics which are a superposition of at least two precessional modes [13]. The particular trace $M(\Delta\tau)$ from Fig. 6.1(b) is further analyzed in Fig. 6.2(a) by a Fourier transformation. The two modes' frequencies can then be determined. By changing the external field, these frequencies shift and the resulting experimental dispersions $\omega_i(H_{\text{ext}})$ can be fitted by (6.2) as seen in Fig. 6.2(b). All data presented in this chapter are plotted in a gray scale, where the (meta)material's response (i.e., the excited spin-wave mode's Fourier power) is indicated by a black peak as a function of externally applied magnetic field and frequency in Fourier space. The maximum time delay between pump and probe pulses of 1 ns consequently limits the resolution to be $(1\text{ ns})^{-1} = 1$ GHz, in the plots represented by the peak line width.

On the reference film, two modes of precession are observed. These are the uniform Kittel mode (white circles and line) as well as the first order perpendicular standing spin wave (PSSW, gray circles and line). Due to the intriguingly low Gilbert damping in ferromagnetic CoFeB, standing spin waves up to high orders $n \leq 5$ are observed in continuous films. While the spin-wave modes carry information on the structural properties of the film material [8], we will in the following focus on periodically structured materials.

6.3 Bloch-Like Modes in CoFeB Antidot Lattices

As described earlier, a metamaterial cannot simply be defined by a relation between the wavelengths and the parameters of the imprinted periodic structure. Instead, it is the emerging effects that make the name metamaterial necessary. For example, in a structured CoFeB film a new – magnonic – spin-wave mode is observed [16]. An SEM image of the two-dimensional square lattice of antidots in an otherwise

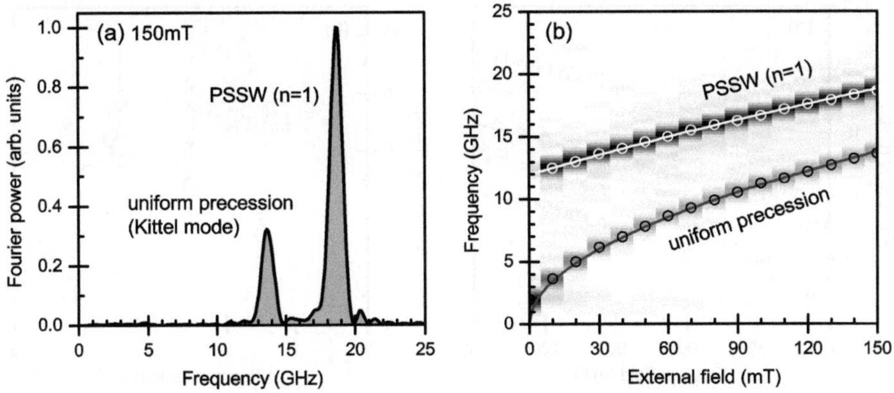

Fig. 6.2 Fourier analysis of the time-resolved magnetization dynamics of a continuous CoFeB film. In (**a**), the Fourier spectrum of the single measurement already shown in Fig. 6.1 is displayed ($\mu_0 H_{\text{ext}} = 150$ mT). Two precessional modes are observed the frequency of which is dependent on the external magnetic field (**b**). *Circles* represent experimentally determined peak positions, the *gray* and *white solid lines* are fits of (6.2) for the case of $k_\perp = 0$ and $k_\perp \neq 0$, respectively

continuous film is included in the inset of Fig. 6.3(b). It depicts the antidot diameter d and the lattice parameter a. Compared to the reference data set from Fig. 6.2, the data recorded on a structured film which are plotted in Fig. 6.3(a) reveal another precessional mode not previously observed (black diamonds and line). With the Damon–Eshbach dispersion $\omega_{\text{DE}}(H_{\text{ext}})$ the magnonic character is verified: Fitting (6.1) to the experimental peak positions yields (as the only fitting parameter) the wave vector to be $k_{\text{DE}} = \pi/a$. From that value and the observed frequency, the spin-wave propagation length can be approximated via the phase velocity: Together with the damping time constant determined in the TRMOKE data, one calculates length scales of about 100 µm which equal approximately 30 magnonic unit cells. This leads to an instructive picture on Bloch-mode excitation in real space: the periodic modulation of the spin-wave potential landscape (i.e., the effective internal magnetic field) imposes its periodicity as a condition on the spin waves propagating away from the spot of excitation. In other words, prerequisites for the observation of spin-wave Bloch states are the low damping in CoFeB and the resulting large propagation length of the spin waves. By changing the periodicity a of the antidot lattice per definition, a new metamaterial is created, in the sense that the magnonic properties are changed. Namely, the DE wave vector can be tuned according to the above-stated relation over a range of 1.5 µm $\leq a \leq$ 3.5 µm. For the measurements in Fig. 6.3(b), the filling fraction $f = \pi d^2/(4a^2) = 12$ % has been held constant. For small antidot separations, an additional effect comes into play: besides the Bloch-like extended modes, also localized ones are excited. These will be further detailed in Sect. 6.5.

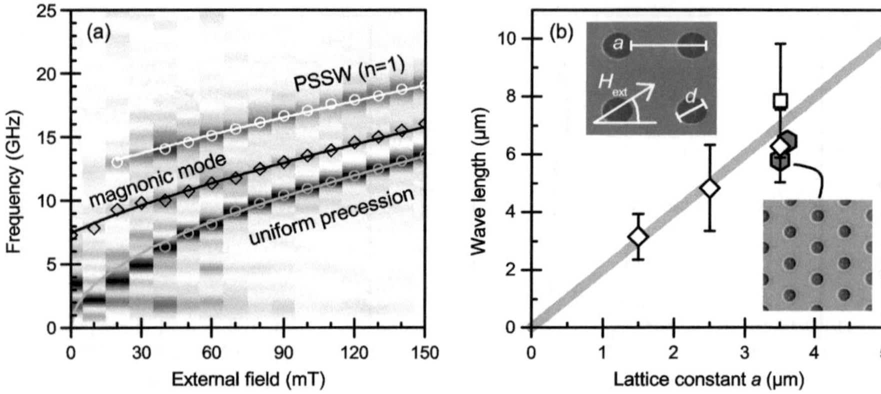

Fig. 6.3 Magnetization dynamics on structured CoFeB films. The Fourier spectrum in (**a**) shows an optically excited magnonic mode (*black diamonds*) additional to the modes also observed on a continuous film (*gray* and *white circles and lines*, see Fig. 6.2). The *solid black line* is a fit of (6.1) which yields the magnonic mode's wave vector k further detailed in (**b**): A change in the antidots' periodicity a shifts the Bloch-like resonance according to $k = \pi/a$ (*gray line*). Included are points for propagation along different directions with respect to H_{ext}, namely 90° (□) and 45° (◇) in a square lattice, as well as 30° and 60° in a hexagonal lattice (⬡). The *insets* of (**b**) show SEM images of a square and hexagonal antidot-lattice unit cell, respectively

6.3.1 Effects of Antidot-Lattice Symmetry

Included in Fig. 6.3(b) are the experimental results obtained in configurations deviating from $k_{\text{DE}} \perp H_{\text{ext}}$. For example, the antidot lattice can be rotated by 45° around the film normal. In that case, the Damon–Eshbach dispersion (6.1) cannot be fitted to satisfyingly represent the experimentally observed magnonic mode dispersion. This deviation can be accounted for by choosing the angle between propagating surface waves and H_{ext} to match 45°. The dispersion then reads [10]

$$\left(\frac{\omega_{45}}{\gamma\mu_0}\right)^2 = H_x^2 + \frac{H_x M_S}{2}\left[1 + \frac{1}{k_{45}t}\left(1 - e^{-k_{45}t}\right)\right] + \frac{M_S^2}{2k_{45}t}\left(1 - e^{-k_{45}t}\right)\left[1 - \frac{1}{k_{45}t}\left(1 - e^{-k_{45}t}\right)\right]. \tag{6.3}$$

Respective fitting results for k_{45} are included in Fig. 6.3(b) as white diamonds. One finds the relation $k_{45} = \pi/a$ to be satisfied. This means that by rotation of the sample the propagation direction of the Bloch-like surface waves is not changed – they still preferably propagate along the nearest-neighbor directions of the antidot lattice.

This remains true for lattices with other than square symmetry. Also in films structured with hexagonal lattices magnonic spin-wave modes are observed. Here, the application of the external field along a high-symmetry axis and assuming a propagation direction in the nearest-neighbor direction (under 30° and 60°, respectively) yields the wave vector $k_{\text{hex}} = \pi/a$. Respective data points are represented in Fig. 6.3(b) by the large filled hexagons.

The population of the magnonic modes suggests the creation of flattened bands that increase the spin-wave density of states (DOS). For spin-waves, the situation is analogous to electrons in a crystal, where a periodic potential is created by the atoms. More precisely, the spin-wave spectrum is modified by the antidots such that bandgaps at the Brillouin zone boundary are introduced. Simultaneously, a flattening of bands takes place which in turn leads to an increase of the density of states. Keeping in mind the condensation-like excitation of precessional modes after optical excitation, this increase of the spin-wave DOS can be considered the reason for the observation of the Bloch-like modes. The broad-band excitation by the pump pulses provides the range of spin waves initially necessary to populate the modes discussed here. The close analogy to electronic crystals will be extended in Sect. 6.4 where a calculation of the spin-wave spectrum is performed via the plane-wave method.

6.4 Spin-Wave Spectra from Plane-Wave Calculations

An analytical approach to theoretically access the creation of bandgaps and the flattening of bands is via the plane-wave method (PWM). It has been developed by Puszkarski et al. and can be used to study the appearance of spin-wave bandgaps in two-dimensional [17], as well as three-dimensional systems [18]. For the present case of a thin film with thickness $t = 50$ nm and wavelengths λ in the micron range, the assumption $\lambda \gg t$ holds. Thus, the lower modes' profiles can be assumed to be uniform in the z-direction, i.e., perpendicular to the thin film. As opposed to the full theory in [10], the exchange interaction can then be neglected for sufficiently small k and the uniform-mode analysis applied [19]. The respective spin waves are confined in a periodic potential which is constructed by a periodic modulation of the magnetization in terms of a Fourier synthesis:

$$M_S(\vec{r}) = \sum_{\vec{G}} M_S(\vec{G}) e^{i\vec{G}\vec{r}},$$

where $\vec{G}$ is a two-dimensional vector of the reciprocal lattice. Using this formulation, the linearized Landau–Lifshitz–Gilbert equation can be solved. Namely, plane waves of a given wave vector are used to solve the eigenvalue problem stemming from the dynamic field and $M_S(\vec{r})$ [1]. The resulting frequency for each mode depends on the position and direction in the reciprocal space of the antidot lattice. This ansatz is similar to the Bloch theorem applied to electrons and photons and yields the magnonic band structure of a periodic ferromagnetic system.

Respective results are plotted in Fig. 6.4. Therein, a calculation for a continuous CoFeB film is compared to a structured film.[1] In the continuous film, the folding of the spin-wave bands at the Brillouin zone boundary is introduced to illustrate the difference when compared to the structured film. For the latter, at the point X' in

[1]Continuous means $d \to 0$ while $a = 3.5$ μm is held constant.

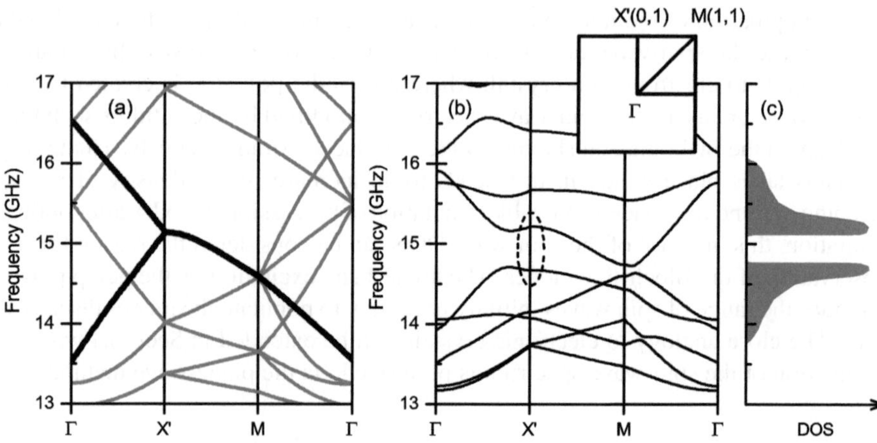

Fig. 6.4 Spin-wave band structures of (**a**) a continuous CoFeB film and (**b**) an antidot lattice with $a = 3.5$ μm and $d = 1$ μm in a CoFeB matrix ($\mu_0 H_{ext} = 130$ mT). In (**a**), the first and second band are marked by the *thicker black lines*. In (**b**), the bandgap created by the periodic structure is highlighted by the *dashed black ellipsis*. In the *inset*, the first Brillouin zone of the square lattice is depicted. The reciprocal directions along X' correspond to $k \perp H_{ext}$ in real space. Schematically depicted in (**c**) is the spin-wave density of states (DOS) for the 1st and 2nd band in this direction, which is populated by the laser pump pulses

reciprocal space, a bandgap as marked by the dashed ellipsis opens up around a frequency of 15 GHz. This corresponds well to the experimentally observed magnonic mode in Fig. 6.3(b). Hence, there is a close relation between the flattening of the spin-wave band at the zone boundary and the experimental finding of the wave vector being equal to π/a. The situation is further illustrated in Fig. 6.4(c). Therein, the density of states as a function of the spin-wave frequency is schematically depicted for the 1st and 2nd band in the $\Gamma - X'$ direction. The peaks introduced by the periodic structure correspond to the experimentally observed spin-wave modes.

Details on the calculations and the application to different ferromagnets can be found in [1]. Concerning size and position of the band gap, tuning is possible by the material parameters of the constituent materials [18].

6.5 Localized Modes in Nickel Antidot Lattices

TRMOKE experiments on magnonic crystals with a nickel matrix do not show any Bloch modes as observed in CoFeB in Sect. 6.3. Instead, a field-independent mode occurs as shown in Fig. 6.5. Its frequency is independent of H_{ext} and remains constant up to $\mu_0 H_{ext} = 150$ mT. Moreover, its amplitude is larger than the one of the uniform mode. Only if the external field is large enough, the excitation efficiency for the Kittel mode is high. This already points to an inhomogeneity of the internal field: If H_{ext} is sufficiently high, regions of constant internal field exist, where the Kittel mode can be populated. Below the critical field, the internal field undergoes

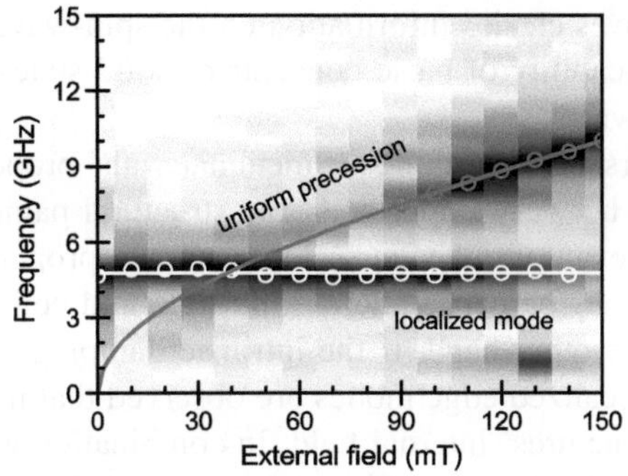

Fig. 6.5 Magnetization dynamics on a structured nickel film with $a = 3.5$ μm and $d = 1$ μm. The strong intrinsic damping leads to the observation of only a localized magnetic mode (*white circles* and *line*). Its amplitude is largest around the antidot edges, and its frequency does not depend on the external field. Only at high $\mu_0 H_{\text{ext}} \geq 70$ mT, the uniform precession is excited by the pump pulses (*gray circles* and *line*)

strong spatial changes and localization sites for spin waves are formed [20]. This is consistent with the experimental finding that the filling fraction f merely changes the amplitude of the localized mode, but does not alter its field-independence.

It has to be considered that the observed effects remain a combination of matrix material and superimposed periodic structure. In ferromagnetic nickel, the intrinsic Gilbert damping is by a factor of 4 larger than in CoFeB. This leads to a strongly decreased propagation length of spin waves, which is not more than 10 μm in nickel (as opposed to 100 μm in CoFeB; see Sect. 6.3). Consequently, the spin waves excited by laser pulses do not propagate far enough to largely interact with the periodic structure of antidots. However, also in CoFeB localized modes are observed, if the periodic distortion becomes too strong. For an antidot separation in the order of the internal field inhomogeneity, additional to the propagating magnonic mode a non-dispersive localized mode appears [16].

The frequency of the localized modes can be shifted by virtue of the antidots' separation. This points to a dipolar coupling of individual locations, i.e., regions of strong field inhomogeneity. The same effect can be achieved by rotation of the external field with respect to the antidot lattice. The dipolar coupling between neighboring localization sites changes and hence, the field-independent modes' frequency is shifted, see Fig. 27 in Ref. [1].

6.6 Outlook: Magnonic Control over Spin Waves

The different effects of a periodic antidot structure in two dimensions have been described. The respective metamaterial(s) illustrate the large potential of artificially structured media in general and magnonics in particular. Namely, they allow the control over spin waves in that the two rivaling effects of localization and Bloch-like extension can be tuned. The selective excitation of magnonic states from a broader continuum of spin waves carries information on the spin-wave density of states. In that respect, the applicability of basic concepts of solid state physics to magnonic crystals has been shown.

Relevant parameters that determine the metamaterials' properties are not only the constituting material, but very importantly the structural parameters of the antidot lattice inside. If, for example, the intrinsic spin-wave propagation length is long enough, information may be carried across multiple unit cells in form of a Bloch-like spin wave. On the other hand, if the intrinsic damping is larger or the filling fraction is too high, localized edge modes are observed that may be used to collect information on the structures' internal field. In combination with other approaches described in the present book, e.g., different material combinations, the perspectives are promising: if taken beyond the proof of principle, spin-wave computation and spin-wave data transmission may be combined to provide a high data throughput and energy efficiency in such devices. One example for a distinct modification of the magnonic periodic structure is a line defect that can function as a wave guide for spin-wave frequencies inside the magnonic gap region, to be functionalized in the future.

Acknowledgements We would like to thank Jakob Walowski for the contribution of the experiment's schematic shown in Fig. 6.1(a) and Georg Herink for careful proof-reading of the manuscript.

References

1. B. Lenk, H. Ulrichs, F. Garbs, M. Münzenberg, Phys. Rep. **507**(4–5), 107 (2011). doi:10.1016/j.physrep.2011.06.003
2. V.V. Kruglyak, S.O. Demokritov, D. Grundler, J. Phys. D, Appl. Phys. **43**(26), 260301 (2010). doi:10.1088/0022-3727/43/26/260301
3. J. Podbielski, F. Giesen, D. Grundler, Phys. Rev. Lett. **96**(16), 167207 (2006). doi:10.1103/PhysRevLett.96.167207
4. S. Neusser, G. Duerr, H. Bauer, S. Tacchi, M. Madami, G. Woltersdorf, G. Gubbiotti, C. Back, D. Grundler, Phys. Rev. Lett. **105**(6), 067208 (2010). doi:10.1103/PhysRevLett.105.067208
5. S. Tacchi, F. Montoncello, M. Madami, G. Gubbiotti, G. Carlotti, L. Giovannini, R. Zivieri, F. Nizzoli, S. Jain, A. Adeyeye, N. Singh, Phys. Rev. Lett. **107**(12), 127204 (2011). doi:10.1103/PhysRevLett.107.127204
6. C.W. Sandweg, M.B. Jungfleisch, V.I. Vasyuchka, A.A. Serga, P. Clausen, H. Schultheiss, B. Hillebrands, A. Kreisel, P. Kopietz, Rev. Sci. Instrum. **81**(7), 73902 (2010). doi:10.1063/1.3454918
7. M. Djordjevic, M. Münzenberg, Phys. Rev. B **75**(1), 12404 (2007). doi:10.1103/PhysRevB.75.012404
8. B. Lenk, G. Eilers, J. Hamrle, M. Münzenberg, Phys. Rev. B **82**(13), 134443 (2010). doi:10.1103/PhysRevB.82.134443
9. R.W. Damon, J.R. Eshbach, J. Appl. Phys. **31**(5), S104 (1960). doi:10.1063/1.1984622
10. B.A. Kalinikos, A.N. Slavin, J. Phys. C **19**(35), 7013 (1986). http://stacks.iop.org/0022-3719/19/7013
11. C. Herring, C. Kittel, Phys. Rev. **81**(5), 869 (1951). doi:10.1103/PhysRev.81.869

12. G. Eilers, H. Ulrichs, M. Münzenberg, A. Thomas, K. Thiel, M. Seibt, J. Appl. Phys. **105**(7), 073701 (2009). doi:10.1063/1.3100044
13. M. Djordjevic, G. Eilers, A. Parge, M. Münzenberg, J.S. Moodera, J. Appl. Phys. **99**(8), 08F308 (2006). doi:10.1063/1.2177141
14. U. Atxitia, O. Chubykalo-Fesenko, Phys. Rev. B **81**(17), 174401 (2010). doi:10.1103/PhysRevB.81.174401
15. E. Beaurepaire, J.C. Merle, A. Daunois, J.Y. Bigot, Phys. Rev. Lett. **76**(22), 4250 (1996). doi:10.1103/PhysRevLett.76.4250
16. H. Ulrichs, B. Lenk, M. Münzenberg, Appl. Phys. Lett. **97**(9), 092506 (2010). doi:10.1063/1.3483136
17. J. Vasseur, L. Dobrzynski, B. Djafari-Rouhani, H. Puszkarski, Phys. Rev. B **54**(2), 1043 (1996). doi:10.1103/PhysRevB.54.1043
18. M. Krawczyk, H. Puszkarski, Phys. Rev. B **77**(5), 54413 (2008). doi:10.1103/PhysRevB.77.054437
19. M. Hurben, J. Magn. Magn. Mater. **139**(3), 263 (1995). doi:10.1016/0304-8853(95)90006-3
20. M.J. Pechan, C. Yu, R.L. Compton, J.P. Park, P.A. Crowell, J. Appl. Phys. **97**(10), 10J903 (2005). doi:10.1063/1.1857412

Chapter 7
Probing Magnons by Spin-Polarized Electrons

K. Zakeri and J. Kirschner

Abstract High wave-vector magnon excitations in itinerant ferromagnets can be investigated by electron scattering experiments. In such an experiment, a spin-polarized beam with a well-defined energy is scattered from the sample surface and the energy distribution of the scattered electrons is measured. Since magnons possess a total angular momentum of $1\hbar$, they can only be excited (annihilated) by incidence of minority (majority) electrons. This fact leads to a peak in the energy loss (gain) region of the intensity spectra when minority (majority) electrons are incident. The scattering itself is elastic and the observed energy loss (or gain) is due to the fact that the ejected electron stems from a lower (or higher) energy level of the excited solid. Such a process is mediated by exchange interaction that is of a pure Coulomb nature and no explicit spin–spin interaction is involved.

We review the recent experimental attempts to probe and investigate the high wave-vector magnons in ultrathin ferromagnetic films by using spin-polarized electrons. Experimental results obtained by spin-polarized electron energy loss spectroscopy will be presented. The focus will be on the basic concepts and the nature of the different types of excitations probed by electrons. A possibility to distinguish between magnon- and phonon-excitations without the need of spin selective detection will be discussed.

7.1 Introduction

Low dimensional magnetic objects such as ultrathin magnetic films and nanostructures possess novel properties that have had a major impact on the innovation of the information and magneto-electronic technology [1]. Many static magnetic properties of this class of materials, which lead to their desired functionalities, are now well-known [2]. Since the speed of operation is important for faster information processing, nowadays the magneto-electronic technology pushes toward high speed operation. This fact implies that a better knowledge of the dynamic magnetic properties of magnetic thin films and nanostructures is required. Hence the processes

K. Zakeri (✉) · J. Kirschner
Max-Planck-Institut für Mikrostrukturphysik, Weinberg 2, 06120 Halle, Germany
e-mail: zakeri@mpi-halle.de

S.O. Demokritov, A.N. Slavin (eds.), *Magnonics*, Topics in Applied Physics 125,
DOI 10.1007/978-3-642-30247-3_7, © Springer-Verlag Berlin Heidelberg 2013

which are linked to spin excitations and ultrafast dynamics in solid are of great importance.

Besides its technological impact, spin dynamics in ferromagnets is of great fundamental interest because it lies in the central understanding of many physical phenomena. For instance, the temperature induced fluctuation of spins at low temperatures can be treated as collective spin excitations called spin waves. Spin waves are also important to understand the electron–electron interactions in high-temperature superconductors [3, 4]. Moreover, they are associated with the newly discovered phenomena like current-induced magnetization switching or current-induced domain wall motion [5–7].

The long wavelength spin wave excitations can be well-described using semi-classical models. However, when the wavelength of excitations approaches the nanometer scale, they are governed by the microscopic exchange interaction. In this case, the excitations are strongly confined in both time and space (their lifetime is mainly governed by the dissipation into the single particle excitations). Hence, they cannot be simply treated as in the case of classical spin waves. This implies that a quantum mechanical description of the system is demanded, and the experimental study of such excitations would provide a true microscopic view of the system.

This chapter is a review of the recent experimental results of high wave-vector magnon excitations by using spin-polarized electrons. The authors realized that the basic concepts of the processes involved in the magnon excitations by spin-polarized electrons are not well-interpreted by many groups in various places; therefore, the basic concepts are emphasized. The magnon dispersion relation and its differences in various systems will not be addressed in this chapter. The nature of the different type of excitations probed by electrons will be discussed in detail. Section 7.2 provides the basic information needed to follow this chapter. The magnetic excitations in ferromagnets are classified and explained in Sects. 7.2.1–7.2.3. The experimental tool, that is, spin polarized electron energy loss spectroscopy (SPEELS), used to investigate such excitations with a relatively high momentum and energy resolution is explained in Sect. 7.3. Moreover, the possibility of using scanning tunneling microscopy to study the surface excitations is briefly discussed. Section 7.4 will provide experimental results of high wave-vector magnon excitations in ultrathin ferromagnetic films. A possibility to distinguish between magnon- and phonon-excitations without the need of spin selective detection is discussed in Sect. 7.4.2.

7.2 Basic Concepts

Ferromagnetism is characterized by a spontaneous magnetization without applying any external magnetic field. In a ferromagnetic solid, all magnetic moments are aligned parallel at $T = 0$ K. It is caused by the direct exchange interaction of the electrons, which leads to a strong coupling of the spins to each other. The excited sate of such a spin system can be described by creation of wavy like motions of the spins over the entire crystal called spin waves [8, 9]. The same picture can also be

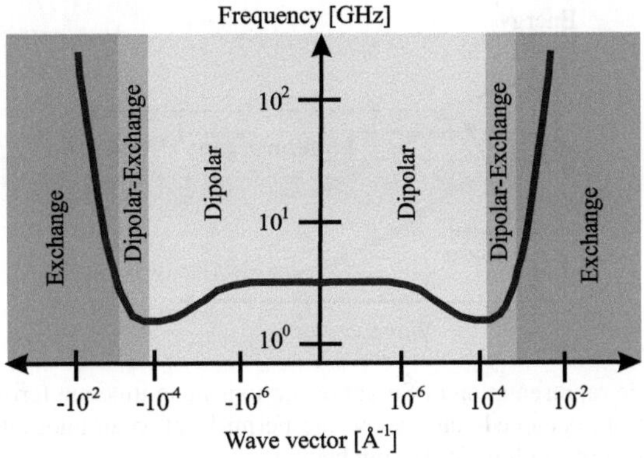

Fig. 7.1 Different types of spin waves within the energy–momentum space. At the very low momentum values, the dominating magnetic energy is the long-range dipolar interaction, hence the spin waves are called dipolar spin waves. For the large momentum values, the dominating magnetic energy is the exchange energy, and thus it determines the energies of the spin waves. An intermediate regime is separating these two classes of spin waves [20]

applied to the antiferromagnets in which the atomic moments are aligned antiparallel at $T = 0$ K. The spin waves in antiferromagnets can be understood in analogy to the ones in ferromagnets.

The classical picture mentioned above is only valid when the magnetic moments are localized. Since in 3*d* ferromagnets the magnetism is governed by conduction electrons, this description cannot provide a clear picture of the system. In addition, another important questions arises: How can the classical spin waves be understood from a microscopic point of view, i.e., by looking at the electronic band structure? This fact was pointed out by Stoner [10, 11] at the beginning of developing of the quantum theory of magnetism. In fact, a unique and satisfactory answer to this question is still missing. The elementary magnetic excitations in metallic ferromagnets can be classified as spin waves and Stoner excitations. Spin waves may be described similar to the excitons (electron–hole pairs). The wave function is a linear superposition of particle hole states (spin triplet particle hole excitations–Stoner excitations) [12]. In the following, we shall explain both excitations in detail.

7.2.1 Spin Waves

Spin waves are collective excitations of a spin system. They are well defined for the magnetic systems with localized moments, e.g., ferromagnetic insulators. In the itinerant ferromagnets, the low energy spin excitations can also be classified as spin waves. Spin waves may be described by Heisenberg Hamiltonian (see Sect. 7.2.3). In this description, the quantized spin waves are called magnons. The word "*magnon*" is chosen in analogy to the word "*phonon*", which describes the quantized lattice vibrations. There is a large variety of magnons over the momentum and energy space. Spin waves caused by the coherent precession of the moments

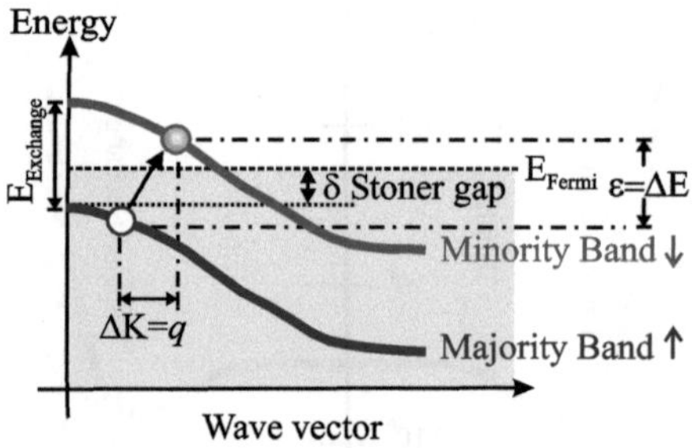

Fig. 7.2 A schematic representation of Stoner excitations in an itinerant ferromagnet. A majority electron jumps from an occupied state below the Fermi-level to an unoccupied state above the Fermi-level, leaving a hole with majority spin behind

($q = 0$) possess energies on the order of magnetic anisotropy energy of the system (few tens of micro-electron-volts) [13]. The lifetime of such spin waves reaches a value of a few tenths of a microsecond [14–17]. Figure 7.1 shows a schematic representation of different types of low-energy and low-wave-vector magnons. As the wave vector increases, the energy of spin waves increases. In the low wave-vector regime, where the magnetic anisotropy and long-range dipolar interactions are dominating, spin waves can be treated classically [18, 19]. At larger wave vectors (larger than 10^{-3} Å^{-1}) the exchange interaction becomes the dominating magnetic interaction and will determine the energy of spin waves.

7.2.2 Stoner Excitations

The electron–electron interactions in ferromagnets lead a to separation of the electronic bands for spin-up (majority) and spin-down (minority) electrons. In a simplified picture, this effect can be described as perturbations of a degenerate, paramagnetic band structure from which the energy degeneracy is lifted by the "*exchange splitting*". The exchange splitting depends on the spin and the wave vector of the electrons within the system. If the Fermi-level lies between the bands derived from a single degenerate paramagnetic band, the lower energy (majority) states are occupied, while the higher energy (minority) states are unoccupied. This fact leads to a spin polarization of the conduction electrons and can explain the magnetic state of the system [11]. In such a case, the spin split bands across the Fermi-level would lead to the possibility of a unique single-particle excitation in the system. In principle, an electron can jump from an occupied majority band, undergo spin reversal, and occupy a state in the minority band. This process leads to a remaining hole of majority spin character in the majority band. The resulting electron–hole pairs (known as "*Stoner excitations*") carry a total spin of $1\hbar$ and a wave vector of q, which is the momentum difference of the electron and hole in the minority and majority bands, respectively. Stoner excitations are spread over the whole Brillouin zone (see Fig. 7.2).

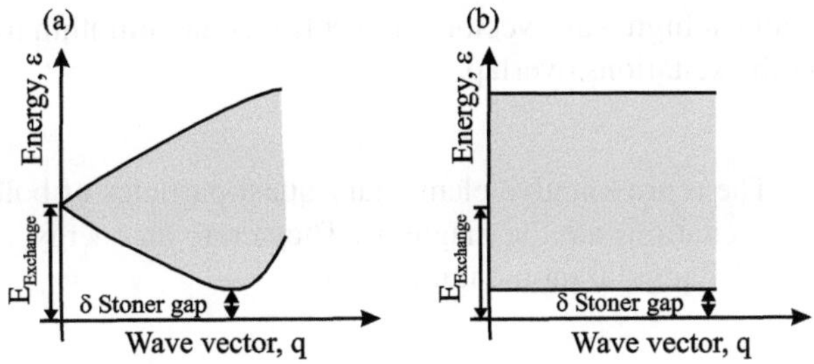

Fig. 7.3 Stoner density of states for a bulk (**a**) and an ultrathin (**b**) metallic ferromagnet with a non-zero Stoner gap δ

A schematic representation of the Stoner excitations is provided in Fig. 7.2. The bands are separated by the exchange splitting. The parameter δ, known as Stoner gap, is defined as the energy between the highest occupied state below the Fermi-level and the Fermi-level, E_{Fermi}. Ferromagnets with non-zero δ are usually called "*strong ferromagnets*" and the ones with $\delta = 0$ are called "*weak ferromagnets*"). The energy, ε, and momentum, q, of the Stoner excitations are given by the energy and momentum difference of the electron and hole in the system, respectively. The distribution of the Stoner excitations in the momentum and energy space is usually called as Stoner density of states or Stoner states. It is given by the joint density of states for occupied and unoccupied states of appropriate spins with the momentum transfer $\Delta K = q$. In the case of bulk metallic ferromagnets, the conservation of the wave vector in three dimensions opens a gap for long-living-magnons with small energy and wave vector. This basically means that there are no Stoner states available for small q and ε ($q \ll$ and $\varepsilon < E_{\mathrm{Exchange}}$). However, in the case of ultrathin metallic ferromagnets only the momentum parallel to the surface is conserved. This implies that the states should be projected into the energy axis meaning that there will be a very small gap left at $q = 0$ (only the Stoner gap). Figure 7.3 represents the Stoner density of states for a strong ferromagnet with a non-zero Stoner gap δ.

Stoner excitations describe a spin reversal in the system and are the elementary single-particle magnetic excitations in itinerant-electron ferromagnets. As it is mentioned above, at low energy and wave vector region, the configuration of the electronic bands does not allow single-particle Stoner excitations (see Fig. 7.3). In this case, the coherent superposition of "*virtual*" Stoner excitations of wave vector q can produce a collective magnetic excitation (or spin wave) of wave vector q at low energies. In regions of energy–momentum space where individual Stoner excitations are allowed, the spin wave and Stoner excitations are strongly coupled so that the excitations are heavily damped. In the case of two-dimensional ferromagnets (thin films and surfaces), the Stoner states cover the whole energy and momentum space [21–23]. It remains only a very tiny space for spin waves with very small energies ($\varepsilon < \delta$, see Fig. 7.3). Stoner excitations play a fundamental role in understanding of the excited states as well as the ground state of itinerant-electron ferromagnets. The band model of ferromagnets leads to the spin waves as the long-living excitations. However, in addition to spin waves, a spectrum of single particle Stoner excitation can also be obtained [24–34]. In principle, a proper theory must provide an account of both the spin waves and the Stoner excitations within a single framework. Especially at high wave vectors or at surfaces and ultrathin metallic ferromagnets, where both excitations overlap.

The representative elementary quasi-particles of both spin waves and Stoner excitations are the magnons. They carry an energy, ε, a momentum, q, and a total angular momentum of $1\hbar$.

7.2.3 Heisenberg Description of Magnons

The magnetically relevant part of the Hamiltonian for a ferromagnetic system in the absence of any external magnetic field is given by:

$$H = -\sum_{i \neq j} J_{ij} \mathbf{S}_i \cdot \mathbf{S}_j. \tag{7.1}$$

Here J_{ij} represents the exchange interaction between spins $\mathbf{S}_i$ and $\mathbf{S}_j$. In the Heisenberg spin Hamiltonian given above, only the symmetric exchange interaction is taken into consideration. We note that in the presence of the magnetic anisotropy in the system, an additional term has to be added to (7.1). The term caused by magnetic anisotropy is directly proportional to the magnetic anisotropy energy of the system and basically acts as a Zeeman term which lifts the energy of the system up by a value of $2K_{\text{eff}}S$, where K_{eff} denotes the effective magnetic anisotropy energy and S is the magnitude of spins. Moreover, in the presence of the antisymmetric exchange interaction known as "*Dzyaloshinskii–Moriya interaction*", an additional term which is proportional to the cross-product of the spins has to be added to the spin Hamiltonian [35, 36]. According to the formalism developed by Kittel [37], one can derive an expression for the magnon dispersion relation. In a classical picture, one may consider the exchange interaction as the source of a torque acting on each magnetic moment. By this consideration, the equation of motion reads as

$$\hbar \frac{d\mathbf{S}_i}{dt} = \tau_i = 2\sum_j J_j (\mathbf{S}_i \times \mathbf{S}_j). \tag{7.2}$$

After a simple mathematical step, (7.2) can be written as

$$i\hbar \frac{d\mathbf{S}_i^+}{dt} = 2JS \sum_j [S_i^+ - S_j^+], \tag{7.3}$$

where $S^+ = S_x + iS_y$ is the raising operator. Usually, for surface magnons a wave form solution in the form of $S_i^+ = A_i \exp[i(\mathbf{q} \cdot \mathbf{R}_j - \omega t)]$ is assumed. A_i is the amplitude of the magnon with the wave vector $\mathbf{q}$ and angular frequency of ω at position $\mathbf{R}_i$. The dot product between $\mathbf{q}$ and $\mathbf{R}_j$ implies that only the in-plane component of the wave vector is important here. Putting this solution into (7.3) results in the following expression:

$$\hbar\omega A_i = 2S\sum J_j\{A_i - A_j \exp[i\mathbf{q}\cdot(\mathbf{R}_j - \mathbf{R}_i)]\}. \quad (7.4)$$

Equation (7.4) is usually used to derive an equation for the magnon dispersion relation which connects the magnons' energy to their propagating wave vector. For an infinitely long cubic crystal and considering only the nearest neighbor interaction, the dispersion relation can be written as this simple form:

$$\varepsilon = \hbar\omega = 2nJS\left[1 - \frac{1}{n}\sum_{\delta}\cos(\mathbf{q}\cdot\mathbf{r})\right], \quad (7.5)$$

where n is the number of nearest neighbors, J is the exchange coupling constant between the neighbors, and $\mathbf{r}$ represents the position vector of the respective neighbor. Since the dispersion relation is not the main subject of the present chapter, we will not discuss it further. No prediction concerning the magnon lifetime can be made by the Heisenberg spin Hamiltonian given in (7.1). Within this description, magnons shall live forever, which is not certainly true for the case of itinerant electron ferromagnets. This classical picture even fails to describe the magnon dispersion relation in this class of materials [24–34, 38, 39].

7.2.4 Spin Dependence of Electron Scattering

Generally, the inelastic scattering of electrons results from the so-called "*dipolar*" and "*impact*" scattering processes [40]. Dipolar scattering refers to the scattering of electrons by the electric dipolar field generated at the surface and may be described using the dielectric dipolar theory [40]. In such a description, detailed knowledge of the electrons and the sample surface is not needed. The dipolar scattering mainly happens for a scattering geometry very close to the specular geometry, where the angle of the incoming beam is the same as the one of the outgoing beam. In the case of off-specular geometries, the dominating process is the impact scattering. In contrast to the dipolar scattering, the inelastically scattered electrons are distributed over a large range of angles. The impact scattering is still poorly described by the theory due to lack of the detailed information about the interaction between the electrons and the sample surface [40].

As mentioned above, the inelastic scattering of spin-polarized electrons is a rather complicated process. The complete description of the processes involved in the scattering of spin-polarized electrons is out of the scope of the present chapter. More information may be found, for example, in Chap. 4 of [41] or in [42, 43]. Let us assume an ideal situation; one sends an electron with a given energy, E_i, and spin orientation (up $\uparrow$ or down $\downarrow$) onto a ferromagnetic surface at a certain geometry and performs a full energy and spin resolved detection of the scattered electron. Now the question would be: What are the possible scattering processes in such a situation? Figure 7.4 illustrates all the possible processes that may occur in an experiment where a primary electron with energy E_i and particular spin orientation

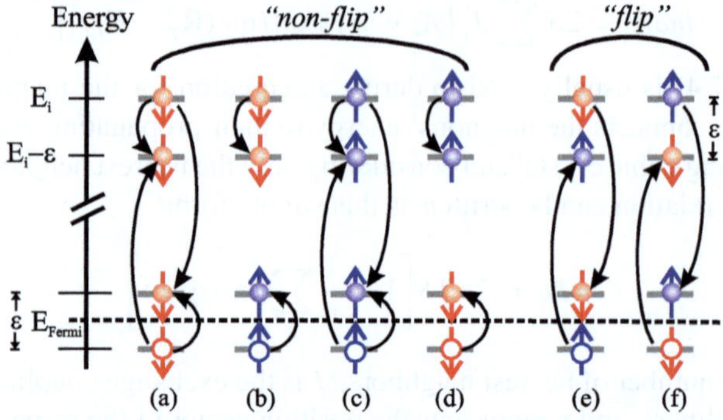

Fig. 7.4 A schematic representation of the possible processes taking place in the scattering of spin-polarized electrons from a magnetic surface. Processes marked as (**a**), (**b**), (**c**), and (**d**) are referred to as "*non-flip*" processes, since the spin of the scattered electron is the same as the one of the primary electron. Processes in which the spin of the scattered electron is opposite to the one of the incoming electron are referred to as "*flip*" processes [(**e**) and (**f**)]

is launched onto a magnetic surface and an electron with energy $E_i - \varepsilon$ is observed after scattering. During the scattering process, electrons in the sample are excited above the Fermi-level, and consequently holes are left in the system. The spin of the electron during the transition is conserved; therefore, only transitions from the states with the same spin orientation are allowed. A primary spin-down electron may experience two loss processes, each resulting in a scattered electron with the same spin orientation as the incident one (these processes will be referred to as "*non-flip*" processes). The incident spin-down electron either excites a minority electron from a state below the Fermi-level to a state above the Fermi-level, or it excites an electron of majority character to the same energy. Therefore, there are two "*spin-down non-flip amplitudes*". These processes are marked as (a) and (b) in Fig. 7.4. The first one has an exchange contribution which partly cancels the direct amplitude. A primary spin-up electron may also experience two loss processes, each resulting in a scattered electron with the same spin orientation as the incident one ("*non-flip*" processes). The incident electron either excites a majority electron from a state below the Fermi-level to a state above the Fermi-level, or it excites an electron of minority character to the same energy. Therefore, there are two "*spin-up non-flip amplitudes*". The corresponding processes are marked as (c) and (d) in Fig. 7.4. Again here the first one (process (c)) has an exchange contribution which partly cancels the direct amplitude. Note that the "*non-flip intensities*" result from the same spin system. A similar behavior is expected in an optical absorption experiment, e.g., photoemission, because the electric and magnetic fields involved are far too small to reverse the spin of an electron. The fact is that, disregarding exchange processes, electron energy loss spectroscopy and optical absorption are basically equivalent. Differences come into play by the possibility of electron exchange processes marked as (e) and (f) in Fig. 7.4. In the "*spin-flip*" process described by (e), the primary spin-down electron transfers its whole energy to a majority spin electron and occupies an empty state above the Fermi-level after the excitation. The excited majority electron, coming from a state below the Fermi-level, emerges with an energy equal to the primary energy minus the energy of the electron–hole pair (Stoner excitation) around the Fermi-level. This process leads to creation of a magnon with the total angular momentum of $1\hbar$ and energy ε. We will come to this point later in Sects. 7.3 and 7.4. The partial intensities for the "*spin-flip*" process with primary spin-up electron (process (f)) may be obtained analogously. The primary spin-up electron transfers its whole energy to a minority spin electron in a state below the Fermi-level and occupies an empty state above the Fermi-level after the excitation. The excited minority electron, coming from a state below the Fermi-level, leaves the sample with an energy equal to the primary energy minus the energy of the electron–hole pair around the Fermi-level. In addition to this process, there is another possibility (not shown in Fig. 7.4): the incoming spin-up electron excites an electron from a state above the Fermi-level and annihilates a hole in a state below the Fermi-level. The excited minority electron, coming from a state above the Fermi-level, leaves the sample with an energy equal to the primary energy plus the energy of the electron–hole pair around the Fermi-level. It looks like that the scattered electron has gained energy. The apparent energy gain is due to the fact that the electron stems from a higher energy level of the system (a state above the Fermi-level). Such a process would lead to annihilation of an available magnon with a total angular momentum of $1\hbar$ and an energy of ε (see Sect. 7.4).

The relative rates of the processes mentioned in Fig. 7.4 for ferromagnetic surfaces have been measured first by Kirschner using a spin-polarized electron beam and a simultaneous spin-polarized detection [44]. It turned out that the "*flip*" rate of spin-down electrons is larger then the one of the spin-up electrons. This is mainly due to the fact that in normal ferromagnets there are more minority states available above the Fermi-level and consequently more majority electrons below the Fermi-level. This implies that the probability of having pairs of majority-hole and minority-electron is higher than the ones of minority-hole and majority-electron [44]. Interestingly, the experiment showed that the relative contributions of different rates depend on the wave vector of the excitations and could also be confirmed by other experiments [45–47].

At this point, we would like to emphasize that the processes referred to as "*flip*" processes above are due to the exchange scattering mechanism and there is no "*direct spin flip*" process involved via the scattering. The "*direct spin flip*" process can, in principle, happen under some circumstances. For instance, when electrons are scattered from a surface with a large spin-orbit coupling or when they experience a magnetic field, they may flip their spin via the well-known Larmor precession. However, these processes are very much slower than the exchange process and happen on the nano-/pico-second time-scale. The explanation of their physical origin is out of the scope of the present chapter (fruitful information may be found, for example, in [42, 43, 48, 49]). These processes are not relevant for the electron scattering from ferromagnetic surfaces with a small spin-orbit coupling. The electric and magnetic fields involved are far too small.

Since magnons carry a total angular momentum of $1\hbar$, they are created by minority electrons in an exchange scattering process [50]. The magnon creation process is mediated by the Coulomb interaction between the electrons and no explicit spin–spin interaction is involved. It takes place within a few tens of attoseconds.

7.3 Spin-Polarized Electron Energy Loss Spectroscopy

The spin-polarized electron energy loss spectroscopy (SPEELS) is based on the scattering of spin polarized electrons from a magnetic surface [51]. Spin polarized electrons are created by using spin-polarized photoemission from a GaAs photocathode. A circularly polarized laser beam is incident into the photocathode. According to the selection rules, the photo-emitted electrons will have their spins either parallel or antiparallel to the incident direction of the laser beam, depending on the helicity of the light. Since a normal GaAs has a total polarization of 50 % (based on the spin-dependent photoemission selection rules), usually a so-called strained semiconductor heterostructure is used for this purpose. Taking the advantage of hereroepitaxy one can grow a semiconductor heterostructure with a large lattice strain [52]. The large strain would modify the band structure so that a large spin polarization (as large as 90 %) can be achieved [52]. The spin-polarized electrons are monochromatized to get a certain energy and are focused onto the sample surface. The spin of the incoming electrons is either parallel or antiparallel to the majority electrons of the sample. The former type of electrons are usually called as spin up and the latter ones are called spin-down electrons. The scattered electrons are collected by a channeltron at a given scattering geometry and their energy is analyzed [51]. Note that no spin-resolved detection is involved in this experiment. The experiments are usually performed for both spin channels, simultaneously.

In Fig. 7.5, the scattering geometry is shown. If one assumes that the energy and momentum in initial and final states before and after scattering are E_i, $\mathbf{k}_i$ and E_f, $\mathbf{k}_f$, respectively, the energy ε and the wave vector q of the excitations can be given by the following expressions:

$$\varepsilon = E_f - E_i; \qquad q = -|\Delta \mathbf{k}_{\parallel}| = |\mathbf{k}_i| \sin\theta - |\mathbf{k}_f| \sin(\theta_0 - \theta). \tag{7.6}$$

In addition to the conservation of energy and momentum, the total angular momentum has to also be conserved. Hence magnon excitations are allowed only when minority electrons are incident. This implies that the magnon peak will appear only in the minority spin channel. Magnon excitations are forbidden when majority electrons are incident onto the sample surface. An example is given in Fig. 7.6, where typical SPEELS spectra measured on a Co film are shown. The peak in the minority spin channel ($I_{\downarrow}$) is due to the long-living magnon excitations and the raising background is attributed to the single-particle electron–hole pair excitations [53].

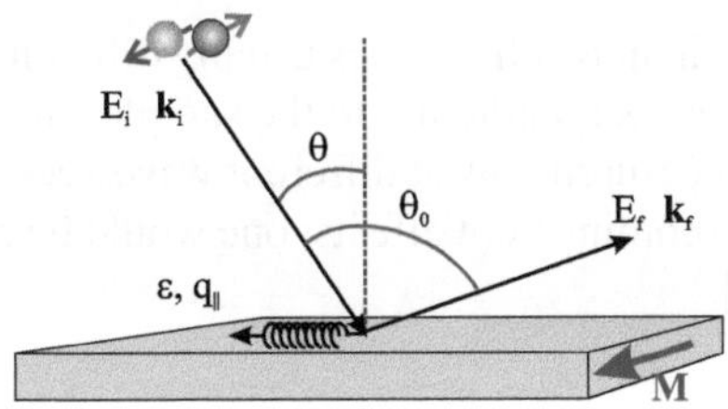

Fig. 7.5 A schematic representation of the scattering geometry used in SPEELS experiments. A monochromatic beam with a well-defined energy and momentum is focused onto the sample surface and the energy distribution of the scattered beam is measured at a given geometry for both spin orientations

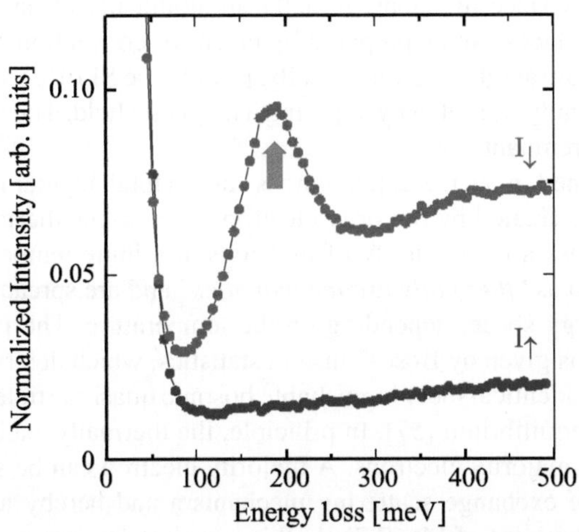

Fig. 7.6 Typical SPEELS spectra recorded on an epitaxial Co(100) film, with a thickness of 8 ML, grown on Cu(100). The energy of incident beam and the in-plane wave vector transfer were about $E_i = 6.7$ eV and $|\Delta \mathbf{k}_{\parallel}| = 0.87$ Å^{-1}, respectively. The peak in the minority spin channel ($I_{\downarrow}$) is due to the magnon excitations and the raising background is attributed to the single-particle electron–hole pair excitations [53]

In Sect. 7.4, it will be explained in detail how the magnons are created and annihilated in SPEELS experiments.

Magnons can also be excited within the tunneling process in scanning tunneling spectroscopy experiments [54–56]. Basically, the process is similar to the one in the SPEELS experiments. The tunneling electrons interact with the ones of the sample, and when they have enough kinetic energy to create an excitation, the tunneling current is enhanced. The excitation process leads to a step in the differential conductivity that is dI/dU and consequently a peak in d^2I/dU^2. The main difference here is that the excitation may happen in the forward and backward tunneling directions. The excitations can be of magnetic and non-magnetic nature. By using a magnetic tip and changing the magnetization direction of the tip, one may confirm that the excitations are of magnetic origin. The main disadvantage of this technique is that the control of the tip magnetization is usually difficult. Due to the fact that the tunneling current is almost perpendicular to the sample surface, the technique is not capable of performing measurements at different wave vectors. Only when standing waves are formed due to confinement effects, one would be able to observe different magnon modes [54–56].

7.4 Recent Experimental Achievements

7.4.1 Magnon Excitations in Ferromagnetic Thin Films

For studying the surface magnons, initially an atomically clean surface has to be obtained. Such surfaces can be prepared by means of conventional thin film growth techniques under ultrahigh vacuum. Usually, prior to the SPEELS measurements the films are magnetically saturated by applying a magnetic field. The measurements are performed in the remnant state.

As it is mentioned in Sect. 7.2.4, magnons carry a total angular momentum of $1\hbar$, therefore, they are created by minority electrons. Due to the thermal fluctuations, a spin system possess a large variety of magnons at a finite temperature. These are usually referred to as "*thermally excited magnons*" and are spread over a large momentum and energy space, depending on the temperature. The population of this class of magnons is given by Bose–Einstein statistics, which determines the statistical distribution of identical indistinguishable bosonic quasi-particles over the energy states in thermal equilibrium [57]. In principle, the thermally excited magnons can be annihilated by majority electrons. A majority electron can be scattered to a minority one via the exchange scattering mechanism and hereby a magnon will be annihilated (see the inset of Fig. 7.7). In this process, the energy of the electron in the final state is larger then the one in the initial state, and hence such a process can be observed in the energy gain region. The intensity of the peak associated with the magnon annihilation process depends on the population of the thermally excited magnons at the measurement temperature. The magnon population is given by a Boltzmann factor, which is about 0.17 for the magnons with an excitation energy of 46 meV at $T = 300$ K. It is nearly zero for $T = 10$ K. This means that at low temperatures the magnon annihilation process is expected to be suppressed and at high temperatures it is pronounced. Here we would like to point out that such a behavior is also expected for phonons that are also classified as bosons. An example is provided in Fig. 7.7 where the SPEELS spectra measured at different temperatures are presented. The spectra are recorded on a 2 ML thick Fe(110) film grown on W(110) with a wave vector of 0.5 Å^{-1}. The magnon propagation direction in this particular experiment is along the [001]-direction of the Fe(110) surface ($\bar{\Gamma}$–$\bar{H}$-direction of the surface Brillouin zone). For each case the spectra for spin-up ($I_{\uparrow}$) and spin-down ($I_{\downarrow}$) electrons are measured. In addition, the sum ($I_{\downarrow} + I_{\uparrow}$), difference ($I_{\downarrow} - I_{\uparrow}$), and asymmetry $[(I_{\downarrow} - I_{\uparrow})/(I_{\downarrow} + I_{\uparrow})]$ spectra are also presented. Let us first start with Fig. 7.7(a). It is apparent that the spectra are dominated by presence of the quasi-elastic peak at $E = 0$ meV. The magnon creation and annihilation peaks are located

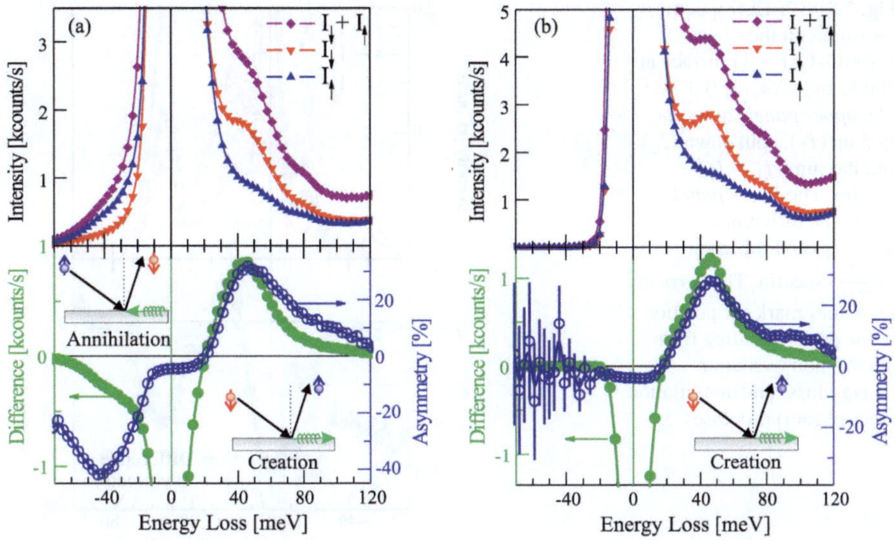

Fig. 7.7 SPEELS spectra measured at $|\Delta \mathbf{k}_{\|}| = 0.5$ Å^{-1} on an ultrathin Fe(110)/W(110) film with a thickness of 2 ML at (**a**) 300 K and (**b**) 10 K. The *upper panels* show the spin up ($I_{\uparrow}$), spin down ($I_{\downarrow}$) and the sum ($I_{\uparrow} + I_{\downarrow}$) spectra. The *lower panels* show the difference ($I_{\downarrow} - I_{\uparrow}$) and asymmetry ($\frac{I_{\downarrow}-I_{\uparrow}}{I_{\downarrow}+I_{\uparrow}}$) spectra. The magnon creation and annihilation processes are schematically sketched in the *insets*

beside the elastic peak in the energy loss and gain region, respectively. These two processes are sketched schematically in the insets of Fig. 7.7. The asymmetry curve shows a change in the sign from negative to positive, when going from gain to loss region. The minima (in the gain region) and the maxima (in the loss region) are the places where the excitation and annihilation processes take place. If one neglects the spin–orbit coupling effects in the system the maxima and minima should be located at the same energies (one negative and the other positive). As it was mentioned earlier, at low temperature where the population of the thermally excited magnons is very low, the magnon annihilation peak supposed to be strongly suppressed. This fact can be clearly seen in Fig. 7.7(b), where the measurements at 10 K, performed on the same sample, are presented. No trace of magnon annihilation could be detected neither in difference nor in asymmetry spectrum.

In order to obtain the magnon dispersion relation, usually the spectra are recorded for different wave vectors [38, 39, 53, 58–61]. The desired wave vector is achieved by changing the scattering geometry, i.e., the angle between incident beam and the sample normal (θ_i, see (7.6) and Fig. 7.5).

7.4.2 Distinguishing Between Magnons and Phonons

In principle, both magnons and phonons can be excited by electrons. Since both kinds of excitations may show very similar energies, it is rather difficult to distin-

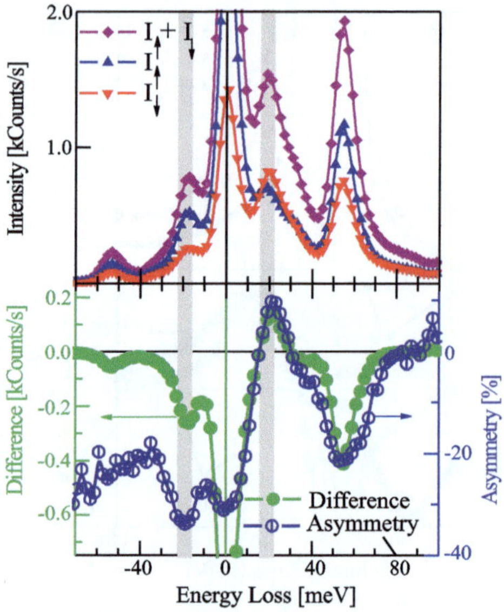

Fig. 7.8 SPEELS spectra measured on the Fe(001)–O(1 × 1) surface at 300 K and $|\Delta \mathbf{k}_{\parallel}| = 0.3$ Å^{-1}. The *upper panel* shows the spin up ($I_{\uparrow}$), spin down ($I_{\downarrow}$) and the sum ($I_{\uparrow} + I_{\downarrow}$) spectra. The *lower panel* shows the difference ($I_{\downarrow} - I_{\uparrow}$) and the asymmetry ($\frac{I_{\downarrow}-I_{\uparrow}}{I_{\downarrow}+I_{\uparrow}}$) spectra. The *vertical gray lines* mark the position of the peaks resulting from the magnon excitation (energy loss) and annihilation (energy gain) processes

guish between them experimentally. However, using the spin degree of freedom of spin-polarized electrons opens a possibility to separate magnons from phonons. As magnons carry a total angular momentum of 1 $\hbar$, they can only be excited by incidence of minority electrons. The time reversal process happens for incidence of majority electrons, which leads to the magnon annihilation. This fact would lead to a sign change in the asymmetry curve in gain and loss regions (see the lower panel of Fig. 7.7(a)). In the case of phonons, the situation is different. Since phonons are spin-independent quasi-particles, they can be created and annihilated by incidence of electrons with any spin direction. We note that the process which leads to phonon creation can also be mediated by the exchange process (the exchange of the electrons with the same spin, see Fig. 7.4). The particular dependence of magnon creation and annihilation on the spin of the incident electrons, as it is different from phonon excitations, is a fundamental feature, which can be used to distinguish between magnons and phonons in spin-polarized electron scattering experiments. The best way to identify the nature of an excited quasi-particle (magnon or phonon), is to compare the signs of asymmetry curves in loss and gain regions. An example is provided in Fig. 7.8, where the spectra of an oxygen passivated Fe(001), measured at $T = 300$ K and $|\Delta \mathbf{k}_{\parallel}| = 0.3$ Å^{-1}, are presented. The magnon and phonon excitations coincide within the same energy window. Interestingly, the intensity of all excitations depends on the spin orientation of the incoming beam. In such a case, the fundamental question is: How can one distinguish between magnons and phonons? As it is discussed above, the identification of phonons and magnons can be done by looking at their different spin nature. Asymmetries of loss and gain spectra of the peaks marked by the vertical gray lines in Fig. 7.8 (at energies of ±19 meV) are of opposite signs, and hence they are associated with magnon excitations. Asymmetry of the other excitations have the same sign and almost identical magnitudes; therefore, they are caused by phonon excitations.

It is interesting to mention that in the case of the Fe(001)–O(1 × 1) surface the asymmetries of phonon induced peaks are always negative. A complementary experiment showed that the asymmetry of the phonon peaks follows the one of the elastic peak. This fact implies that very likely the asymmetry of the phonon peaks has the same origin as the one of the elastic peak [62]. However, the details of the effect are expected to depend on many parameters like the energy of the incident beam, excitation energy, and the scattering geometry. Even, the sign of the spin asymmetry may change while changing the electron energy. Recently, the dispersion relation of magnons and phonons could be measured simultaneously by varying the wave vector and the dispersion branches could be separated based on their different spin nature [62].

7.5 Conclusion

We discussed the processes involved in high wave-vector magnon excitations in itinerant ferromagnets by using spin-polarized electron scattering experiments. Since magnons carry a total angular momentum of $1\hbar$, they can only be excited by incidence of minority electrons and annihilated by incidence of majority ones. The excitation process is commonly called as an inelastic electron scattering process. However, we think that the process itself is elastic and the observed energy loss (or gain) is due to the fact that the ejected electron stems from a lower (or higher) energy level of the excited solid. The process is extremely fast (taking place within a few tens of attoseconds). It is mediated by the exchange interaction, i.e., the Coulomb interaction between electrons and no explicit spin–spin interaction is needed to be taken into consideration.

A possibility to distinguish between magnon- and phonon-excitations based on their spin nature is discussed. Magnons can only be excited by incidence of minority electrons and annihilated by incidence of majority ones whereas, phonons can be excited and annihilated by incidence of electrons with any spin orientation. This fact would lead to a sign change in the asymmetry curve in the gain and loss regions where the magnons are excited. For the case of phonons, no change in the sign of the asymmetry curve is expected. This means that the magnons and phonons can be distinguished by comparing the sign of the symmetry curves in loss and gain regions and there is no need of further spin resolved detection of the scattered electrons.

Acknowledgements We thank all the present and former co-workers: H. Ibach, R. Vollmer, M. Etzkorn, P.S. Anil Kumar, W.-X. Tang, J. Prokop, I. Tudosa, Y. Zhang, T.R.F. Peixoto, and T.-H. Chuang who have contributed to the various experiments discussed here.

References

1. S.A. Wolf, D.D. Awschalom, R.A. Buhrman, J.M. Daughton, S. von Molnar, M.L. Roukes, A.Y. Chtchelkanova, D.M. Treger, Science **294**, 1488 (2001)

2. C.A.F. Vaz, J.A.C. Bland, G. Lauhoff, Rep. Prog. Phys. **71**, 056501 (2008)
3. J.P. Carbotte, E. Schachinger, D.N. Basov, Nature **401**, 354 (1999)
4. T. Dahm, V. Hinkov, S.V. Borisenko, A.A. Kordyuk, V.B. Zabolotnyy, J. Fink, B. Büchner, D.J. Scalapino, W. Hanke, B. Keimer, Nat. Phys. **5**, 217 (2009)
5. K.-J. Lee, A. Deac, O. Redon, J.-P. Nozieres, B. Dieny, Nat. Matters **3**, 877 (2004)
6. D.-S. Han, S.-K. Kim, J.-Y. Lee, S.J. Hermsdoerfer, H. Schultheiss, B. Leven, B. Hillebrands, Appl. Phys. Lett. **94**, 112502 (2009)
7. Y. Le Maho, J.-V. Kim, G. Tatara, Phys. Rev. B **79**, 174404 (2009)
8. F. Bloch, Z. Phys. **61**, 206 (1930)
9. C. Herring, C. Kittel, Phys. Rev. **81**, 869 (1951)
10. E.C. Stoner, Proc. R. Soc. A **154**, 656 (1936)
11. E.C. Stoner, Proc. R. Soc. A **165**, 372 (1938)
12. D.M. Edwards, Proc. R. Soc. A **235**, 305 (1962)
13. M. Farle, Rep. Prog. Phys. **61**, 755 (1998)
14. C.E. Patton, C.H. Wilts, F.B. Humphrey, J. Appl. Phys. **38**, 1358 (1967)
15. C.E. Patton, Dynamic processes in magnetic thin films. Domain wall motion and ferromagnetic resonance, Ph.D. thesis, California Institute of Technology (1967)
16. D.L. Mills, S.M. Rezende, *Spin Dynamics in Confined Magnetic Structures II* (Springer, Berlin, 2003)
17. K. Zakeri, J. Lindner, I. Barsukov, R. Meckenstock, M. Farle, U. von Hörsten, H. Wende, W. Keune, J. Rocker, S.S. Kalarickal, K. Lenz, W. Kuch, K. Baberschke, Z. Frait, Phys. Rev. B **76**, 104416 (2007)
18. J.A.C. Bland, B. Heinrich, *Ultrathin Magnetic Structures III* (Springer, Berlin, 2005)
19. P.A. Grünberg, Surf. Sci. **18**, 1 (1985)
20. B. Hillebrands, K. Ounadiela (Eds), in *Spin Dynamics in Confined Magnetic Structures III*. Topics in Applied Physics, vol. 83 (Springer, Berlin, 2003)
21. J. Kirschner, D. Rebenstorff, H. Ibach, Phys. Rev. Lett. **53**, 698 (1984)
22. J. Kirschner, S. Suga, Surf. Sci. **178**, 327 (1986)
23. J. Kirschner, E. Langenbach, Solid State Commun. **66**, 761 (1988)
24. R.B. Muniz, D.L. Mills, Phys. Rev. B **66**, 174417 (2002)
25. R.B. Muniz, A.T. Costa, D.L. Mills, J. Phys., Condens. Matter **15**, S495 (2003)
26. A.T. Costa, R.B. Muniz, D.L. Mills, Phys. Rev. B **68**, 224435 (2003)
27. A.T. Costa, R.B. Muniz, D.L. Mills, Phys. Rev. B **74**, 214403 (2006)
28. R.B. Muniz, A.T. Costa, D.L. Mills, IEEE Trans. Magn. **44**, 1974 (2008)
29. A.T. Costa, R.B. Muniz, J.X. Cao, R.Q. Wu, D.L. Mills, Phys. Rev. B **78**, 054439 (2008)
30. E. Şaşioglu, A. Schindlmayr, C. Friedrich, F. Freimuth, S. Blügel, Phys. Rev. B **81**, 054434 (2010)
31. P. Buczek, A. Ernst, P. Bruno, L.M. Sandratskii, Phys. Rev. Lett. **102**, 247206 (2009)
32. P. Buczek, A. Ernst, L.M. Sandratskii, Phys. Rev. Lett. **105**, 097205 (2010)
33. P. Buczek, A. Ernst, P. Bruno, L.M. Sandratskii, J. Magn. Magn. Mater. **322**, 1396 (2010)
34. P. Buczek, A. Ernst, L.M. Sandratskii, Phys. Rev. Lett. **106**, 157204 (2011)
35. I.E. Dzyaloshinskii, Sov. Phys. JETP **5**, 1259 (1957)
36. T. Moriya, Phys. Rev. **120**, 91 (1960)
37. C. Kittel, *Introduction to Solid State Physics* (Wiley, New York, 1996)
38. W.X. Tang, Y. Zhang, I. Tudosa, J. Prokop, M. Etzkorn, J. Kirschner, Phys. Rev. Lett. **99**, 087202 (2007)
39. J. Prokop, W.X. Tang, Y. Zhang, I. Tudosa, T.R.F. Peixoto, K. Zakeri, J. Kirschner, Phys. Rev. Lett. **102**, 177206 (2009)
40. H. Ibach, D.L. Mills, *Electron Energy Loss Spectroscopy and Surface Vibrations* (Academic Press, New York, 1982)
41. J. Kessler, *Polarized Electrons* (Springer, Berlin, 1985)
42. J. Kirschner, *Polarized Electrons at Surfaces*. Springer Tracts in Modem Physics, vol. 106 (Springer, Heidelberg, 1985)

43. J. Kirschner, Inelastic electron scattering by ferromagnets, in *Polarized Electrons in Surface Physics*, ed. by R. Feder (World Scientific, Singapore, 1985)
44. J. Kirschner, Phys. Rev. Lett. **55**, 973 (1985)
45. R. Feder, J. Kirschner, Solid State Commun. **40**, 547 (1981)
46. D. Venus, J. Kirschner, Phys. Rev. B **37**, 2199 (1988)
47. M.S. Hammond, G. Fahsold, J. Kirschner, Phys. Rev. B **45**, 6131 (1992)
48. J. Kirschner, R. Feder, J.F. Wendelken, Phys. Rev. Lett. **47**, 614 (1981)
49. D. Venus, H.L. Johnston, Phys. Rev. B **50**, 15787 (1994)
50. M. Plihal, D.L. Mills, J. Kirschner, Phys. Rev. Lett. **82**, 2579 (1999)
51. H. Ibach, D. Bruchmann, R. Vollmer, M. Etzkorn, P.S.A. Kumar, J. Kirschner, Rev. Sci. Instrum. **74**, 4089 (2003)
52. P. Drescher et al., Appl. Phys. A **63**, 203 (1996)
53. R. Vollmer, M. Etzkorn, P.S. Anil Kumar, H. Ibach, J. Kirschner, Phys. Rev. Lett. **91**, 147201 (2003)
54. T. Balashov, A.F. Takacs, W. Wulfhekel, J. Kirschner, Phys. Rev. Lett. **97**, 187201 (2006)
55. C.L. Gao, A. Ernst, G. Fischer, W. Hergert, P. Bruno, W. Wulfhekel, J. Kirschner, Phys. Rev. Lett. **101**, 167201 (2008)
56. T. Balashov, A.F. Takacs, M. Däne, A. Ernst, P. Bruno, W. Wulfhekel, Phys. Rev. B **78**, 174404 (2008)
57. R. Kobu, M. Toda, N. Hashitsume, *Statistical Physics II Nonequilibrium Statistical Mechanics* (Springer, Heidelberg, 1992)
58. M. Etzkorn, P.S. Anil Kumar, W. Tang, Y. Zhang, J. Kirschner, Phys. Rev. B **72**, 184420 (2005)
59. M. Etzkorn, P.S. Anil Kumar, J. Kirschner, High-energy surface spin waves studied by spin-polarized electron energy loss spectroscopy, in *Handbook of Magnetism and Advanced Magnetic Materials*, ed. by H. Kronmüller, S. Parkin, Novel Techniques for Characterizing and Preparing Samples, vol. 3 (Wiley, New York, 2007)
60. Y. Zhang, P. Buczek, L.M. Sandratskii, W.X. Tang, J. Prokop, I. Tudosa, T.R.F. Peixoto, K. Zakeri, J. Kirschner, Phys. Rev. B **81**, 094438 (2010)
61. K. Zakeri, Y. Zhang, J. Prokop, T.-H. Chuang, N. Sakr, W.-X. Tang, J. Kirschner, Phys. Rev. Lett. **104**, 137203 (2010)
62. Y. Zhang, P.A. Ignatiev, J. Prokop, I. Tudosa, T.R.F. Peixoto, W.X. Tang, K. Zakeri, V.S. Stepanyuk, J. Kirschner, Phys. Rev. Lett. **106**, 127201 (2011)

Chapter 8
Micromagnetic Simulations in Magnonics

M. Dvornik, Y. Au, and V.V. Kruglyak

Abstract We review the use of numerical micromagnetic simulations ("micromagnetics") for investigations in magnonics, the study of spin waves and their quanta – magnons. We argue that, when used with suitable post-processing tools, micromagnetics provide the power and flexibility necessary both for interpretation of complex magnonic phenomena observed in realistic magnetic structures and devices and for prediction of novel effects. We foresee that the development of multiscale and multiphysics extensions of micromagnetic solvers will broaden both the scope of micromagnetic simulations in magnonics and the field of magnonics itself. For example, the extension of micromagnetics to solvers based on atomistic spin models will underpin application of the developed methodology to studies of phenomena involving both magnons and other fundamental excitations of the solid state. In a more distant perspective, it is highly intriguing to study spin waves in non-stationary conditions (i.e. in structures with time dependent material properties), such as those realized in experiments with samples under ultrafast optical pumping.

8.1 Introduction

Technically, the age of magnonics [1] should be counted from the discovery of spin waves more than 80 years ago [2]. Yet, the term "magnonics" itself emerged only recently [3–5], mainly to emphasize the field's obvious similarities with those of electronics, photonics and plasmonics. At the same time, the term "magnonic crystals" [6] was introduced for periodic magnetic media [7–9], exploiting the same analogy with photonic crystals in photonics. These two events in semantics played a major role in triggering the currently observed renaissance of magnonics as the study of spin waves and their quanta – magnons, often interpreted as emergence of the field's new identity [1, 10–16].

This process has also been underpinned by objective factors, the major one being the revolutionary development of nanotechnology. In particular, the advances in

M. Dvornik · Y. Au · V.V. Kruglyak (✉)
School of Physics, University of Exeter, Stocker Road, Exeter EX4 4QL, UK
e-mail: V.V.Kruglyak@exeter.ac.uk

S.O. Demokritov, A.N. Slavin (eds.), *Magnonics*, Topics in Applied Physics 125,
DOI 10.1007/978-3-642-30247-3_8, © Springer-Verlag Berlin Heidelberg 2013

nanofabrication have brought into spotlight magnetic nanostructures, in which the confinement effects and the exchange interaction both play significant roles defining the spectrum [17, 18] and dispersion [10, 19] of spin waves. The investigations have led the development of experimental tools able to detect the magneto-dynamical phenomena, with the major roles currently played by the Brillouin Light Scattering (BLS) [10, 17, 19], the time resolved scanning Kerr microscopy (TRSKM) [16, 18, 20], and the cavity and vector network analyzer based ferromagnetic resonance (FMR) measurements [21, 22]. It has soon been realized that the interpretation of the experimental findings at the nanoscale requires development of advanced numerical methods for solving the equations of motion of the magnetization [23], leading to adaptation of micromagnetic solvers [24–26] to dynamical problems.

Other factors brought about by nanotechnology were the development of advanced computing hardware and the continuous progress in miniaturization of data processing architectures, currently dominated by complementary metal-oxide-semiconductor (CMOS) technology [27]. The latter progress inspired magnonics researchers and engineers to interpret their results in terms of not only the well-established applications in microwave signal processing [28] but also in spin wave based logic computation – magnonic logic [1, 12, 13, 29–31] – with prospects of integration into monolithic microwave integrated circuits (MMIC). At the same time, the approaching scaling limit of CMOS has stimulated search for viable alternative data processing technologies, with magnonic logic considered as one of the promising candidates [27].

The growth of the available computing power has made numerically solvable problems closer to those faced by realistic experiments and envisaged applications. Moreover, the numerical modeling ("micromagnetic simulations" or simply "micromagnetics") has become remarkably popular, currently practiced in probably every magnetics lab. Hence, it appears timely to review here the micromagnetic methods that have been developed specifically in the context of magnonics. Our experience shows that the use of right methodology can not only broaden the range of solvable problems but also save valuable computational time for the more common ones. Furthermore, we describe our software ("semargl" [32]) that was developed specifically to address problems arising in the field of magnonics and associated technical difficulties connected with the need to efficiently handle large amounts of simulated data.

8.2 Real-Space–Time Domain Analysis: Magnonic Devices

The output from micromagnetic simulations belongs to the real-space–time domain, making them most suitable for analysis of phenomena occurring in the same domain and hence minimizing the need for additional data post-processing. Such problems include, e.g. magnetic switching [33, 34], vortex and domain wall motion [12, 35] and magnetization dynamics in spin transfer torque devices [36, 37]. In magnonics, the time and real space domain simulations have been used to study spin wave phase dependent phenomena [12, 29] and to demonstrate action of magnonic logic devices [12, 31]. The simulations also confirmed the important property of spin waves to follow the bending of magnonic waveguides, provided that the static magnetization is aligned along their length [38, 39]. This property is vital for reducing reflection loss in magnonic logic and signal processing devices. It also distinguishes magnonic waveguides from those of photonics. Indeed, spin waves are naturally confined to and (in this geometry) guided within magnetic materials, while in photonics one has to learn to confine and guide light prior to using it for data processing.

Figure 8.1 demonstrates the propagation of spin waves along a curved magnonic waveguide. Figure 8.1(a) shows the ground magnetic state of a U-shaped waveguide of rectangular cross-section with width and thickness of 100 and 10 nm, respectively. A 150×50 nm^2 "antenna" with thickness of 30 nm is placed on top of the right arm of the waveguide (also shown in Fig. 8.1(a)) at a spacing of 10 nm. In the simulations, we assume that both the waveguide and antenna are made of a Permalloy-like material.[1] The damping constant is set to $\alpha = 0.005$. The cell size in the simulations is $5 \times 5 \times 5$ nm^3, i.e. about the exchange length in the material. First, the system is relaxed at zero applied field with carefully pre-aligned magnetization direction so as to obtain the ground state shown in Fig. 8.1(a). Due to the shape anisotropy, the nanomagnet antenna is magnetized along its long axis (the x axis defined in Fig. 8.1(b)), which is orthogonal to the magnetization in the waveguide underneath the antenna (the negative y direction defined in Fig. 8.1(b)).

Subsequently, a global uniform rf field (covering both the waveguide and the antenna) of 0.1 Oe amplitude and 11.5 GHz frequency is applied in the y direction. Following the concept demonstrated experimentally in [40, 41], the frequency is selected to match that of the antenna's uniform FMR mode, as determined in a separated simulation using a pulsed excitation. As a result, the antenna is resonantly driven into uniform precession. The precessing magnetization of the antenna produces a rotating stray magneto-dipole field in its neighborhood. In contrast to the global driving field, the latter rotating field is highly non-uniform. Figure 8.1(b)–(e) shows snapshots of the out-of-plane component of the waveguide's magnetization (m_z) at moments of time equal to 0.25 T, 0.5 T, 0.75 T and 1 T, where T is the oscillation period of the rf field (i.e. $1/11.5$ GHz $= 86.96$ ps), after the system has achieved a steady dynamic state.

One can see that the stray field produced by the antenna couples efficiently to a spin wave of about 100 nm wavelength propagating inside the waveguide. The **k** vector of the spin wave aligned parallel to the local magnetization, which corresponds to the so-called "backward volume" geometry [23]. However, because of the short wavelength, the frequency of the spin wave is exchange dominated. So, it is moving forward (in terms of the wave front). This behavior is opposite to that

[1] The saturation magnetization is $M_S = 8.0 \times 10^5$ A/m, the exchange stiffness is $A = 1.3 \times 10^{-11}$ J/m, and the gyromagnetic ratio is $\gamma = 2.32 \times 10^5$ m/As. The corresponding exchange length is $\sqrt{\frac{2A}{\mu_0 M_S^2}} \approx 5$ nm. No crystal anisotropy is assumed. The value of the damping constant is specified separately for each simulation.

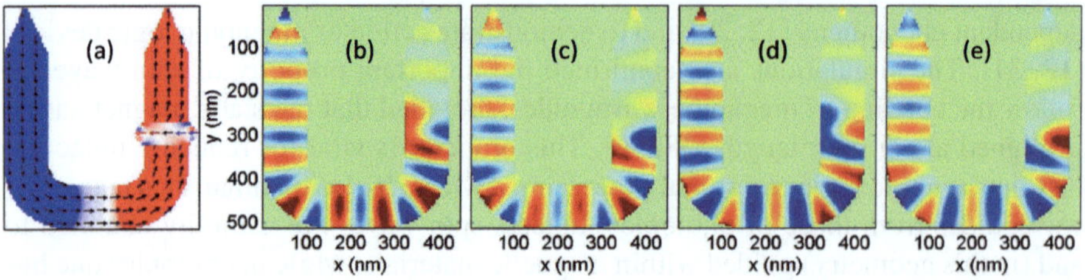

Fig. 8.1 (**a**) The ground magnetic state configuration of the U-shaped magnonic waveguide is shown together with that of the nanomagnet "antenna" sitting on its top, at remanence. (**b**) to (**e**) Simulated snapshots of the dynamic magnetization in the magnonic waveguide from (**a**) are shown for different moments relative to the phase of the global microwave magnetic field exciting the structure: $0.25T$, $0.5T$, $0.75T$ and $1T$, respectively, where T is the oscillation period of the incident microwave. The static background has been subtracted from the dynamic snapshots

observed for magnetostatic spin waves in this geometry. The spin wave initially propagates along the positive y direction. However, it is also capable of changing the direction of propagation at the 90 degree turns without any noticeable reflection, thereby following the physical bending of the U-shaped waveguide. After changing traveling direction twice, the spin wave enters the opposite arm on the U-shaped waveguide. Please note that the damping parameter in the $y < 150$ nm region of the waveguide is increased to 0.05 so as to dump the spin wave near and hence to minimize the reflection from the physical end of the waveguide.

The simulation demonstrates that, in this particular case, exchange dominated spin waves are capable of efficiently adapting their propagation direction as defined by the waveguide, which will prove extremely helpful in designing compact magnonic circuitries. On an integrated circuit chip, the space is precious and is often saved by folding different components together. The magnonic circuitry readily conforms to this demand. Furthermore, the system functions at zero applied field. At the same time, traditional and even recently proposed magnonic applications require application of an external magnetic field from few hundred oersteds up to several kOe, either in-plane or out-of-plane relative to the chip's surface. This requires the magnonic chip to be accompanied by a bulky external magnet, which is against the spirit of miniaturization aimed by the magnonic circuitry. The circuitry demonstrated in Fig. 8.1 at remanence eliminates the need for such a magnet. Instead, the system relies totally on the shape anisotropy of the waveguide and antenna to stabilize itself magnetically, yet allowing operation at a high frequency (11.5 GHz).

This particular study demonstrates how micromagnetic simulations can provide insights into magnonic circuitry designs that go beyond of what could be conjured by purely analytical methods. The system is geometrically non-symmetric, and the effect of the antenna's rotating stray field on the waveguide is hardly expected intuitively, given that a back action of the magnetization dynamics inside the waveguide on the antenna could also be important. The behavior of the spin wave encountering the 90 degree corner of the U-shaped waveguide is highly desired but its rigorous analytical treatment is formidable. However, micromagnetic simulations are definitely capable to help elucidate the way in which the radius of curvature at such corners governs the scattering of spin waves, albeit at the expense of computation time.

8.3 Real-Space–Frequency Domain Analysis: Magnonic Normal Modes

There are, however, cases when results of micromagnetic simulations are awkward to analyze in the real space and time domain. A typical example of this is the response of finite sized magnetic elements to excitation by a transient magnetic field [18, 20, 42–44]. To disentangle the response which consists of several oscillatory components corresponding to different normal modes of precession confined in the element, the time resolved signal is converted into the frequency domain. For the latter purpose as well as other needs of magnonics addressed in the following sections of this chapter, we have developed specialized high performance software *semargl* [32]. The software accepts data in the output format of Object Oriented Micromagnetic Framework (OOMMF) [24] and generates its own output in the user specified form (magnonic spectrum, dispersion, etc.) and format (graphical or numerical), as schematically depicted in Fig. 8.2. In the following, we present examples of the software's application to some typical problems of magnonics.

Let us begin by analyzing the response of a $100 \times 50 \times 10\ \text{nm}^3$ stadium shaped Permalloy-like (see footnote 1) element subject to a bias magnetic field of 1.5 kOe and excited by a transient magnetic field with the "sinc" temporal profile

$$H_{\text{ex}} = H_{\text{ex},0}\frac{\sin(2\pi f_{\max}(t-t_0))}{2\pi f_{\max}(t-t_0)}$$

and amplitude $H_{\text{ex},0}$ of 10 Oe. $f_{\max} = 20$ GHz is the cut-off frequency of the excitation [12]. The center of the excitation field is shifted by time t_0 spanning 20 zeroes of the sinc function. The dynamical states $\mathbf{M}(\mathbf{r},t)$ of the sample are recorded within 80 ns after the excitation with step dT of 10 ps. The subsequent Fourier analyses are therefore limited to frequencies up to $f_{\text{lim}} = \frac{1}{2dT} = 50$ GHz. The ground magnetic state is subtracted from the transient states, so that only the dynamic magnetization is analyzed:

$$\mathbf{m}(\mathbf{r},t) = \big(\mathbf{M}(\mathbf{r},t) - \mathbf{M}_0(\mathbf{r})\big)M_s^{-1}, \tag{8.1}$$

where $\mathbf{M}(\mathbf{r},t)$, $\mathbf{M}_{\text{S}}$ and M_{S} are the dynamic magnetization, the magnetization in the ground state and the saturation magnetization, respectively. $\mathbf{r}$ is the spatial coordinate (the radius-vector). This procedure corresponds to the "real-space–time domain analysis" box in Fig. 8.2, which essentially represents mapping of $\mathbf{m}(\mathbf{r},t)$ onto the color space, e.g. as shown in Fig. 8.1.

The "frequency domain analysis" box in Fig. 8.2 refers to different methods by which spectra of spin waves in the sample can be calculated. The spectra are generally calculated using a Discrete Fourier transform (DFT) of the dynamic magneti-

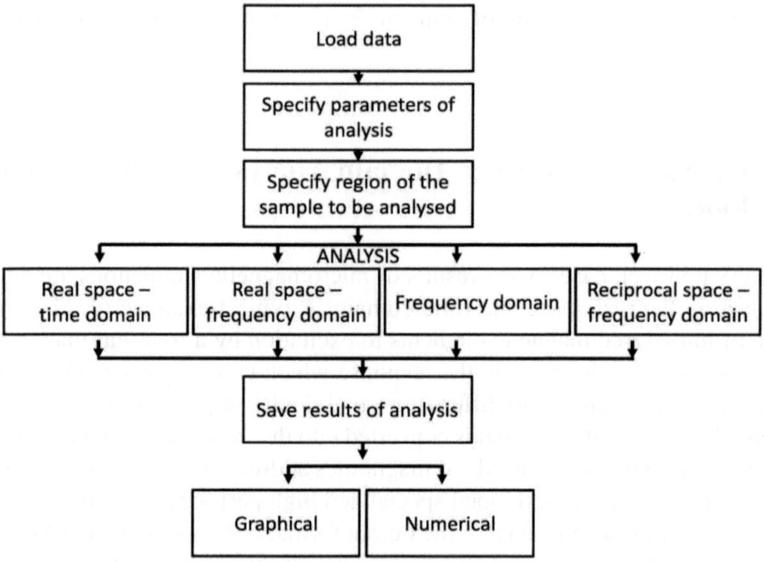

Fig. 8.2 The block diagram of *semargl* software is shown

zation

$$\mathbf{m}(\mathbf{r},\omega) = \mathrm{DFT}_t\big[w(t)\mathbf{m}(\mathbf{r},t)\big], \tag{8.2}$$

where DFT_t and $w(t)$ represent the DFT operator and the windowing function, respectively. However, the frequency domain analysis could be performed in three different ways that could be described as

$$\mathbf{m}(\omega) = \mathrm{DFT}_t\left[\sum_V p(\mathbf{r})\mathbf{m}(\mathbf{r},t)\right], \tag{8.3a}$$

$$\mathbf{m}(\omega) = \sum_V \big|p(\mathbf{r})\mathbf{m}(\mathbf{r},\omega)\big|, \tag{8.3b}$$

$$\mathbf{m}(\omega) = \sum_{d\mathbf{k}} \big|\mathrm{DFT}_{\mathbf{r}}\big[p(\mathbf{r})\mathbf{m}(\mathbf{r},\omega)\big]\big|, \tag{8.3c}$$

where the summation is performed either over a specified volume of the sample (see (8.3a) and (8.3b)) or over a specified volume in the reciprocal space (see (8.3c)). $\mathrm{DFT}_{\mathbf{r}}$ represents the spatial DFT operator, and $p(\mathbf{r})$ is a spatial weighting factor that is used to mimic various experimental conditions, such as the profile of the optical probe spot or the skin depth. The method of analysis represented by (8.3a) suppresses contributions from spatially non-uniform modes, e.g. canceling antisymmetric modes altogether. Methods (8.3b) and (8.3c) are free from this limitation, and so all normal modes contribute to the spectrum, provided that they are excited by the selected type of excitation. Importantly, method (8.3c) also allows one to flexibly restrict the consideration to modes with longer wavelengths by excluding contribu-

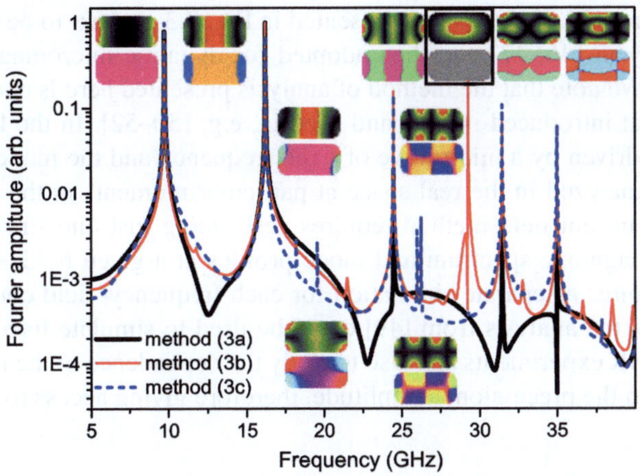

Fig. 8.3 The spectra of magnonic normal modes confined in the $100 \times 50 \times 10\ \text{nm}^3$ stadium shaped magnetic nanoelement are shown together with the spatial profiles of the dominant modes. The *top* and *bottom* images in each pair represent spatial profiles of the DFT amplitude and phase, respectively. A bias magnetic field of 1.5 kOe is applied along the major axis of the element. The different spectra shown are calculated using the different analysis methods as described in the text

tions from modes at the edge of the Brillouin zone of the discrete micromagnetic model. As we show below, the latter modes, which correspond to relative oscillations of neighboring cells and are presumably excited due to numerical errors, can lead to artificial peaks in the lower frequency part of the spectrum due to aliasing associated with the discrete temporal sampling.

The "real-space–frequency domain analysis" box in Fig. 8.2 refers to characterization of spatial profiles of the normal modes identified in the calculated magnonic spectra. For example, frequencies ω_i, of the normal modes can be extracted by fitting peaks in $\mathbf{m}(\omega)$ to the Lorentzian functions. The spatial profiles of the modes are then obtained by mapping $\mathbf{m}(\mathbf{r}, \omega_i)$ onto a color space, as in case of the real-space–time domain analysis.

Methodologically, the mode profiles allow us to identify the aliasing artefacts in the spectra. The results of application of the different analysis methods to the time resolved response of the magnetic nanoelement are presented in Fig. 8.3. Expectedly, the spectrum obtained using method (8.3b) contains several additional peaks as compared to that obtained using method (8.3a). However, inspection of the amplitude and phase maps (those in the black frame) corresponding to the major of such peaks shows that the precession uniformly occupies the majority of the element while the phase alternates between neighboring cells. The frequency of such precession would be exchange dominated and lie well in the THz range. Moreover, such a mode would have extremely weak coupling to the uniform transient field used in the simulations and hence could not be excited with the observed high amplitude. The considerations point to the artificial origin of the mode.

One of the early examples of the use of micromagnetic modeling to study spectra of spin waves in small magnetic elements is found in [45], while the sort of mapping of magnonic normal modes as that presented in Fig. 8.3 appears to be demonstrated experimentally in [46, 47] and then adopted for dynamic micromagnetic simulations in [48]. We note that the method of analysis presented here is computationally superior to that introduced in [49] and used in, e.g. [50–52]. In the latter method, the sample is driven by a microwave of fixed frequency and the mode spatial character is then analyzed in the real space at particular moments in the time domain. Indeed, the transient field method requires performing just one simulation to obtain the full magnonic spectrum and mode profiles at a given field, while the one from [49] requires a separate simulation for each frequency–field combination. At the same time, the analysis from [49] could be used to simulate fixed frequency – swept field FMR experiments and also to study the dependence of the magnetization dynamics upon the precessional amplitude, therefore giving access to the nonlinear regime.

8.4 Reciprocal-Space–Frequency Domain: Magnonic Dispersion and Scattering Parameters

The wave dispersion relation offers the most convenient way to characterize properties of waves. In particular, the great majority of analytical spin wave theories yield the magnonic dispersion as their main result [3–9, 14, 23, 53–55]. For properly formulated problems, the results of micromagnetic simulations can also be presented in this important form by performing spatial Discrete Fourier transform $\mathrm{DFT}_{\mathbf{r}}[p(\mathbf{r})\mathbf{m}(\mathbf{r},\omega)]$ [38, 56–59]. For an illustration, let us consider the case of spin waves propagating in a waveguide representing a straight 10.5 μm × 100 nm × 10 nm Permalloy (see footnote 1) nanowire with a 100 nm long cobalt (Co) layer placed in its center [60], as shown in the inset of Fig. 8.4. The cell size of the rectangular mesh is equal $s_x \times s_y \times s_z = 1 \times 100 \times 10\ \mathrm{nm}^3$, and so, the width and thickness of the mesh cell coincide with the corresponding dimensions of the sample. Hence, the model is one dimensional (1D). The Co layer plays role of a scatter center ("defect"), which could be used in, e.g. the design of magnonic interferometers [29–31]. The end regions of the waveguide (marked as "D") are damped to prevent reflections of spin waves and associated interference effects. The spin waves are excited by a sinc magnetic pulse of 50 Oe localized at the boundary between the left damped and center regions of the waveguide. The cut-off frequency is set to $f_{\max} = 4$ THz [12]. Each simulation is run for 8 ns and the data are recorded every $dT = 120$ fs, which corresponds to the excitation frequency bandwidth $f_{\lim}$ of 4.17 THz. So, condition $f_l < f_{\lim}$ necessary to prevent aliasing is satisfied.

The results of applying $\mathrm{DFT}_t[\mathbf{m}(\mathbf{r},t)]$ and then $\mathrm{DFT}_{\mathbf{r}}[p(\mathbf{r})\mathbf{m}(\mathbf{r},\omega)]$ procedures to the data from the non-damped part of the left half of the waveguide are shown in Fig. 8.4 together with those from identical simulations performed neglecting the magneto-dipole interaction ("exchange approximation"). The greyscale intensity is proportional to the spin wave amplitude at particular values of the wave number and frequency. At high frequencies, the simulated dispersion curve has a characteristic

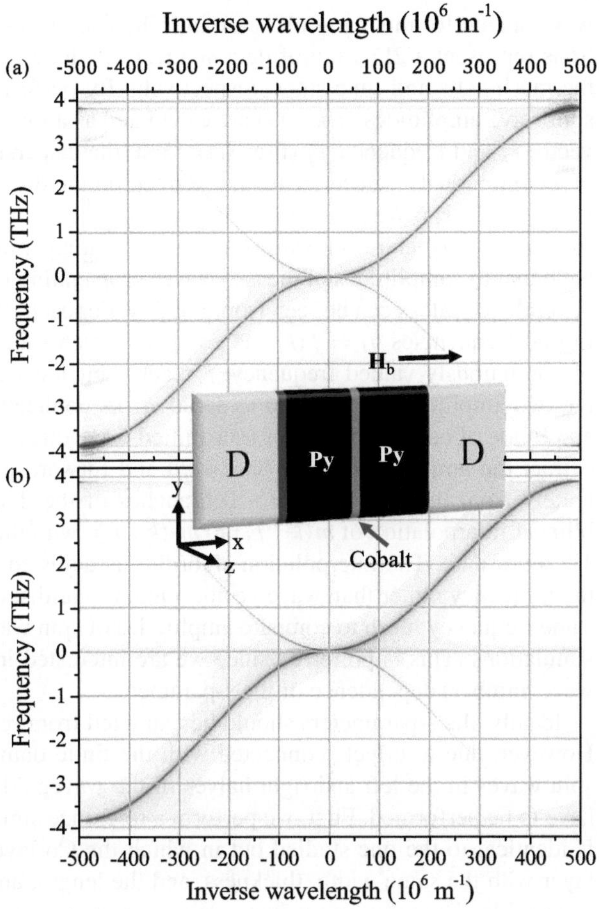

Fig. 8.4 The dispersion of spin waves in the "input" waveguide is shown for simulations performed in exchange (**a**) and dipole-exchange (**b**) approximations. The greyscale intensity represents the absolute value of the spin wave amplitude

downward curvature owing to the discrete nature of the numerical problem [60]. Indeed, the highest frequency and wave number accessible in the simulations correspond to the edge of the Brillouin zone of the spectrum of the 1D chain of spins with a period equal to the cell size. Hence, the frequency range of agreement between the theory and simulations could be expanded by reducing the cell size. Physically, the cell size could not be reduced below the characteristic atomic scale. This hints to the need to adopt atomistic spin models within micromagnetic solvers in order to study magnonic phenomena at THz frequencies. Such solvers might facilitate exploration of quantum properties of magnons, such as Bose–Einstein condensation, quantum entanglement and Berry phase.

The signs of the wave number and slope of the dispersion curve show the direction of the magnonic phase and group velocities, respectively, allowing one to isolate the amplitudes of the incident and reflected waves. The access to the amplitude information suggests a means for calculation of the values of the magnonic scattering coefficients, i.e. the reflection and transmission coefficients or "s-parameters", as was realized for this particular case in [60]. The results of the time domain simulations represent a 2D array of data $m_z(x_i, t_j)$, where i and j are integer indices of the mesh cells and time steps, respectively. By performing a 2D DFT of the data, spin wave amplitudes $m(k_i, f_j)$ are calculated as a function of discrete valued wave vector k_i and frequency f_j (Fig. 8.4). First, the dispersion is calculated in the form of a continuously valued frequency defined on the discrete mesh of the wave numbers, $f_i = f(k_i)$. We assume that dispersion $f(k)$ of spin waves is equivalent for the forward ($k > 0$) and backward ($k < 0$) propagating spin waves, i.e. $f(k) = f(-k)$. Note that the amplitudes of the two waves propagating in opposite directions are not generally equal, as can be seen from Fig. 8.4. For each $|i|$, we use cubic interpolation to find frequencies $f_i = f(k_i)$ as points at which functions $m(k_i, f) + m(k_{-i}, f)$ of continuously valued frequency f reach their local maxima. The summation of the two amplitudes is required to avoid errors connected with the reduction of the amplitude of either reflected or transmitted wave due to, e.g. interference. Then, we extract the amplitudes of the backward and forward propagating spin waves separately from the $k < 0$ and $k > 0$ branches of the dispersion, respectively, using bilinear interpolation of $m(k_i, f_j)$ to $m_i(k_i, f_i)$, with the latter now being a discrete 1D set of data. The interpolation algorithm is adjusted so that the discretization of the frequency rather than wave vector remains equidistant. This allows us to use the same frequency mesh to compare amplitudes of spin waves extracted from different simulations. This is preferred since we are interested in the frequency (rather than wave number) dependence of the s-parameters.

Ideally, the s-parameters should be extracted from results of a single simulation. However, due to effects connected with the finite damping and group velocity of spin waves in the left and right halves of the waveguide, two different simulations have to be performed. First, we perform a reference simulation for a waveguide that is identical to the one studied but in which the Co layer is replaced by a uniform layer with the same width, thickness, and the length, and with the properties of the waveguide. The reference waveguide is therefore magnetically uniform. Reference amplitudes $m^f_{R,I}(f)$ and $m^f_{R,III}(f)$ of the forward spin waves in the input and output waveguides, respectively, are then calculated as functions of the frequency. After that, simulations for the waveguide with the Co layer are performed, and amplitudes $m^b_I(f)$ and $m^f_{III}(f)$ of the reflected from and transmitted through the Co layer spin waves are obtained for the left and right halves of the waveguide, respectively. Finally, the values of s-parameters S_{11} and S_{21} are calculated as

$$\begin{aligned} S_{11} &= \frac{m^b_I}{m^f_{R,I}} = R(f), \\ S_{21} &= \frac{m^b_{III}}{m^f_{R,III}} = T(f) \end{aligned} \tag{8.4}$$

where $T(f)$ and $R(f)$ are the transmission and reflection coefficients, respectively. The damped regions must of course be excluded from the analysis. For each particular problem (i.e. magnonic device tested), the length of the input (left) and output

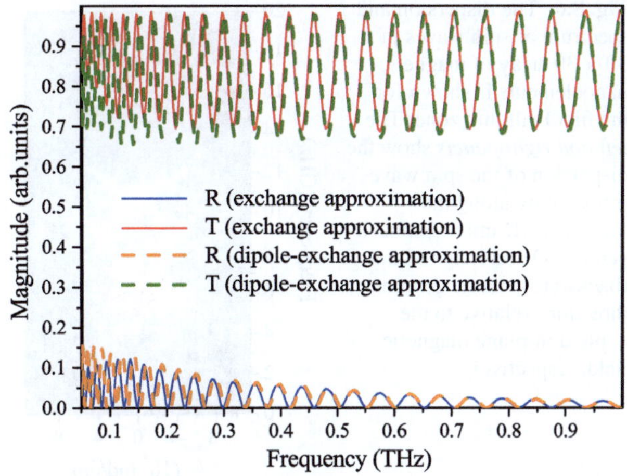

Fig. 8.5 The frequency dependence of the squared reflection and transmission coefficients of spin waves from the Co layer is shown for simulations performed in the exchange (*solid lines*) and dipole-exchange (*dashed lines*) approximations

(right) waveguides has to be optimized so as to obtain the required spectral resolution in the subsequent Fourier analyses and at the same time to allow spin waves with the smallest group velocities to reach the device-under-test well within the simulation time. In the present case, the wave number bandwidth k_{lim} of the simulations is $k_{\mathrm{lim}}/2\pi = 0.5/s_x = 0.5 \times 10^9\ \mathrm{m}^{-1}$.

From Fig. 8.4, one could conclude that the dispersions calculated in the dipole-exchange and exchange approximations are virtually identical. To verify this impression, Fig. 8.5 compares the corresponding frequency dependences of the reflection and transmission coefficients. Both reflection and transmission curves agree well at frequencies above about 200 GHz. This demonstrates the applicability of the exchange approximation in this frequency range and therefore allows one to exploit the analogy existing between the exchange spin waves and the motion of an electron in a non-uniform potential, pointed out and exploited in, e.g. [61, 62]. However, the agreement is significantly worse at low frequencies, emphasizing the need to take into account the magneto-dipole interaction in that case. Similarly to the comparison of the exchange dominated magnonic dispersion with its analytically calculated form [60], the difference in wavelength calculated here in the exchange and dipole-exchange approximations is effectively compared against the thickness of the Co layer playing the role of a Fabry–Perot resonator for spin waves.

Let us now look again at the spectrum shown in Fig. 8.3. Even for an isolated nanoelement, the spectrum is quite complex, featuring a number of peaks. When two or more such elements are brought close enough to experience significant magneto-dipole interaction, each of the resonances of the isolated element splits into several peaks, the number of which is generally equal (neglecting degeneracy) to the number of interacting elements. The case of two elements was considered in, e.g. [63]. The complexity of the magnonic spectra and associated profiles of the collective modes grows rapidly as the number of interacting elements increases. Very soon, the character of the collective modes becomes difficult to analyze in the real space, and the reciprocal space analysis becomes preferable. As an example, Fig. 8.6 presents

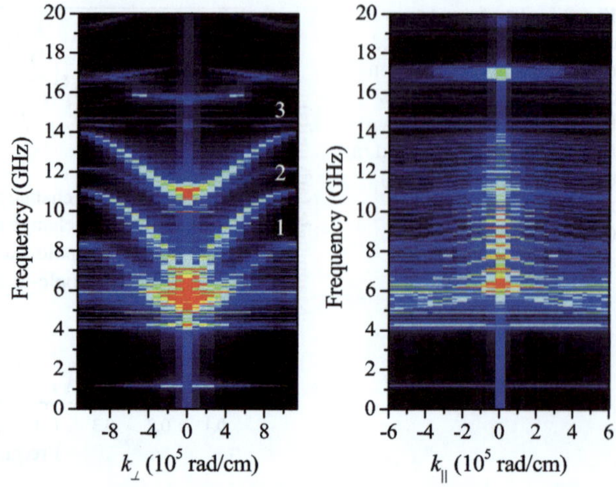

Fig. 8.6 The dispersion and spectrum of spin waves in a 30 × 30 array of magnetic nanoelements is shown for the first Brillouin zone. The *left and right panels* show the dispersion of the spin waves propagating along the transverse (Damon–Eshbach geometry) and longitudinal (backward volume geometry) directions relative to the applied in-plane magnetic field, respectively

magnonic dispersion of a 2D magnonic crystal formed by a 30 × 30 array of magnetic nanoelements that are identical to those in Fig. 8.3. As expected, the majority of resonances fall onto regular dispersion branches characterized by either positive or negative slope, as determined by the symmetry of the magneto-dipole interaction between elements [23]. The horizontal branches signify localized modes, e.g. confined by the profile of the demagnetizing field in the array as a whole [58].

The sample's response at zero wave vector describes its effectively continuous properties in relation to long wavelength electromagnetic waves. With respect to the latter, the array represents a so-called "magnonic meta-material" [18, 64] – an effectively continuous material properties of which are defined not only by those of the magnetic material of its constituent elements ("meta-atoms") but also by their geometry, e.g. their shape or aspect ratio, and associated spin wave resonances. By tweaking the properties of the meta-atoms, it is possible, e.g. to design negative permeability [64] of or to tailor magnonic dispersion [58] in the metamaterial. We note however that, depending on the classification assumed, magnonic crystals also represent a class of magnonic meta-materials – the so-called magnonic bandgap meta-materials [1].

8.5 Semargl

The presented examples demonstrate a remarkable strength of micromagnetic simulations in relation to magnonics. The ability to analyze their output in both real space and time domain and reciprocal domains appears particularly valuable. However, both the simulations and their analysis have a considerable scope for improvement. For example, we are not aware of any calculation of 3D magnonic dispersions from micromagnetic simulations, while the presented methodology of calculation of the magnonic scattering coefficients has only been demonstrated for 1D models.

The limitations are due to not only methodological reasons but also the computational load associated with the size of modeled samples. Indeed, the resolution in the reciprocal space is determined by the total size of the sample while the cell size is dictated by the exchange length in the material and geometry of the sample which, e.g. could be patterned [58]. As a result, the number of cells in the problem easily exceeds limits of what could be comfortably handled by available computing hardware and micromagnetic solvers. Moreover, the available micromagnetic packages do not offer advanced post-processing tools. Instead, commercial packages, such as Origin [65] or MatLab [66], are often used for analysis which, however, does not guaranty that large amounts of data could be processed on reasonable time scales. Custom implementations based on relatively low-level programming languages, e.g. C/C++ or FORTRAN, could offer reduced memory footprint and faster performance, but their development requires significant time.

The analyses presented in this paper have been performed using *semargl* software [32], specifically intended to perform analysis of output produced by OOMMF [24], which has been used for simulations here. The software offers both graphical user interface and outstanding computational performance, and is designed to accept OOMMF's OVF 2.0 files of either "binary 4" or "binary 8" formats as described in [24]. The parallel processing is realized by means of the OpenMP framework [67] and tuned to be Non-Uniform Memory Access (NUMA) aware [68], so that the performance of the software can also scale efficiently across multiple processors. The DFT routines are called from the Core Math Library of AMD [69]. To reduce the memory footprint, the data is stored in memory in the single precision format. However, this is not expected to lead to accumulation of significant errors, since only few computational steps are involved in the analysis.

8.6 Conclusions and Outlook

In this paper, we have demonstrated several methods in which micromagnetic simulations could be used to study magnonic phenomena in nanoscale samples. The remarkable complexity of precessional dynamics observed in realistic magnetic nanostructures and devices leads us to believe that micromagnetic simulations will remain the tool of choice for interpretation of experimental data and performance of magnetic devices, predicting unknown effects and testing analytical theories. We predict that the development of multiscale and multiphysics extensions of dynamic micromagnetic solvers will broaden the scope of both micromagnetic simulations in magnonics and magnonics itself, with the process being assisted by the continuing progress in and availability of computing hardware. For example, the extension of micromagnetics to solvers based on atomistic spin models will underpin application of the developed methodology to studies of phenomena involving both magnons and other fundamental excitations of the solid state. In a more distant perspective, it is highly intriguing to study spin waves in non-stationary conditions (i.e. in structures with time dependent material properties), such as those realized in experiments with samples under ultrafast optical pumping.

Acknowledgements The authors gratefully acknowledge funding received from the European Community's Seventh Framework Programme (FP7/2007-2013) under Grant Agreements no 233552 (DYNAMAG) and 228673 (MAGNONICS) and from the Engineering and Physical Sciences Research Council (EPSRC) of the United Kingdom under project EP/E055087/1.

References

1. V.V. Kruglyak, S.O. Demokritov, D. Grundler, J. Phys. D, Appl. Phys. **43**, 264001 (2010), and references therein
2. F. Bloch, Z. Phys. **61**, 206 (1930)
3. M. Krawczyk, J.-C. Lévy, D. Mercier, H. Puszkarski, Phys. Lett. A **282**, 186 (2001)
4. V.V. Kruglyak, A.N. Kuchko, Physica B **339**, 130 (2003)
5. V.V. Kruglyak, A.N. Kuchko, J. Magn. Magn. Mater. **272–276**, 302 (2004)
6. S.A. Nikitov, P. Tailhades, C.S. Tsai, J. Magn. Magn. Mater. **236**, 320 (2001)
7. C. Elachi, IEEE Trans. Magn. **MAG-11**, 36 (1975)
8. R.L. Stamps, B. Hillebrands, Phys. Rev. B **44**, 5095 (1991)
9. E.L. Albuquerque, M.G. Cottam, Phys. Rep. **376**, 225 (2003), and references therein
10. G. Gubbiotti, S. Tacchi, M. Madami, G. Carlotti, A.O. Adeyeye, M. Kostylev, J. Phys. D, Appl. Phys. **43**, 264003 (2010), and references therein
11. A.A. Serga, A.V. Chumak, B. Hillebrands, J. Phys. D, Appl. Phys. **43**, 264002 (2010), and references therein
12. S.K. Kim, J. Phys. D, Appl. Phys. **43**, 264004 (2010), and references therein
13. A. Khitun, M. Bao, K.L. Wang, J. Phys. D, Appl. Phys. **43**, 264005 (2010), and references therein
14. H. Al-Wahsh, A. Akjouj, B. Djafari-Rouhani, L. Dobrzynski, Surf. Sci. Rep. **66**, 29 (2011), and references therein
15. B. Lenk, H. Ulrichs, F. Garbs, M. Münzenberg, Phys. Rep. **507**, 107 (2011), and references therein
16. P.S. Keatley, V.V. Kruglyak, P. Gangmei, R.J. Hicken, Philos. Trans. R. Soc., Math. Phys. Eng. Sci. **369**, 3115 (2011), and references therein
17. J. Jorzick, S.O. Demokritov, B. Hillebrands, M. Bailleul, C. Fermon, K.Y. Guslienko, A.N. Slavin, D.V. Berkov, N.L. Gorn, Phys. Rev. Lett. **88**, 47204 (2002)
18. V.V. Kruglyak, P.S. Keatley, A. Neudert, R.J. Hicken, J.R. Childress, J.A. Katine, Phys. Rev. Lett. **104**, 027201 (2010)
19. V.L. Zhang, H.S. Lim, C.S. Lin, Z.K. Wang, S.C. Ng, M.H. Kuok, S. Jain, A.O. Adeyeye, M.G. Cottam, Appl. Phys. Lett. **99**, 143118 (2011)
20. W.K. Hiebert, A. Stankiewicz, M.R. Freeman, Phys. Rev. Lett. **79**, 1134 (1997)
21. J. Ding, M. Kostylev, A.O. Adeyeye, Phys. Rev. Lett. **107**, 047205 (2011)
22. S. Neusser, H.G. Bauer, G. Duerr, R. Huber, S. Mamica, G. Woltersdorf, M. Krawczyk, C.H. Back, D. Grundler, Phys. Rev. B **84**, 184411 (2011)
23. A.G. Gurevich, G.A. Melkov, *Magnetization Oscillations and Waves* (CRC Press, Boca Raton, 1996)
24. M. Donahue, D.G. Porter, OOMMF User's guide, Version 1.0, Interagency Report NISTIR 6376, NIST, Gaithersburg, MD 1999 at http://math.nist.gov/oommf/
25. http://www.micromagus.de/
26. http://www.soton.ac.uk/~fangohr/nsim/nmag/
27. The International Technology Roadmap for Semiconductors, http://www.itrs.net/Links/2009ITRS/Home2009.htm
28. M. Pardavi-Horvath, J. Magn. Magn. Mater. **215–216**, 171 (2000)
29. R. Hertel, W. Wulfhekel, J. Kirschner, Phys. Rev. Lett. **93**, 257202 (2004)
30. S.V. Vasiliev, V.V. Kruglyak, M.L. Sokolovskii, A.N. Kuchko, J. Appl. Phys. **101**, 113919 (2007)
31. K.S. Lee, S.K. Kim, J. Appl. Phys. **104**, 053909 (2008)

32. http://www.magnonics.org/semargl/
33. B.C. Choi, Y.K. Hong, J. Rudge, E. Girgis, J. Kolthammer, G.W. Donohoe, Phys. Status Solidi, B Basic Solid State Phys. **244**, 4486 (2007)
34. L. Kaganovskiy, D. Litvinov, S. Khizroev, S. Wilcox, J. Appl. Phys. **110**, 043901 (2011)
35. S. Lepadatu, J.S. Claydon, C.J. Kinane, T.R. Charlton, S. Langridge, A. Potenza, S.S. Dhesi, P.S. Keatley, R.J. Hicken, B.J. Hickey, C.H. Marrows, Phys. Rev. B **81**, 020413 (2010)
36. D.V. Berkov, N.L. Gorn, J. Phys. D, Appl. Phys. **41**, 164013 (2008)
37. F. Montoncello, L. Giovannini, F. Nizzoli, R. Zivieri, G. Consolo, G. Gubbiotti, J. Magn. Magn. Mater. **322**, 2330 (2010)
38. S. Choi, K.-S. Lee, S.-K. Kim, Appl. Phys. Lett. **89**, 062501 (2006)
39. S. Bance, T. Schrefl, G. Hrkac, A. Goncharov, D.A. Allwood, J. Dean, J. Appl. Phys. **103**, 07E735 (2008)
40. Y. Au, T. Davison, E. Ahmad, P.S. Keatley, R.J. Hicken, V.V. Kruglyak, Appl. Phys. Lett. **98**, 122506 (2011)
41. Y. Au, E. Ahmad, O. Dmytriiev, M. Dvornik, T. Davison, V.V. Kruglyak, Appl. Phys. Lett. **100**, 182404 (2012)
42. C.S. Lin, H.S. Lim, Z.K. Wang, S.C. Ng, M.H. Kuok, A.O. Adeyeye, J. Nanosci. Nanotechnol. **11**, 2615 (2011)
43. C. Ragusa, M. Carpentieri, F. Celegato, P. Tiberto, E. Enrico, L. Boarino, G. Finocchio, IEEE Trans. Magn. **47**, 2498 (2011)
44. R. Huber, D. Grundler, Proc. SPIE **8100**, 81000D (2011)
45. O. Gerardin, H. Le Gall, M.J. Donahue, N. Vukadinovic, J. Appl. Phys. **89**, 7012 (2001)
46. J.P. Park, P. Eames, D.M. Engebretson, J. Berezovsky, P.A. Crowell, Phys. Rev. Lett. **89**, 277201 (2002)
47. M. Buess, R. Höllinger, T. Haug, K. Perzlmaier, U. Krey, D. Pescia, M.R. Scheinfein, D. Weiss, C.H. Back, Phys. Rev. Lett. **93**, 77207 (2004)
48. R.D. McMichael, M.D. Stiles, J. Appl. Phys. **97**, 10J901 (2005)
49. S. Jung, B. Watkins, L. DeLong, J.B. Ketterson, V. Chandrasekhar, Phys. Rev. B **66**, 132401 (2002)
50. V.V. Kruglyak, P.S. Keatley, R.J. Hicken, J.R. Childress, J.A. Katine, Phys. Rev. B **75**, 024407 (2007)
51. C.C. Dantas, L.A. de Andrade, Phys. Rev. B **78**, 024441 (2008)
52. C.C. Dantas, Physica E **44**, 675 (2011)
53. H.T. Nguyen, M.G. Cottam, J. Phys. D, Appl. Phys. **44**, 315001 (2011)
54. H. Yang, G.H. Yun, Y.J. Cao, J. Phys. D, Appl. Phys. **44**, 455001 (2011)
55. I.P. Coelho, M.S. Vasconcelos, C.G. Bezerra, J. Magn. Magn. Mater. **323**, 3162 (2011)
56. V.V. Kruglyak, R.J. Hicken, J. Magn. Magn. Mater. **306**, 191 (2006)
57. F.S. Ma, H.S. Lim, Z.K. Wang, S.N. Piramanayagam, S.C. Ng, M.H. Kuok, Appl. Phys. Lett. **98**, 153107 (2011)
58. M. Dvornik, V.V. Kruglyak, Phys. Rev. B **84**, 140405 (2011)
59. F.S. Ma, H.S. Lim, Z.K. Wang, S.N. Piramanayagam, S.C. Ng, M.H. Kuok, IEEE Trans. Magn. **47**, 2689 (2011)
60. M. Dvornik, A.N. Kuchko, V.V. Kruglyak, J. Appl. Phys. **109**, 07D350 (2011)
61. E. Schlömann, J. Appl. Phys. **35**, 159 (1964)
62. E. Schlömann, J. Appl. Phys. **35**, 167 (1964)
63. M. Dvornik, P.V. Bondarenko, B.A. Ivanov, V.V. Kruglyak, J. Appl. Phys. **109**, 07B912 (2011)
64. R.V. Mikhaylovskiy, E. Hendry, V.V. Kruglyak, Phys. Rev. B **82**, 195446 (2010)
65. http://www.originlab.com/
66. http://www.mathworks.co.uk/
67. R. Gerber, *Getting Started with OpenMP* (Intel Corporation, Santa Clara, 2008)
68. N. Soveiko, M.S. Nakhla, R. Achar, IEEE Trans. Comput.-Aided Des. Integr. Circuits Syst. **29**, 65 (2010)
69. AMD Core Math Library Redistribution Agreements (2010). http://download2-developer.amd.com/amd/ACML/ACML_June_24_2010.pdf, as of 24 June 2010

Part III
Magnon Spintronics: Spin Currents, Spin Pumping and Magnonic Spin-Torque Devices

Chapter 9
Spin Waves, Spin Currents and Spin Seebeck Effect

Hiroto Adachi and Sadamichi Maekawa

Abstract The spin Seebeck effect is now established as a universal aspect of ferromagnets that enables thermal injection of spin currents from a ferromagnet into attached nonmagnetic metals over macroscopic scale of several millimeters. We show that the spin-wave degrees of freedom play a crucial role in the spin Seebeck effect.

9.1 Introduction

The interplay of heat transport and spin waves has been discussed for a long time. In 1955, Sato [1] pointed out that theoretically at low enough temperatures the spin-wave thermal conductivity in an insulating ferromagnet can dominate over the phonon thermal conductivity. The spin-wave contribution to the thermal conductivity was experimentally observed in yttrium iron garnet (YIG) [2, 3], and these works stimulated further investigations of the dynamical and thermal behaviors of spin waves [4]. Recently, there has been a renewed interest in this old topic because of the discovery of the spin Seebeck effect. In 2008, Uchida et al. [5] demonstrated that when a temperature gradient is applied to a $Ni_{19}Fe_{81}$ film, spin currents are injected into attached nonmagnetic metals. This phenomenon, termed the spin Seebeck effect, is now established as a universal aspect of ferromagnets since this phenomenon is observed in a variety of materials ranging from metallic ferromagnets $Ni_{81}Fe_{19}$ [5] and Co_2MnSi [6], a semiconducting ferromagnet (Ga,Mn)As [7], to insulating magnets $LaY_2Fe_5O_{12}$ [8] and $(Mn,Zn)Fe_2O_4$ [9].

The discovery of the spin Seebeck effect surprised the community because the length scale seen in experiments was extraordinarily longer than the spin-flip diffusion length of conduction electrons. A key in resolving this mystery was revealed

H. Adachi · S. Maekawa (✉)
Advanced Science Research Center, Japan Atomic Energy Agency, Tokai 319-1195, Japan
e-mail: maekawa.sadamichi@jaea.go.jp

H. Adachi
e-mail: adachi.hiroto@jaea.go.jp

H. Adachi · S. Maekawa
CREST, Japan Science and Technology Agency, Sanbancho, Tokyo 102-0075, Japan

S.O. Demokritov, A.N. Slavin (eds.), *Magnonics*, Topics in Applied Physics 125,
DOI 10.1007/978-3-642-30247-3_9, © Springer-Verlag Berlin Heidelberg 2013

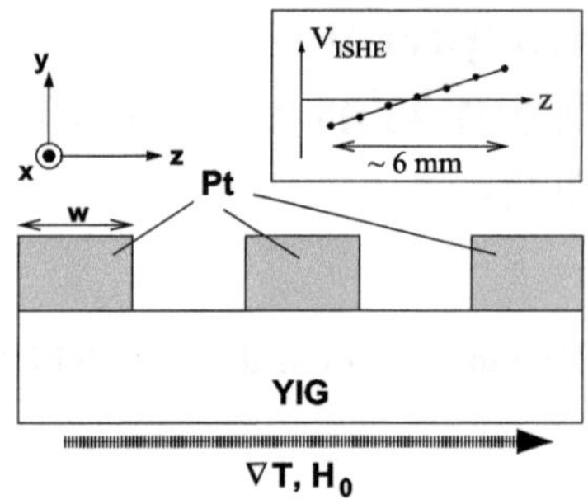

Fig. 9.1 Side-view schematic of the experimental setup to observe the spin Seebeck effect. (*Inset*) Spatial dependence of the observed voltage

in an experiment on electric signal transmission through a ferromagnetic insulator YIG [10], demonstrating that spin currents can be carried by low-lying collective excitations of localized spins, i.e., spin waves or magnons. Subsequently, it was reported that despite the absence of conduction electrons the spin Seebeck effect can be observed in $LaY_2Fe_5O_{12}$, a magnetic insulator [8]. These experiments suggest that, contrary to the conventional wisdom gathered over the last two decades that the spin current is carried by *conduction* electrons [11], the magnon is a promising candidate as a carrier for the spin Seebeck effect.

In this article, we give a theoretical framework to understand the spin Seebeck effect in terms of magnon excitations. In addition, we discuss the active role of nonequilibrium phonons in the spin Seebeck effect.

9.2 Local Picture of Thermal Spin Injection

Let us begin with a brief introduction to the spin Seebeck effect. To simplify the argument, we focus on the "spin Seebeck insulator" [8]. Since the spin Seebeck effect is observed in the insulating magnet of yttrium iron garnet (YIG), the conduction electrons are safely neglected in our discussion. Figure 9.1 shows a side view of the experimental setup, where several Pt strips are attached on top of a YIG film in a static magnetic field $H_0\hat{z}$ ($\gg$ anisotropy field) which aligns the localized magnetic moment along $\hat{z}$. A temperature gradient ∇T is applied along the z-axis, and it induces a spin voltage across the YIG/Pt interface. This spin voltage then injects a spin current I_s ($\parallel \hat{y}$) into the Pt strip (or ejects it from the Pt strip). A fraction of the injected/ejected spin current I_s is converted into a charge voltage V_{ISHE} ($\parallel \hat{x}$) through the so-called inverse spin Hall effect [12]:

$$V_{\text{ISHE}} = \Theta_H \big(|e| I_s\big) \left(\frac{\rho}{w}\right), \tag{9.1}$$

where $|e|$, Θ_H, ρ, and w are respectively the absolute value of electron charge, spin-Hall angle, electrical resistivity, and width of the Pt strip along the z direction (see Fig. 9.1). Therefore, the observed charge voltage V_{ISHE} is a measure of the

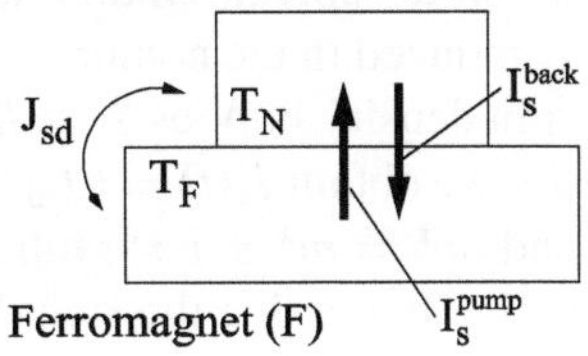

Fig. 9.2 Ferromagnet(F)/nonmagnetic-metal(N) junction. $F(N)$ has local temperature T_F (T_N)

injected/ejected spin current I_s. Experimentally, we can measure the spatial distribution of the thermal spin injection by changing the position of the Pt strip. The observed signal is schematically shown in the inset of Fig. 9.1. Quite surprisingly, the observed signal shows a near linear spatial dependence covering several millimeters.

To obtain an intuitive insight into the spin Seebeck effect, we first consider a model shown in Fig. 9.2 for the thermal spin injection. In this model, a ferromagnet (F) and a nonmagnetic metal (N) are interacting weakly through s–d coupling J_{sd} at the interface. Since we focus here on the local picture of thermal spin injection, we neglect any spatial variations of physical quantities, and assume that the size of the localized spin is unity. We also assume that each segment is initially in thermal equilibrium with no interactions with its neighbors. We then switch on the s–d interaction and calculate the resultant nonequilibrium dynamics of the system.

In the ferromagnet F, the physics is described by the localized moment $\boldsymbol{M}$, for which the dynamics is modeled by the Landau–Lifshitz–Gilbert equation:

$$\frac{d}{dt}\boldsymbol{M} = \left[\gamma(\boldsymbol{H}_{\text{eff}} + \boldsymbol{h}) - \frac{J_{\text{sd}}}{\hbar}\boldsymbol{s}\right] \times \boldsymbol{M} + \frac{\alpha}{M_s}\boldsymbol{M} \times \partial_t \boldsymbol{M}, \tag{9.2}$$

where $\boldsymbol{H}_{\text{eff}}$ is the effective field, γ the gyromagnetic ratio, α the Gilbert damping constant, and M_s the saturation magnetization. In the above equation, the noise field $\boldsymbol{h}$ represents thermal fluctuations in F, and through the fluctuation–dissipation theorem it is assumed to obey the following Gaussian ensemble [13]: $\langle h^i(t)\rangle = 0$ and $\langle h^i(t)h^j(t')\rangle = \frac{2k_B T_F \alpha}{\gamma a_S^3 M_s}\delta_{ij}\delta(t-t')$, where $a_S^3 = \hbar\gamma/M_s$ is the cell volume of the ferromagnet.

In the nonmagnetic metal N, the physics is described by the itinerant spin density $\boldsymbol{s}$, and its dynamics is modeled by the Bloch equation:

$$\frac{d}{dt}\boldsymbol{s} = -\Gamma_{\text{sf}}\delta\boldsymbol{s} - \frac{J_{\text{sd}}}{\hbar}\frac{\boldsymbol{M}}{M_s} \times \boldsymbol{s} + \boldsymbol{l}, \tag{9.3}$$

where Γ_{sf} is the spin-flip relaxation rate, and $\delta\boldsymbol{s} = \boldsymbol{s} - s_0\boldsymbol{M}/M_s$ the nonequilibrium spin accumulation with local equilibrium spin density [14] $s_0 = \chi_N J_{\text{sd}}$ and paramagnetic susceptibility χ_N in N. In this equation, the noise source $\boldsymbol{l}$ is described as a Gaussian ensemble satisfying the fluctuation–dissipation theorem [15], i.e., $\langle l^i(t)\rangle = 0$ and $\langle l^i(t)l^j(t')\rangle = 2k_B T_N \chi_N \Gamma_{\text{sf}}\delta_{ij}\delta(t-t')$.

From now on we focus on the spin-wave region where the magnetization $\boldsymbol{M}$ fluctuates only weakly around the ground state value $M_s\hat{\boldsymbol{z}}$, and we set $\boldsymbol{M}/M_s =$

$\hat{z} + \boldsymbol{m}$ to separate small fluctuations $\boldsymbol{m}$ from the ground state value. The spin current I_s induced in the nonmagnetic metal N can be calculated as the rate of change of the spin density in N as $I_s = \frac{d}{dt}\langle s^z(t)\rangle$. Performing a perturbative approach in terms of $J_{\rm sf}$, we obtain $I_s(t) = (J_{\rm sd}/\hbar)\Im{\rm m}\langle s^+(t)m^-(t')\rangle_{t'\to t}$ from (9.3), where $s^\pm = s^x \pm {\rm i}s^y$ and $m^\pm = m^x \pm {\rm i}m^y$. Introducing the Fourier representation $f(t) = \int \frac{d\omega}{2\pi} f_\omega e^{-{\rm i}\omega t}$ and noticing that the right hand side of I_s is a function of $t - t'$ in the steady state, we obtain

$$I_s = \frac{J_{\rm sd}}{\hbar}\int_{-\infty}^{\infty}\frac{d\omega}{2\pi}\langle\!\langle s_\omega^+ m_{-\omega}^-\rangle\!\rangle, \tag{9.4}$$

where the average $\langle\!\langle \cdots \rangle\!\rangle$ is defined by $\langle s_\omega^+ m_{\omega'}^-\rangle = 2\pi\delta(\omega+\omega')\langle\!\langle s_\omega^+ m_{-\omega}^-\rangle\!\rangle$.

To evaluate the right-hand side of (9.4), we linearize the transverse components of (9.2) and (9.3) with respect to $s_\pm$ and $m_\pm$. Then, to the lowest order in $J_{\rm sd}$, we obtain

$$s_\omega^+ = \frac{l_\omega^+}{-{\rm i}\omega + \Gamma_{\rm sf}} + s_0\Gamma_{\rm sf}\frac{\gamma h_\omega^+}{(\omega_0+\omega-{\rm i}\alpha\omega)(-{\rm i}\omega+\Gamma_{\rm sf})}, \tag{9.5}$$

$$m_\omega^- = \frac{\gamma h_\omega^-}{\omega_0-\omega-{\rm i}\alpha\omega} + J_{\rm sd}\frac{l_\omega^-}{(\omega_0-\omega-{\rm i}\alpha\omega)(-{\rm i}\omega+\Gamma_{\rm sf})}, \tag{9.6}$$

where $\omega_0 = \gamma H_0$, $h^\pm = h^x \pm {\rm i}h^y$, and $l^\pm = l^x \pm {\rm i}l^y$. From the above equations, it is seen that $\boldsymbol{s}$ and $\boldsymbol{m}$ are affected by both the noise field $\boldsymbol{h}$ in F and the noise source $\boldsymbol{l}$ in N through the s–d interaction $J_{\rm sd}$ at the interface. Substituting the above equations into (9.4), we can express the spin current injected into N as follows:

$$I_s = I_s^{\rm pump} - I_s^{\rm back}, \tag{9.7}$$

where $I_s^{\rm pump}$ and $I_s^{\rm back}$ are defined by

$$I_s^{\rm pump} = -G_s\int_{-\infty}^{\infty}\frac{d\omega}{2\pi}\frac{\omega}{|\omega-\omega_0+{\rm i}\alpha\omega|^2}\langle\!\langle \gamma h_\omega^+\gamma h_{-\omega}^-\rangle\!\rangle, \tag{9.8}$$

$$I_s^{\rm back} = -G_s\int_{-\infty}^{\infty}\frac{d\omega}{2\pi}\frac{\omega}{|\omega-\omega_0+{\rm i}\alpha\omega|^2}\left(\frac{\alpha\Gamma_{\rm sf}}{\chi_N\hbar}\right)\langle\!\langle l_\omega^+ l_{-\omega}^-\rangle\!\rangle. \tag{9.9}$$

Here $G_s = J_{\rm sd}^2\chi_N/\Gamma_{\rm sf}\hbar$, and we used the condition $\Gamma_{\rm sf} \gg \omega_0$. Note that the minus sign in front of G_s arises from defining the positive direction of I_s. The expression for $I_s^{\rm pump}$ can be readily seen to represent the spin current pumped into N by the thermal noise field $\boldsymbol{h}$ in F (the so-called pumping component [16]), whereas $I_s^{\rm back}$ represents the spin current coming back into F by the thermal noise source $\boldsymbol{l}$ in N (the so-called backflow component [17]). Employing the two fluctuation–dissipation relations and after performing the ω integration, the pumping and backflow components reduce to a single expression [18, 19]

$$I_s = -G_s\frac{k_{\rm B}}{\hbar}(T_F - T_N), \tag{9.10}$$

where we have used $a_S^3 M_s = \gamma\hbar$. Interestingly, in calculating the z-component of $\langle \boldsymbol{m} \times \frac{d}{dt}\boldsymbol{m} \rangle$ from (9.2), the pumping component can be expressed as $I_s^{\mathrm{pump}} = -G_s\langle [\boldsymbol{m} \times \frac{d}{dt}\boldsymbol{m}]^z \rangle$. This shows that we can calculate the thermal spin injection into N based on the Landau–Lifshitz–Gilbert equation (see (9.2)). The backflow component I_s^{back}, which must be subtracted from I_s^{pump} to evaluate the net spin injection, can be obtained by calculating I_s^{pump} at *local thermal equilibrium*. This procedure was used in [20] to perform the numerical simulation on the spin Seebeck effect.

Equations (9.7) and (9.10) tell us that if both F and N are in local thermal equilibrium (i.e., $T_F = T_N$), there is no net spin injection into the attached nonmagnetic metal N. However, conversely, that means that if the ferromagnet deviates from local thermal equilibrium for some reason, a finite spin current is injected into (or ejected from) the attached nonmagnetic metal N. This consideration leads us to the following simple picture for the spin Seebeck effect: The essence of the spin Seebeck effect is that the localized spins in the ferromagnet are excited by the heat current flowing through the ferromagnet, which then generates finite spin injection because of the imbalance between the pumping component I_s^{pump} and the backflow component I_s^{back}. It is important to note here that the heat current which excites localized spins has two contributions: the magnon heat current and the phonon heat current. Accordingly, there are two relevant processes to the spin Seebeck effect. The first, in which the localized spins are excited by a magnon heat current, corresponds to the magnon-driven spin Seebeck effect discussed in [18] and [19]. The second, in which the localized spins are excited by phonon heat current, corresponds to the phonon-drag spin Seebeck effect discussed in [21].

9.3 Magnon-Driven Spin Seebeck Effect

In the magnon-driven spin Seebeck effect, the localized spins in the ferromagnet are excited by the magnon heat current flowing through the ferromagnet, thereby producing a nonzero spin injection into the attached nonmagnetic metal. So far, two approaches to this process have been proposed: one based on scattering theory [18] and the other based on linear-response theory [19]. While both approaches describe the same physics in general, there are a few differences in detail, especially in the definition and interpretation of the effective temperature. This difference mainly stems from the assumptions of the two approaches: the linear-response approach considers the deviation from the *local equilibrium* conditions following the standard argument of the linear response [22], whereas the scattering approach considers the deviation from the *global equilibrium* condition with uniform temperature distribution [23]. Here we discuss the basic idea along within the linear-response formulation given in [19].

We consider a model shown in Fig. 9.3, where ferromagnet (F) and nonmagnetic metals (N) are divided into the three temperature domains of F_1/N_1, F_2/N_2, and F_3/N_3. Here, there is no temperature difference between the ferromagnet and attached nonmagnetic metals, i.e., $T_{N_1} = T_{F_1} = T_1$, $T_{N_2} = T_{F_2} = T_2$,

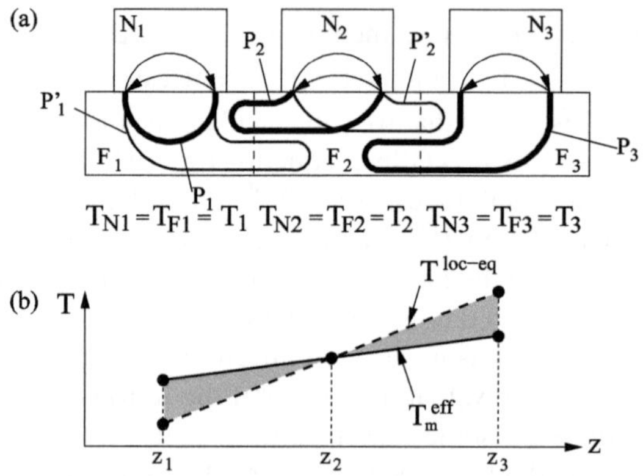

Fig. 9.3 (**a**) System composed of ferromagnet (F) and nonmagnetic metals (N) divided into three temperature domains of F_1/N_1, F_2/N_2, and F_3/N_3, with local temperatures T_1, T_2, and T_3, respectively. The *thin solid lines* with arrows (*bold lines without arrows*) represent electron propagators (magnon propagators). (**b**) Spatial profile of the "effective" magnon temperature $T_{\mathrm{m}}^{\mathrm{eff}}$ and local equilibrium temperature $T^{\mathrm{loc-eq}}$. Because the spin accumulation in the nonmagnetic metal senses the local equilibrium temperature $T^{\mathrm{loc-eq}}$, the spin Seebeck effect picks up the temperature difference between $T^{\mathrm{loc-eq}}$ and $T_{\mathrm{m}}^{\mathrm{eff}}$ (the *shaded area*)

and $T_{N_3} = T_{F_3} = T_3$. As before, we assume that each domain is initially in thermal equilibrium without interactions with neighboring domains, then switch on the interactions among domains and calculate the resultant nonequilibrium dynamics of the system. Hereon, we use the diagrammatic representation of each elementary process [24].

Consider first the process P_1 shown in Fig. 9.3(a), where the magnons travel around the ferromagnet F_1 without sensing the temperature difference between F_1 and F_2. This corresponds to the process discussed in the previous section, and from the condition $T_{N_1} = T_{F_1}$ we get zero spin injection (see (9.10)). We next consider the process P_1' in Fig. 9.3(a) in which magnons sense the temperature difference between F_1 and F_2. In this case, magnons deviate from the local thermal equilibrium condition and produce finite spin injection, because they sense not only the temperature T_1 but also the temperature T_2. Evaluating the diagram P_1' by making use of the procedure as in the previous subsection, the spin Seebeck signal is calculated to be [19]

$$I_s = G_s \omega_0 \tau_{\mathrm{m}} \frac{k_{\mathrm{B}}(T_1 - T_2)}{\hbar(\Lambda/a_S)} = G_s \omega_0 \tau_{\mathrm{m}} \frac{k_{\mathrm{B}}}{\hbar} a_S \nabla T. \tag{9.11}$$

Note that the signal is proportional to the magnon lifetime $\tau_{\mathrm{m}} = 1/(\alpha\omega_0)$ since the carriers of the heat current in this process are magnons. The spin current injected into the right terminal N_3 can be calculated in the same manner by considering the process P_3, which gives the same strength as but opposite sign to the spin current injected into N_1 because of the relation $T_1 - T_2 = -(T_3 - T_2)$. The spin current injected into the middle terminal N_2 vanishes because the two relevant processes (P_2 and P_2') cancel out. Note that effects from spatial variations in magnetization

$\boldsymbol{M}[T(\boldsymbol{r})]$ through the local temperatures $T(\boldsymbol{r})$ are taken into account in our treatment because the temperature dependence of $\boldsymbol{M}$ in the magnon region is automatically described by the number of thermal magnons discussed here.

The result obtained can be interpreted by introducing the concept of effective temperatures [25] which characterizes the nonequilibrium state. The key in our discussion is the existence of a local thermal equilibrium temperature $T^{\text{loc-eq}}$ that can be identified, for example, as the temperature of *optical* phonons having a localized nature with a large specific heat but small thermal conductivity. Note that most of the phonon heat current is carried by *acoustic* phonons. The idea of the approach given in [25] is that the distribution function of a nonequilibrium state is mimicked by a distribution function of an approximate equilibrium state with an effective temperature. By considering the process P_1', where the magnons sense not only the temperature T_1 but also the temperature T_2, the effective magnon temperature in F_1 is approximately given by

$$T_{\text{m}}^{\text{eff}}(z_1) = (1 - r)T_1 + rT_2, \tag{9.12}$$

where $r \sim \alpha^{-1}(a_S/\Lambda)$. In the same manner, the effective magnon temperature in F_3 is given by $T_{\text{m}}^{\text{eff}}(z_3) = (1-r)T_3 + rT_2$. Finally, the effective magnon temperature in F_2 is not renormalized (i.e., $T_{\text{m}}^{\text{eff}}(z_2) = T_2$) because of the cancellation of processes P_2 and P_2'.

These results are summarized in Fig. 9.3(b). Here $T_{\text{m}}^{\text{eff}}$ is the effective magnon temperature, and $T^{\text{loc-eq}}$ is the local equilibrium temperature which can be identified as the temperature of localized (optical) phonons modeled by Einstein phonons. The spin accumulation in the nonmagnetic metal senses the local equilibrium temperature $T^{\text{loc-eq}}$ because the nonmagnetic metal is isolated and not extended to the direction parallel to the temperature gradient. Therefore, this effective temperature difference between $T^{\text{loc-eq}}$ and $T_{\text{m}}^{\text{eff}}$ produces the thermal spin injection following (9.10). In other words, the spin Seebeck effect picks up the difference in the effective temperature corresponding to the shaded area in Fig. 9.3(b).

9.4 Phonon-Drag Spin Seebeck Effect

At the end of Sect. 9.2, we discussed that the heat current flowing through the ferromagnet excites localized spins, hence causing the spin Seebeck effect. Moreover, there are two relevant processes underpinning the spin Seebeck effect because both magnon and phonon heat currents can excite localized spins. The latter, in exciting localized spins and producing finite thermal spin injection, gives rise to the phonon-drag contribution to the spin Seebeck effect [21].

Consider process P_1 in Fig. 9.4(a). Here the magnons themselves do not sense the temperature difference between T_1 and T_2, but interact with phonons which sense the temperature T_1 as well as T_2, and fall into nonequilibrium. These phonons disturb the local equilibrium condition of the magnons in F_1, and cause finite spin

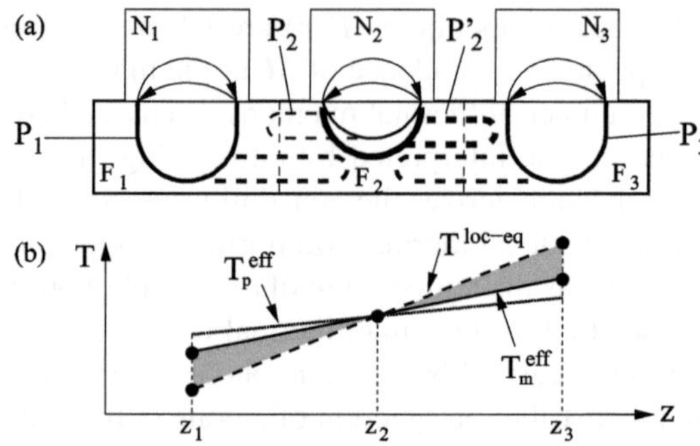

Fig. 9.4 Pictorial representation of the phonon-drag spin Seebeck effect. (**a**) Phonon drag process. The *dashed line* represents a phonon propagator. Meanings of the other lines are the same as in Fig. 9.3. (**b**) Spatial profile of effective magnon temperature $T_{\rm m}^{\rm eff}$, effective phonon temperature $T_{\rm p}^{\rm eff}$, and local equilibrium temperature $T^{\rm loc-eq}$ in the phonon-drag process. The spin Seebeck effect picks up the temperature difference between $T^{\rm loc-eq}$ and $T_{\rm m}^{\rm eff}$ (the *shaded area*)

injection into the nonmagnetic metal N_1. This process gives the thermal spin injection with strength

$$I_s = -G_s \Upsilon_{\rm m-p}^2 \Gamma_{\rm sf} \tau_{\rm p} \mathcal{B} \frac{k_{\rm B}}{\hbar} a_S \nabla T, \tag{9.13}$$

where $\Upsilon_{\rm m-p}$ is the effective magnon–phonon coupling constant and $\tau_{\rm p}$ the phonon lifetime. In the above equation, $\mathcal{B}$ is given by $\mathcal{B} = \mathcal{B}_1 \cdot \mathcal{B}_2$ with $\mathcal{B}_1 = (T/T_D)^{5/4} \int_0^{T_D/T} \frac{du}{4\pi^3} u^6 / \sinh^2(\frac{u}{2})$ and $\mathcal{B}_2 = (T/T_M)^{9/2} \int_0^{T_M/T} \frac{dv}{2\pi^2} v^{7/2}/(e^v - 1)$, where T_D is the Debye temperature and T_M is the temperature corresponding to the high-energy magnon cutoff.

The important point of (9.13) is that the spin Seebeck signal due to phonon drag is proportional to the phonon lifetime $\tau_{\rm p}$ because the carriers of heat current in this process are phonons. Since it is well known that the phonon lifetime is strongly enhanced at low temperatures (typically below 100 K) due to a rapid suppression of the umklapp scattering, (9.13) means that the spin Seebeck effect is enormously enhanced at low temperatures. In contrast, the signal at zero temperature should vanish because of the third law of thermodynamics. Therefore, the phonon-drag spin Seebeck effect must have a pronounced peak at low temperatures, which has been confirmed by recent experiments [26, 27]. Note that the phonon-drag scenario can explain the local nature of the spin Seebeck effect as well [7, 28].

For an intuitive understanding of the phonon-drag spin Seebeck effect, the notion of effective temperature is again quite useful. The strength of the thermal spin injection is proportional to the difference between the effective magnon temperature $T_{\rm m}^{\rm eff}$ and the local thermal equilibrium temperature $T^{\rm loc-eq}$, as before. Now, though, the deviation of $T_{\rm m}^{\rm eff}$ from $T^{\rm loc-eq}$ is caused by nonequilibrium phonons, which pushes magnons away from local thermal equilibrium conditions through the magnon–phonon interaction. Note that the effective phonon temperature $T_{\rm p}^{\rm eff}$ has a smaller slope than the effective magnon temperature $T_{\rm m}^{\rm eff}$ because acoustic phonons have usually a larger thermal conductivity. The spin accumulation in the nonmagnetic metal senses the local equilibrium temperature $T^{\mathrm{loc-eq}}$ and does not equilibrate with the effective phonon temperature $T_{\mathrm{p}}^{\mathrm{eff}}$ because the nonmagnetic metal is isolated and not extended to the direction parallel to the temperature gradient. One must also take note that the discussion by Sanders and Walton [23] makes no account of the phonon-drag process.

9.5 Conclusion

We have discussed the physics of the spin Seebeck effect, and clarified the important role played by magnons. Moreover, we have shown that nonequilibrium phonons also play an active role. With this understanding underlying the spin Seebeck effect, more experimental and theoretical efforts are in order to elucidate the coupling between these two degrees of freedoms.

Acknowledgements We are grateful for fruitful discussions with E. Saitoh, K. Uchida, S. Takahashi, J. Ohe, J. Heremans, and G.E.W. Bauer. This work was supported by Grant-in-Aid for Scientific Research from MEXT, Japan.

References

1. H. Sato, Prog. Theor. Phys. **13**, 119 (1955)
2. B. Lüthi, J. Phys. Chem. Solids **23**, 35 (1962)
3. R.L. Douglass, Phys. Rev. **129**, 1132 (1963)
4. H. Mori, K. Kawasaki, Prog. Theor. Phys. **27**, 529 (1962)
5. K. Uchida, S. Takahashi, K. Harii, J. Ieda, W. Koshibae, K. Ando, S. Maekawa, E. Saitoh, Nature **455**, 778 (2008)
6. S. Bosu, Y. Sakuraba, K. Uchida, K. Saito, T. Ota, E. Saitoh, K. Takanashi, Phys. Rev. B **83**, 224401 (2011)
7. C.M. Jaworski, J. Yang, S. Mack, D.D. Awschalom, J.P. Heremans, R.C. Myers, Nat. Mater. **9**, 898 (2010)
8. K. Uchida, J. Xiao, H. Adachi, J. Ohe, S. Takahashi, J. Ieda, T. Ota, Y. Kajiwara, H. Umezawa, H. Kawai, G.E.W. Bauer, S. Maekawa, E. Saitoh, Nat. Mater. **9**, 894 (2010)
9. K. Uchida, T. Nonaka, T. Ota, H. Nakayama, E. Saitoh, Appl. Phys. Lett. **97**, 262504 (2010)
10. Y. Kajiwara, K. Harii, S. Takahashi, J. Ohe, K. Uchida, M. Mizuguchi, H. Umezawa, H. Kawai, K. Ando, K. Takanashi, S. Maekawa, E. Saitoh, Nature **464**, 262 (2010)
11. S. Maekawa (ed.), *Concepts in Spin Electronics* (Oxford Univ. Press, Oxford, 2006)
12. E. Saitoh, M. Ueda, H. Miyajima, G. Tatara, Appl. Phys. Lett. **88**, 182509 (2006)
13. W.F. Brown Jr., Phys. Rev. **130**, 1677 (1963)
14. S. Zhang, Z. Li, Phys. Rev. Lett. **93**, 127204 (2004)
15. S. Ma, G.F. Mazenko, Phys. Rev. B **11**, 4077 (1975)
16. Y. Tserkovnyak, A. Brataas, G.W.W. Bauer, B.I. Halperin, Rev. Mod. Phys. **77**, 1375 (2005)
17. J. Foros, A. Brataas, Y. Tserkovnyak, G.E.W. Bauer, Phys. Rev. Lett. **95**, 016601 (2005)
18. J. Xiao, G.E.W. Bauer, K. Uchida, E. Saitoh, S. Maekawa, Phys. Rev. B **81**, 214418 (2010)
19. H. Adachi, J. Ohe, S. Takahashi, S. Maekawa, Phys. Rev. B **83**, 094410 (2011)
20. J. Ohe, H. Adachi, S. Takahashi, S. Maekawa, Phys. Rev. B **83**, 115118 (2010)

21. H. Adachi, K. Uchida, E. Saitoh, J. Ohe, S. Takahashi, S. Maekawa, Appl. Phys. Lett. **97**, 252506 (2010)
22. R. Kubo, M. Toda, N. Hashitsume, *Statistical Physics II: Nonequilibrium Statistical Mechanics* (Springer, Berlin, 1991)
23. D.J. Sanders, D. Walton, Phys. Rev. B **15**, 1489 (1977)
24. S.K. Ma, *Modern Theory of Critical Phenomena* (Addison-Wesley, Reading, 1976)
25. P.C. Hohenberg, B.I. Shraiman, Physica D **37**, 109 (1989)
26. C.M. Jaworski, J. Yang, S. Mack, D.D. Awschalom, R.C. Myers, J.P. Heremans, Phys. Rev. Lett. **106**, 186601 (2011)
27. K. Uchida, T. Ota, H. Adachi, J. Xiao, T. Nonaka, Y. Kajiwara, G.E.W. Bauer, S. Maekawa, E. Saitoh, J. Appl. Phys. **111**, 103903 (2012)
28. K. Uchida, H. Adachi, T. An, T. Ota, M. Toda, B. Hillebrands, S. Maekawa, E. Saitoh, Nat. Mater. **10**, 737 (2011)

Chapter 10
Spin Pumping at Ytrium Iron Garnet Interfaces

Capucine Burrowes and Bretislav Heinrich

Abstract A brief review is presented of spin pumping theory at ferromagnetic metal (FM)/normal metal (NM) interfaces or at magnetic insulator (MI)/normal metal (NM) interfaces. Spin transport in a NM is discussed using either spin diffusion theory or spin current decoherence via a spin fluctuation mechanism. These concepts were applied to our experimental results on spin pumping at YIG/Au and YIG/Pd interfaces. Thin YIG films (between 9 and 20 nm thick) were prepared by means of pulsed laser deposition (PLD) on a gallium gadolinium garnet substrate, GGG(111). Quantitative studies of spin pumping at the YIG/Au interface was carried out using magnetic double layers, YIG/Au/Fe, where the Fe layer acted as a spin sink. Spin pumping was first investigated at untreated YIG surfaces where the spin mixing conductance $g_{\uparrow\downarrow}$ at the YIG/Au interface was found to be $\sim 1 \times 10^{14}$ cm^{-2}. Removing the YIG surface contamination using Ar gun sputtering increased the spin mixing conductance to 5×10^{14} cm^{-2} which is 70 % of that predicted by theory and 50 % of that found for the Fe/Au interface. In the YIG/Pd system, the spin transport in Pd is affected by spin fluctuations caused by a large Stoner enhancement factor in this material. The spin mixing conductance was found to be 2.9×10^{14} cm^{-2}, and the spin current decoherence length was found to be about 5 nm.

10.1 Introduction

Recently attention has turned to developing ideas and systems in which a Spin Transfer Torque (STT) can be achieved by means of pure spin currents. A newly emerging field called spin caloritronics [1] addresses the generation of spin currents by thermal gradients. In his pioneering work, Slonczewski has shown that a higher efficiency in STT devices can be achieved by using magnetic insulator/normal metal structures [2]. Magnetic insulators, yttrium iron garnet (YIG) in particular, exhibit very low magnetic losses and a difference between magnon and phonon temperatures can create large pure spin currents.

C. Burrowes · B. Heinrich (✉)
Physics Department, Simon Fraser University, Burnaby, BC, Canada
e-mail: bheinric@sfu.ca

S.O. Demokritov, A.N. Slavin (eds.), *Magnonics*, Topics in Applied Physics 125,
DOI 10.1007/978-3-642-30247-3_10,

In this chapter, we shall review the spin pumping mechanism at metallic ferromagnet (FM) and magnetic insulator (MI)/normal metal (NM) interfaces. In this work, F will refer both to FM and MI. The chapter is organized as follows: In Sect. 10.2, the General Theory of Spin Pumping at FM/NM and MI/NM interfaces will be described. The spin transport in the NM spacer is considered in two useful limits: (a) Spin diffusion in the NM (Sect. 10.2.1) and (b) paramagnon spin current dephasing (Sect. 10.2.2) – in this case the NM notation will be replaced by NMF. The experimental results will be described in Sect. 10.3. The spin pumping at the YIG/Au and YIG/Pd interfaces will be presented in Sects. 10.3.2 and 10.3.3, respectively. The spin mixing conductances in YIG/Au and YIG/Pd will be compared to those obtained at the Fe/Au(001) and Fe/Pd(001) crystalline interfaces.

10.2 Theory of Spin Pumping

The spin dynamics in the classical limit in a single ultrathin ferromagnetic film can be described by the Landau–Lifshitz–Gilbert (LLG) equation of motion [3]:

$$\frac{1}{\gamma}\frac{\partial \vec{M}}{\partial t} = -[\vec{M} \times \vec{H}_{\text{eff}}] + \frac{\alpha}{\gamma}\left[\vec{M} \times \frac{\partial \vec{u}}{\partial t}\right], \tag{10.1}$$

where $\vec{M}$ is the total magnetization, and is constant across the film thickness, $\vec{u}$ is the unit vector in the direction of the instantaneous magnetization $\vec{M}$, $\gamma = g|e|/2mc$ is the absolute value of the gyromagnetic ratio, and α is the dimensionless Gilbert damping parameter. The first term on the right-hand side represents the precessional torque in the internal field $\vec{H}_{\text{eff}}$, and the second term represents the Gilbert damping torque [3].

Tserkovnyak et al. [4] and Heinrich et al. [5] have shown that an interface Gilbert damping can be generated by pumping the spin current from FM into an adjacent NM layer. The net flow of spin momentum per unit surface area in a FM1/NM/FM2 structure in the ballistic limit is given by:

$$\vec{I}_{\text{sp}} = \frac{\hbar}{4\pi}\,\text{Re}(g_{\uparrow\downarrow})\left[\vec{u} \times \frac{\partial \vec{u}}{\partial t}\right], \tag{10.2}$$

where the real part of spin mixing conductance $g_{\uparrow\downarrow}$ per unit area is dependent on the transmission $t^{\uparrow}, t^{\downarrow}$ and reflection $r^{\uparrow}, r^{\downarrow}$ coefficients for majority ($\uparrow$) and minority ($\downarrow$) spins, respectively,

$$g_{\uparrow\downarrow} = \frac{1}{2}\sum_m \left(|r_{\uparrow,m} - r_{\downarrow,m}|^2 + |t_{\uparrow,m} - t_{\downarrow,m}|^2\right). \tag{10.3}$$

Simanek and Heinrich [6] have shown that spin pumping has its origin in the time retarded response of the interlayer exchange coupling. The static interlayer exchange coupling (leading to RKKY oscillations in the NM) is given by the instantaneous response of electrons in the NM to the interface exchange coupling between the FM and NM layers while the time retarded response (using the full Kubo response theory) leads to spin pumping which is in agreement with the standard spin pumping theory presented by Tserkovnyak et al. [4] and demonstrated by Heinrich et al. [5].

In spin caloritronics, one uses a thermal gradient across the MI to create a non-equilibrium distribution of magnons at the MI/NM interface. The pumped spin current, across the area A, parallel to the saturation magnetization is given by:

$$\vec{I}_{\mathrm{sp}}^{||} = \frac{\hbar \gamma k_{\mathrm{B}}}{2\pi M_s V_a} \mathrm{Re}(g_{\uparrow\downarrow})\big(T_F^m - T_N\big)A, \tag{10.4}$$

where k_{B} is the Boltzmann coefficient, V_a is the temperature dependent magnetic coherence volume in MI, T_F^m is the temperature of the magnons, T_N is the temperature of the NM at the MI/NM interface, and $g_{\uparrow\downarrow}$ is the spin mixing conductance per unit area, see [1].

In the metallic FM/NM interface, the spin mixing conductance $g_{\uparrow\downarrow}$, is given by (10.3), and is governed by the sum $\sum_m (|r_{\uparrow(\downarrow),m}|^2 + |t_{\uparrow(\downarrow),m}|^2)$. Since this expression adds 1 to each channel l, and since the contribution of $r_l^{\uparrow} r_l^{\downarrow *} + t_l^{\uparrow} t_l^{\downarrow *}$ is negligible for the metallic interface [7], the expression becomes:

$$g_{\uparrow\downarrow} = \sum_l \big(1 - \mathrm{Re}\big(r_l^{\uparrow} r_l^{\downarrow *} + t_l^{\uparrow} t_l^{\downarrow *}\big)\big) \approx \frac{k_F^2}{4\pi} = 1.2 n^{\frac{2}{3}}, \tag{10.5}$$

where l is the number of spin channels impinging on the FM/NM interface, k_F is the wave-number at the Fermi surface of the NM, and n is the density of electrons per spin in the NM.

However, the situation at the MI/NM interface is different. The transmission parameters are zero, and the absolute value of the reflectivity parameters is one. Brataas et al. [8] and Xiao et al. [1] pointed out that $g_{\uparrow\downarrow}$ is not zero. Indeed, the reflectivity parameters include the phases, $r_{\uparrow(\downarrow)} = 1 \times e^{i\varphi_{\uparrow(\downarrow),l}}$ which are dependent on the channel number l. The spin mixing conductance is in this case given by:

$$g_{\uparrow\downarrow} = \sum_l \big(1 - \cos(\varphi_{\uparrow,l} - \varphi_{\downarrow,l})\big). \tag{10.6}$$

It turns out that the theoretical predictions for $g_{\uparrow\downarrow}$ for MI/NM and FM/NM interfaces are 0.7×10^{15} and $1.1 \times 10^{15}\ \mathrm{cm}^{-2}$, respectively. This suggests that the phase contributions are partly canceled out [9].

Quantitative Description of Spin Transport in a NM In order to define the chemical potentials for the majority and minority electrons, one needs a globally diffuse system where the momentum distribution of the electrons incident on each interface is randomized, resulting in an accumulated density of spins in the NM, $\vec{s}_N$. Tserkovnyak et al. [7] have shown that such randomization in the momentum space can be provided by a sufficiently large interface diffuse scattering. In that case, the spin current density from F (which is FM or MI) to NM is renormalized by a factor 2 compared to (10.2):

$$\vec{I}_{\mathrm{sp}}^{\mathrm{ren}} = \frac{\hbar}{4\pi} \mathrm{Re}(2 g_{\uparrow\downarrow}) \left[\vec{u} \times \frac{\partial \vec{u}}{\partial t}\right]. \tag{10.7}$$

This spin current at the F/NM interface leads to an accumulated spin momentum density $\vec{s}_N$ inside the NM spacer layer. For small rf excitations, the accumulated spin density is perpendicular to the static magnetization $\vec{M}_S$ and this significantly simplifies the calculation of the flow of spin current in the NM. At the F/NM interface the accumulated spin density partly flows back to the F spin reservoir. The spin current flowing back to F (back-flow) is given by $-\frac{1}{2}v_F\vec{s}_N$, where v_F is the Fermi velocity, and is in the case of small rf excitations entirely absorbed at the F/NM interface [7, 10].

The net flow of spin momentum at the F/NM node is given by:

$$\vec{I}_{\rm sp}^{\rm net} = \vec{I}_{\rm sp}^{\rm ren} - \frac{1}{2}v_F\vec{s}_N. \tag{10.8}$$

Conservation of the spin momentum requires for the macrospin $\vec{M}_{\rm tot}$:

$$\frac{1}{\gamma}\frac{\partial\vec{M}_{\rm tot}}{\partial t} = \vec{I}_{\rm sp}^{\rm net} \tag{10.9}$$

where $\vec{M}_{\rm tot} = \vec{M}d_F$, $\vec{M}$ is the density of magnetization, and d_F is the thickness of the ferromagnetic film.

After simple vectorial algebra, one obtains for the interface spin pumping a contribution to the Gilbert damping parameter $\alpha_{\rm sp}$:

$$\alpha_{\rm sp} = \frac{\gamma\widetilde{I}_{\rm sp}^{\rm net}}{M_S d} \tag{10.10}$$

where $\vec{I}_{\rm sp}^{\rm net} = \widetilde{I}_{\rm sp}^{\rm net}(\vec{u}\times\partial\vec{u}/\partial t)$. Spin transport in a NM will be discussed in two limits: (a) spin diffusion in which the electron mean free path is smaller than the spin diffusion length (Au, Ag, Cu) and (b) spin current dephasing by paramagnon fluctuations caused by strong electron spin correlations (Pt, Pd). Spin transport solutions and the corresponding $\alpha_{\rm sp}$ will be obtained using magnetoelectronic Kirchhoff's laws [8].

10.2.1 Spin Transport by Spin Diffusion in a NM

One can write a full set of magnonic Kirchhoff's equations using the layer F1 as a magnetic terminal providing a spin current source and the layer F2 as a magnetic terminal acting as a spin sink.

The F1/NM, NM/F2, and NM/vacuum interfaces play the role of circuit nodes, and the NM spacer plays the role of a spatially distributed spin conductor.

F1/NM node: Kirchhoff's law gives the net spin momentum flow,

$$\vec{I}_{\rm sp}^{\rm net} = -D\frac{\partial\vec{s}_N}{\partial x}, \tag{10.11}$$

where D is the diffusion coefficient in the NM, given by:

$$D = v_F^2\tau_m/3, \tag{10.12}$$

with τ_m the electron momentum relaxation time. Notice that the terminal F1 acts also as a spin sink for the spin momentum back flow.

NM/F2 node: Continuity of the current at the NM/F2 node, where the F2 terminal acts as a spin sink, is given by:

$$-D\frac{\partial \vec{s}_N}{\partial x} = \frac{1}{2}v_F\vec{s}_N. \tag{10.13}$$

Equation (10.13) is valid only if F2 is off resonance, and therefore contributes negligibly to spin pumping. The magnetic current reaching the NM/F2 interface is then fully absorbed by the F2 layer (spin detector) and acts as a driving torque for the F2 layer and results in its rf excitation, see [11].

NM/vacuum node: The second terminal is a vacuum (ambient) that usually reflects the spin current,

$$\frac{\partial \vec{s}_N}{\partial x} = 0. \tag{10.14}$$

The spin momentum transport across the NM spacer is described by the diffusion equations (equivalent to a distributed conductance) [7, 10]:

$$\frac{\partial \vec{s}_N}{\partial t} = D\frac{\partial^2 \vec{s}_N}{\partial x^2} - \frac{1}{\tau_{\text{sf}}}\vec{s}_N, \tag{10.15}$$

where τ_{sf} is the spin flip relaxation time, x is the coordinate in a direction perpendicular to the interface. Equation (10.15) determines the penetration of $\vec{s}_N$ inside the NM spacer. The spin diffusion length in the limit of low driving frequencies ($\omega \ll 1/\tau_{\text{sf}}$) is given by:

$$\delta_{\text{sd}} = (D\tau_{\text{sf}})^{\frac{1}{2}} = v_F(\tau_{\text{sf}}\tau_m/3)^{\frac{1}{2}}. \tag{10.16}$$

The above diffusion equations can be solved analytically using boundary conditions (10.11), (10.13), and (10.14). For a single magnetic structure F/NM, the additional interface damping in F1, α_{sp}, is given by:

$$\alpha_{\text{sp}}^s = \left[1 - \frac{(1+e^{-2kd})\frac{1}{2}v_F}{(Dk+\frac{1}{2}v_F)+e^{-2kd}(\frac{1}{2}v_F - Dk)}\right] \times \frac{g\mu_{\text{B}}}{2\pi M_s}g_{\uparrow\downarrow}\frac{1}{d_F}, \tag{10.17}$$

where d is the thickness of the NM and $k = 1/\delta_{\text{sp}}$.

For a double magnetic structure F1/NM/F2, the additional interface spin pumping damping is given by:

$$\alpha_{\text{sp}}^d = \left[1 - \frac{((Dk+\frac{1}{2}v_F)+e^{-2kd}(Dk-\frac{1}{2}v_F))\frac{1}{2}v_F}{(Dk+\frac{1}{2}v_F)^2 - e^{-2kd}(Dk-\frac{1}{2}v_F)^2}\right] \times \frac{g\mu_{\text{B}}}{2\pi M_s}g_{\uparrow\downarrow}\frac{1}{d_F}, \tag{10.18}$$

where μ_{B} is the Bohr magneton and g is the Landé g-factor.

In the limit $d \gg \delta_{\text{sp}}$, (10.17) and (10.18) lead to the same result. Notice that the square brackets in (10.17) and (10.18) are only dependent on τ_m and τ_{sf}. Therefore, fitting the data requires the determination of the spin mixing conductance $g_{\uparrow\downarrow}$, the bulk Gilbert damping α_b, and the parameters τ_m and τ_{sf}.

10.2.2 Spin Transport in the Presence of Paramagnons

One can expect that the propagation of a coherent spin current in the NM would be affected by strong spin electron–electron correlation effects which are particularly present in Pd and Pt. Those two materials are known to have a strongly enhanced paramagnetic susceptibility due to their large Stoner enhancement factors compared to Ag, Au, and Cu [12]. Associated local fluctuating magnetic moments (paramagnons) in Pt and Pd (abbreviated NMF) are coupled to the coherent spin current generated by the ferromagnet F, leading to a gradual decrease (decoherence) of the pumped spin current. Indeed, the spin pumping measurements in Pd and Pt show clearly that the spin diffusion theory is not applicable because the length scale of the spin current penetration in the Pt and Pd films is significantly smaller than the electron mean free path for these materials [13, 14].

The spin current pumped from F into NMF (Pt, Pd) can be described by using the ballistic approximation for which $\vec{I}_{\mathrm{ball}} = \frac{1}{2} v_F \vec{s}_{\mathrm{NMF}}$. One can then assume that the accumulated spin density will decrease exponentially with the distance away from the interface due to the interaction with the fluctuating paramagnons: The characteristic decay length is λ_{dec}. In this case, the spin current reaching the outer interface of the NMF film and returning back to the F/NMF interface will be given by:

$$\frac{1}{2} v_F \vec{s}_{\mathrm{NMF}} \left(e^{-2d_{\mathrm{NMF}}/\lambda_{\mathrm{dec}}} \right), \tag{10.19}$$

where d_{NMF} is the thickness of the NMF film and λ_{dec} is the mean decoherence length. By applying Kirchhoff's law at the F/NMF node, one gets for the net flow at the F/NMF interface:

$$\vec{I}_{\mathrm{sp}}^{\mathrm{ren}} - \frac{1}{2} v_F \vec{s}_{\mathrm{NMF}} = \frac{1}{2} v_F \vec{s}_{\mathrm{NMF}} \left(1 - e^{-2d_{\mathrm{NMF}}/\lambda_{\mathrm{dec}}} \right). \tag{10.20}$$

This equation leads to the interface Gilbert damping:

$$\alpha_{\mathrm{NMF}} = \frac{g \mu_B g_{\uparrow\downarrow}}{2\pi M_s d_F} \times \frac{(1 - e^{-2d_{\mathrm{NMF}}/\lambda_{\mathrm{dec}}})}{(2 - e^{-2d_{\mathrm{NMF}}/\lambda_{\mathrm{dec}}})}. \tag{10.21}$$

A similar expression was used in the study by Foros et al. of Pd grown on Fe [13], but it ignored the back flow in (10.20), and therefore it approaches α_{NMF} only in the limit where $2d_{\mathrm{NMF}}/\lambda_{\mathrm{dec}} \gg 1$.

10.3 Experimental Results and Discussion

10.3.1 FMR

Spin injection across the ferrimagnetic insulator (YIG)/normal metal (Au or Pd) interface was studied using ferromagnetic resonance (FMR). The Gilbert damping was obtained from the slope of the FMR linewidth as a function of the microwave frequency:

$$\Delta H = \alpha_{\mathrm{tot}} \frac{\omega}{\gamma} + \Delta H_0, \tag{10.22}$$

where γ is the absolute value of the gyro-magnetic ratio and α_{tot} is the total Gilbert damping parameter. The zero frequency offset ΔH_0 is caused by long range magnetic inhomogeneities. $\alpha_{\mathrm{tot}} = \alpha_b + \alpha_{\mathrm{sp}}$, where α_b is the bulk Gilbert damping parameter [15].

FMR measurements were carried out in microwave cavities at 10, 14, 24, and 36 GHz with the static magnetic field directed along the film surface.

The saturation induction of our YIG films was determined using SQUID magnetometry and was found to be 1.31 kG for 9 nm thick films. The effective perpendicular demagnetizing field $4\pi M_{\mathrm{eff}} = 4\pi M_s - H_{u,\perp}$ ($H_{u,\perp}$ is the uniaxial perpendicular anisotropy field) and the g-factors for our YIG films were determined from the FMR data measured at different microwave frequencies, see [16]. Moreover, for Fe and YIG thin films, the effective perpendicular demagnetizing fields $4\pi M_{\mathrm{eff}}$ are 10.6 and 1.9 kOe, respectively. This large difference in $4\pi M_{\mathrm{eff}}$ separates the Fe and YIG FMR resonance fields, and therefore the Fe will act only as a spin sink and won't contribute to spin pumping.

The bulk damping was obtained from the uncovered YIG (untreated or treated by sputtering). Subtracting α_b from the measured α in YIG/Au/Fe and using (10.18), one can determine the spin mixing conductance $g_{\uparrow\downarrow}$ at the YIG/Au interface. The average parameters for the momentum and spin flip relaxation parameters are $\tau_m = 1.2 \times 10^{-14}\ \mathrm{s}^{-1}$, $\tau_{\mathrm{sf}} = 15 \times 10^{-14}\ \mathrm{s}^{-1}$, and the spin diffusion length is $\delta_{\mathrm{sd}} = 35$ nm [17]. In (10.18), the Au spacer decreases α_{sp} compared to that in the ballistic spin pumping due to the loss of spin momentum in the Au spacer. Indeed, in the ballistic limit,

$$\alpha_{\mathrm{sp}}^{\mathrm{ballistic}} = \frac{g\mu_{\mathrm{B}}}{4\pi M_s} g_{\uparrow\downarrow} \frac{1}{d}. \tag{10.23}$$

For a 6.1 nm thick Au layer, this correction is small and decreases the spin pumping damping from the ballistic limit by 17 %.

10.3.2 YIG/Au

YIG ($Y_3Fe_2(FeO_4)_3$) films grown on (111) $Gd_3Ga_5O_{12}$ substrates were prepared using pulsed laser deposition (PLD) (see [16]). The YIG films were characterized by X-ray photoelectron spectroscopy (XPS) and the Au and Fe films were deposited using molecular beam epitaxy (MBE) at pressures in the low 10^{-10} Torr. The Au, Pd, and Fe films were polycrystalline. The thickness of aYIG sample was determined by means of low angle X-ray diffraction. XPS spectral analysis of our YIG indicated that Fe was deficient at the YIG surface. The surface chemical compositions of thick and thin YIG films were identical and different from the bulk. This difference is caused by the PLD process.

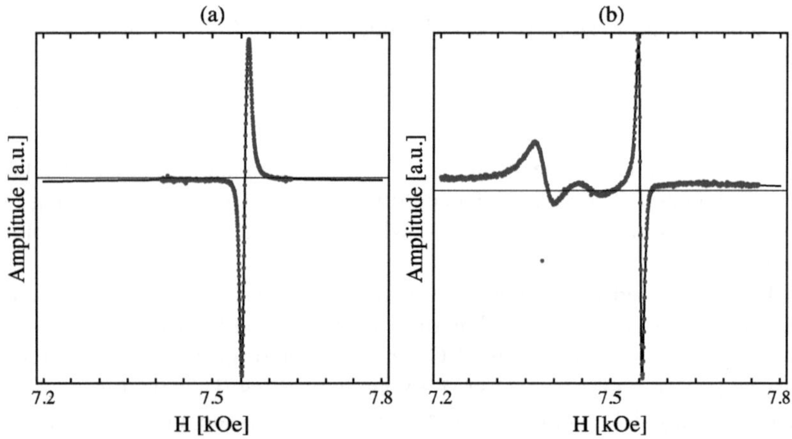

Fig. 10.1 The field derivative of the FMR signal as a function of the external field. (**a**) $f = 23.988$ GHz, bare YIG 9.0 nm thick grown on GGG(111) substrate: The FMR line shape required the sum of 3 Lorenzian absorption lines separated by a few oersteds. (**b**) $f = 23.9751$ GHz, 9.0YIG/2.0Au/4.3Fe/6.1Au. The numerals are in nm. The three peaks observed in the bare sample became well split due to ferromagnetic interlayer exchange coupling between the YIG and the Fe layer. See details in [16]

Spin Pumping Using Untreated YIG Surfaces Important results were obtained on 9.0YIG/2.0Au/4.3Fe/6.1Au, where the numerals are thicknesses in nm. In this sample, the FMR signal was split into three well separated FMR absorption lines [16], see Fig. 10.1. This means that this sample has three different areas, two areas show a coupling through the thin 2.0 nm Au spacer between the YIG and the Fe, and one area is uncoupled. The surface roughness of the YIG films measured using Atom Force Microscope is 0.5 nm which resulted in a negligible magnetostatic coupling of 1×10^{-4} erg/cm^2 between the YIG and Fe films. Therefore, the layers are coupled by a ferromagnetic interlayer exchange coupling parameters $J_{ex} = +(0.08 \text{ and } 0.15)$ erg/cm^2. Even more importantly, the slope of the FMR linewidth as a function of microwave frequency, $\Delta H(f)$, increased for the coupled samples compared to the bare sample while the uncoupled sample showed no increase in this slope. This is in perfect agreement with an alternative theory of spin pumping which was proposed by Simanek and Heinrich [18]. In fact, this is perhaps the first clear experimental demonstration of the relationship between the *s–d* interface exchange coupling and the spin pumping mechanism.

The sample 9.0YIG/6.1Au/4.3Fe/6.1Au does not show any appreciable shift in the FMR field compared to that in the bare YIG; therefore, the static interlayer exchange coupling for the 6.1 nm Au spacer was negligible. However, the slope of $\Delta H(f)$ clearly increased compared to that in the bare YIG, indicating that a long range spin pumping contribution, as expected, remains present [16]. The RKKY induced spin density in the NM oscillates with the Fermi spanning *k* wavectors [19] and decreases with the thickness of the NM spacer layer. Due to interface roughness, this coupling quickly approaches zero and is usually zero for a Au spacer thickness

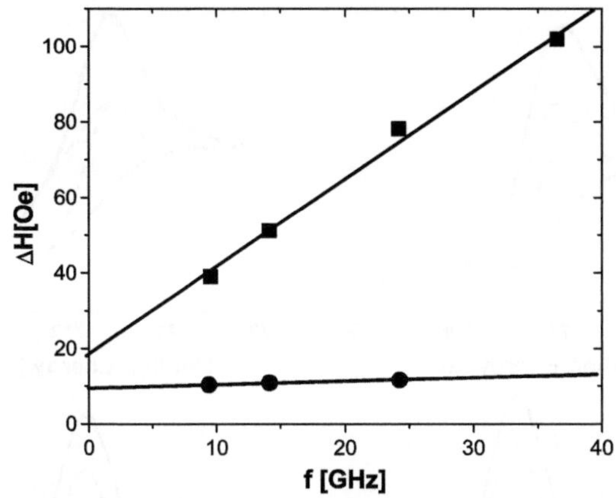

Fig. 10.2 The FMR linewidth (HWHM) as a function of microwave frequency. The *solid lines* were obtained using linear fits to the data points. The *circular symbols* and *straight line* correspond to the single magnetic layer sample 9.0YIG(sputtered). The *square symbols* and *straight line* correspond to the double magnetic layer 9.0YIG/6.1Au/4.3Fe/6.1Au. The difference in the slopes of the *solid lines* results in $g_{\uparrow\downarrow} \simeq 5.0 \times 10^{14}$ cm^{-2}

greater than 3 nm [20]. The spin pumping contribution does not oscillate with the Au spacer thickness and its efficiency only decreases due to the loss of spin momentum by a spin flip relaxation mechanism in Au; see (10.18). For a Au thickness significantly less than the spin diffusion length, this decrease is only minor and (10.23) is nearly correct.

The increase in the Gilbert damping in the above samples was interpreted using the spin pumping (10.18) (taking into account a small spin loss in Au). The spin mixing conductance at the untreated YIG/Au interface was found to be large $g_{\uparrow\downarrow,\mathrm{Fe}} = 1.0 \times 10^{14}$ cm^{-2}, and it was an appreciable fraction of that found for the Fe/Au(001) interface, $g_{\uparrow\downarrow,\mathrm{Fe}} = 1.1 \times 10^{15}$ cm^{-2}. The analysis were carried out assuming that the magnetic moment at the YIG/Au interface was equal to that in the bulk YIG.

Spin Pumping Using a Treated YIG Surface Since the chemical composition of a YIG surface has a deficiency of Fe compared to the bulk, we treated the YIG surface using in situ etching and deposition techniques. We could correlate the spin pumping efficiency with chemical modifications of the YIG film surface.

The surface of the YIG films were treated using low angle ($\sim 25°$) Ar^+ sputtering (ion etching) at energies of 0.8 and 1 kV while the sample was rotated in order to achieve a uniform surface etching. The chemical states of Y, O, Fe, C were monitored by means of XPS spectroscopy. By controlled ion etching (see caption in Fig. 10.3), one was able to increase the spin mixing conductance at the YIG/Au interface to $g_{\uparrow\downarrow} \simeq 5.0 \times 10^{14}$ cm^{-2} (see Fig. 10.2). This result represents a factor of 5 increase compared to the untreated YIG/Au interface and it is close to the value predicted by the recent first principles calculations $g_{\uparrow\downarrow} \simeq 7 \times 10^{14}$ cm^{-2} [21].

The strength of the spin mixing conductance was correlated with changes in the Fe and O X-ray photoelectron spectroscopy (XPS) spectra; see Fig. 10.3. Extensive

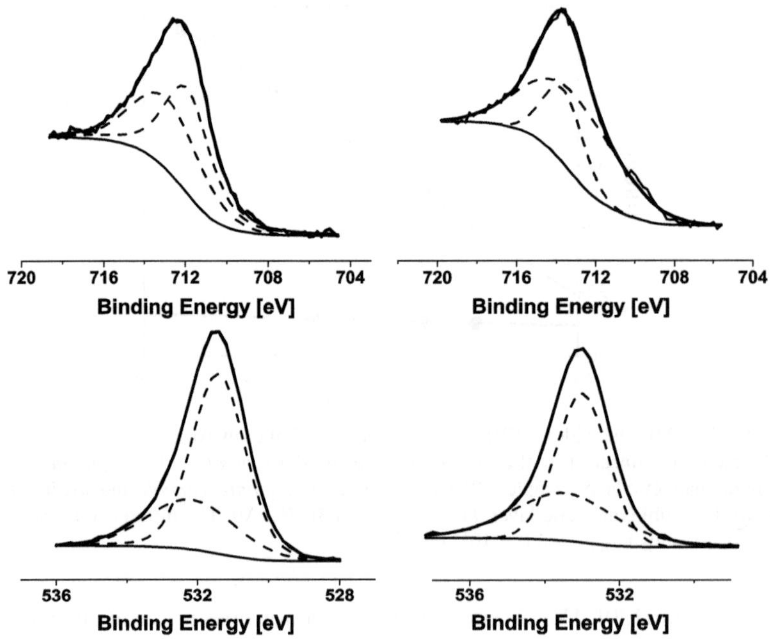

Fig. 10.3 The Fe $2p_{3/2}$ (*top figures*) and O 1s (*bottom figures*) XPS peaks for bare YIG before and after sputtering. *Figures on the left side* represent XPS spectra before sputtering. *Figures on the right side* represent the same spectra after sputtering. Low angle incidence sputtering was carried out at 1 kV for 22 minutes. Each of the Fe $2p_{3/2}$ and O 1s XPS lines consisted of the sum of two nearly coincident XPS lines. After sputtering the higher binding energy of Fe, $E_{h,b}^{2p} = 714.6$ eV, shifted towards the lower binding energy by 0.4 eV and the intensities of the two lines became more nearly equal. For O the higher binding energy line moved toward the lower binding energy $E_{l,b}^{1s} = 533.3$ eV by 0.3 eV but its intensity decreased [22]

sputtering (over 28 minutes in our case) resulted in the appearance of the metallic state of Fe2p, which led to a decrease the spin mixing conductance, $g_{\uparrow\downarrow} \simeq 3.5 \times 10^{14}$ cm^{-2} [22].

Two other modifications were carried out. In one case, we deposited the equivalent of 1 ML of metallic Fe on bare YIG. In the other case, we etched the surface of the YIG using H atoms. In both cases metallic Fe was obtained. The deposited Fe was not chemically reacting to Fe oxide in YIG and chemical etching resulted in a partial reduction of the Fe oxide to the metallic state. Both cases resulted in zero spin mixing conductance. This is contrary to the theoretical predictions of [21] suggesting that the spin pumping can be increased by adding a thin metallic Fe layer between the YIG film and the Au normal metal overlayer. In our view, the discrepancy between the theoretical prediction and our results suggests that the metallic Fe in our case is not only chemically decoupled from the Fe oxide but also exhibits no magnetic coupling to the magnetic Fe in YIG. The metallic Fe in our case kills the spin pumping effect, and therefore acts as a spin reflecting mirror at the YIG/Au interface. One should point out that the metallic Fe in this case is very thin (less than 1 ML), and therefore is in the paramagnetic state. The absence of spin pumping in the presence of the metallic Fe at the YIG/Au interface represents an interesting result on its own and requires further theoretical work. *In fact, a paramagnetic atomic layer of Fe acted as a spin current blockade.*

At this point it is interesting to compare spin pumping efficiency for microwaves, thermal excitations, and spin polarized currents. The dc spin current in YIG (spin momentum oriented along the static magnetization) driven by an rf field is given by:

$$\frac{\hbar}{4\pi} g_{\uparrow\downarrow} \omega \sin^2 \theta \omega. \tag{10.24}$$

Assuming $\theta = \pi/2$ for 10 GHz, the dc spin current is equal to $2 \times 10^{10} \hbar\ \mathrm{nm}^{-2}\,\mathrm{s}^{-1}$. Even for a frequency of 3 MHz, the dc spin pumping current remains appreciable, about $6 \times 10^7 \hbar\ \mathrm{nm}^{-2}\,\mathrm{s}^{-1}$.

The dc spin current driven by thermal excitation is given by:

$$\vec{I}^{\,\|}_{\mathrm{sp}} = \frac{\hbar \gamma k_{\mathrm{B}}}{2\pi M_s V_a} \operatorname{Re}(g_{\uparrow\downarrow}) \big(T_F^m - T_N\big). \tag{10.25}$$

The coherence volume in YIG equals $10^3\ \mathrm{nm}^3$ and considering that $(T_F^m - T_N) = 10$ K, the dc spin current is equal to $1 \times 10^8 \hbar\ \mathrm{nm}^{-2}\,\mathrm{s}^{-1}$.

In spin torque Magnetic Random Access Memory (MRAM) experiments, the current is about $2 \times 10^6\ \mathrm{A\,cm}^{-2}$ [23]. Dividing that by the charge on the electron, one obtains 2×10^{10} electrons $\mathrm{nm}^{-2}\,\mathrm{s}^{-1}$. Assuming that the spin polarization is 0.5, the spin torque generated by a spin polarized current is equal to $6 \times 10^{10} \hbar\ \mathrm{nm}^{-2}\,\mathrm{s}^{-1}$. The difference of 10 K in thermal excitation corresponds to an effective angular frequency of 0.2 GHz.

10.3.3 YIG/Pd

FMR measurements of spin pumping in Pd were obtained using YIG/Pd structures. The spin pumping parameter α_{sp} in Pd as a function of the Pd thickness is shown in the figure (see Fig. 10.4). The experimental data were fit with (10.21) using the fitting parameters $g_{\uparrow\downarrow} = 2.9 \times 10^{14}\ \mathrm{cm}^{-2}$ and $\lambda_{\mathrm{dec}} \simeq 5$ nm.

These results will be compared with those obtained on crystalline Fe/Pd [13] and polycrystalline Py/Pd [14] structures, where Py is permalloy.

A thick Pd film acts as a spin sink (see (10.21) and (10.23)). The $g_{\uparrow\downarrow}$ in YIG/Pd (thick) and in YIG/Au/Fe are equal to $2.9 \times 10^{14}\ \mathrm{cm}^{-2}$ and $5.2 \times 10^{14}\ \mathrm{cm}^{-2}$, respectively. This means that the spin sink in a thick Pd film is 40 % less efficient than the second Fe layer in YIG/Au/Fe. The spin mixing conductance $g_{\uparrow\downarrow}$ in the crystalline Fe/Pd(001) structure [13] is only 18 % lower than that found for a perfect sink in an Fe/Au/Fe(001) structure. Bailey et al. [14] obtained $g_{\uparrow\downarrow} = 0.8 \times 10^{15}\ \mathrm{cm}^{-2}$ in Py/Cu/Pd, which is nearly the same as that for a Fe/Pd(001) structure but is a factor of 3 times larger than that observed for the YIG/Pd structure.

Bailey et al. found that the decoherence length in polycrystalline Pd was $\lambda_{\mathrm{dec}} \simeq 4.7$ nm which is close to that estimated in our study on a YIG/Pd structure.

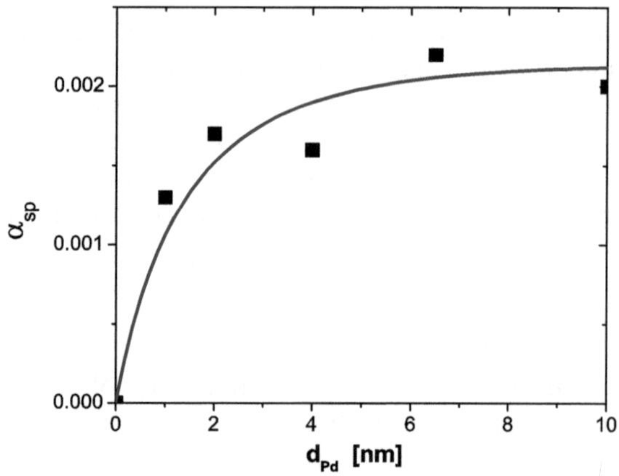

Fig. 10.4 Spin pumping Gilbert damping parameter α_{sp} in 20.0YIG/Pd as a function of the Pd thickness d_{Pd}. The *solid line* was obtained by fitting the data with (10.21) resulting in $g_{\uparrow\downarrow} \simeq 2.9 \times 10^{14}$ cm^{-2} and $\lambda_{dec} \simeq 5$ nm

The fitting parameters in the Foros paper [13] were obtained using a simplified equation which did not include back flow (see (10.20)). The parameters in this case were: $g_{\uparrow\downarrow} = 0.9 \times 10^{15}$ cm^{-2} and $\lambda_{dec} \simeq 9$ nm. We refit these data using the full equation (10.21). The fit is equally good and the parameters are slightly larger (less than 10 %) : $g_{\uparrow\downarrow} = 0.96 \times 10^{15}$ cm^{-2} and $\lambda_{dec} \simeq 9.4$ nm.

Acknowledgements The authors would like to thank their colleagues B. Kardasz, E. Montoya, and E. Girt for their extensive collaboration and stimulating discussions during the study of spin pumping in magnetic heterostructures. We are grateful to Yiyan Su, Young-Yeal Song, and Mingzheng Wu for providing us with high quality YIG films and their collaboration in the YIG spin pumping studies. We would like to thank J.F.C. Cochran for stimulating discussions and his invaluable help in preparation of this manuscript. We would like to thank G.E.W. Bauer, Y. Tserkovnyak, A. Brataas, and K. Xia for extensive discussions on spin pumping which helped greatly to improve the clarity of our chapter. We would like to express our appreciation to W. Bailey for sharing with us the results of his extensive studies on spin pumping in metallic structures.

We would like to thank especially the Natural Science Engineering Research Council (NSERC) of Canada and the Canadian Institute for Advanced Research (CIFAR) for continued research funding which makes our work possible.

References

1. J. Xiao, G.E.W. Bauer, K.U. Uchida, E. Saitoh, S. Maekawa, Phys. Rev. B **81**, 214418 (2010)
2. J. Slonczewski, Phys. Rev. B **82**, 054403 (2010)
3. B. Heinrich, *Ultrathin Magnetic Structures III*, ed. by B. Heinrich, J.F.C. Bland (Springer, Berlin, 2003), Chap. 5
4. Y. Tserkovnyak, A. Brataas, G.E.W. Bauer, Phys. Rev. Lett. **88**, 117601 (2002)
5. B. Heinrich, Y. Tserkovnyak, G. Woltersdorf, A. Brataas, R. Urban, G. Bauer, Phys. Rev. Lett. **90**, 187601 (2003)
6. E. Šimanek, B. Heinrich, Phys. Rev. B **67**, 144418 (2003)
7. Y. Tserkovnyak, A. Brataas, G.E. Bauer, B.I. Halperin, Rev. Mod. Phys. **77**, 1375 (2005)
8. A. Brataas, G.E.W. Bauer, P.J. Kelly, Phys. Rep. **427**, 157 (2006)
9. X. Jia, K. Liu, G. Bauer, Europhys. Lett. **96**, 117601 (2011)
10. B. Kardasz, B. Heinrich, Phys. Rev. B **81**, 94409 (2010)

11. O. Mosendz, G. Woltersdorf, B. Kardasz, B. Heinrich, C.H. Back, Phys. Rev. B **79**, 224412 (2009)
12. M.B. Brodsky, A.J. Freeman, Phys. Rev. Lett. **45**, 133 (1980)
13. J. Foros, G. Woltersdorf, B. Heinrich, A. Brataas, J. Appl. Phys. **97**, 10A714 (2005)
14. W. Bailey et al., Phys. Rev. B, submitted
15. B. Heinrich, J.A.C. Bland (eds.), *Ultrathin Magnetic Structures*, vol. II (Springer, Berlin, 1994), 2005
16. B. Heinrich, C. Burrowes, E. Montoya, B. Kardasz, E. Girt, Y.Y. Song, Y. Sun, M. Wu, Phys. Rev. Lett. **107**, 066604 (2011)
17. B. Kardasz, O. Mosendz, B. Heinrich, M. Freeman, J. Appl. Phys. **103**, 07C509 (2008)
18. E. Simanek, B. Heinrich, Phys. Rev. B **67**, 144418 (2003)
19. P. Bruno, J. Phys. Condens. Matter **11**, 9403 (1999)
20. Z. Celinski, B. Heinrich, J. Magn. Magn. Mater. **99**, L25 (1991)
21. X. Jia, K. Liu, K. Xia, G.E.W. Bauer, Europhys. Lett. **96**, 17005 (2011)
22. C. Burrowes, B. Heinrich, E. Montoya, B. Kardasz, Y.Y. Song, Y. Sun, M. Wu, Appl. Phys. Lett. **98**, 052508 (2012)
23. Y. Huai, AAPPS Bull. **18**(6), 33 (2008)

Chapter 11
Spin-Torque Microwave Detectors

Oleksandr V. Prokopenko, Ilya N. Krivorotov, Thomas J. Meitzler, Elena Bankowski, Vasil S. Tiberkevich, and Andrei N. Slavin

Abstract It is known that the spin-transfer torque (STT) effect provides a new method of manipulation of magnetization in nano-scale devices. According to the STT effect, bias DC current traversing magnetic multilayers can transfer angular magnetic moments from one layer to another, which can give rise to microwave dynamics of magnetization in the layer. However, it is clear that an inverse effect is also possible. This inverse effect leads to the so-called spin-torque diode effect, first originally observed experimentally in 2005. The spin-torque diode effect is a rectification effect of the input microwave current in a magnetoresistive junction. In this case, the resonance oscillations of the junction resistance can mix with the oscillations of the input microwave current and produce a large enough output DC voltage across the junction. The devices which utilize this effect are called the spin-torque microwave detectors (STMD). In this chapter, we review the general properties of STMDs and consider the performance of a STMD in two different dynamic regimes of detector operation: in the well-known traditional in-plane regime of STMD oper-

O.V. Prokopenko
Faculty of Radiophysics, Taras Shevchenko National University of Kyiv, Kyiv, 01601, Ukraine
e-mail: ovp@univ.kiev.ua

I.N. Krivorotov
Department of Physics and Astronomy, University of California, Irvine, CA 92697, USA
e-mail: ikrivoro@uci.edu

T.J. Meitzler · E. Bankowski
Ground Systems Survivability, U.S. Army TARDEC, Warren, MI 48397, USA

T.J. Meitzler
e-mail: thomas.j.meitzler.civ@mail.mil

E. Bankowski
e-mail: elena.n.bankowski.civ@mail.mil

V.S. Tiberkevich · A.N. Slavin (✉)
Department of Physics, Oakland University, Rochester, MI 48309, USA
e-mail: slavin@oakland.edu

V.S. Tiberkevich
e-mail: tyberkev@oakland.edu

S.O. Demokritov, A.N. Slavin (eds.), *Magnonics*, Topics in Applied Physics 125,
DOI 10.1007/978-3-642-30247-3_11,

ation and in the recently discovered novel out-of-plane regime of STMD operation. We analyze the performance of a STMD and consider the typical applications for such detectors in both regimes.

11.1 Introduction

The spin-transfer torque (STT) effect in magnetic multilayers was theoretically predicted in [1–3] and experimentally observed in [4–14]. It provides a new method of manipulation of the magnetization direction in nano-magnetic systems [15]: magnetization switching [4, 5], generation of microwave oscillations under the action of a dc electric current [6–12], and the spin-torque diode effect [13, 14], which can then be used for the development of practical microwave detectors, the so called spin-torque microwave detectors (STMD), and also for quantitatively measuring STT [16, 17].

The spin-torque diode effect is a quadratic rectification effect of the input microwave current $I_{RF}(t)$ in a magnetoresistive junction, which is commonly observed in the traditional regime of operation of a STMD, when the frequency f_s of the current $I_{RF}(t) = I_{RF}\sin(2\pi f_s t)$ is close to the ferromagnetic resonance (FMR) frequency f_0 of the junction. In this case, the induced resonance oscillations of the junction resistance $R(t)$ can mix with the oscillations of the input microwave current $I_{RF}(t)$ and produce a sufficiently large output DC voltage $U_{DC} = \langle I_{RF}(t)R(t)\rangle$ across the junction (here $\langle\ldots\rangle$ denotes averaging over the period of oscillations $1/f_0$).

In the traditional in-plane (IP) regime of operation of an STMD [13, 14, 18], the STT excites a *small-angle in-plane* magnetization precession about the equilibrium direction of magnetization in the "free" layer (FL) of a magnetic tunnel junction (MTJ) (see the red dashed IP-curve in Fig. 11.1). Below we shall refer to this regime of STMD operation as the IP-regime. Using analytical theory, we show that the performance of a passive STMD in the IP-regime is limited by thermal noise. We also show how to calculate the optimal parameters of a STMD operating in this regime. We believe that the developed formalism can be used for optimization of noise-handling properties for construction of a practical STMD.

In contrast to the well-known IP-regime of STMD operation, in this paper we also consider a different regime of operation of an STMD, based on the excitation of *large-angle out-of-plane* (OOP) magnetization precession under the action of an input microwave current $I_{RF}(t)$ (see the blue dashed OOP-curve in Fig. 11.1). Using analytical and numerical calculations, we show that all the major STMD characteristics in the OOP-regime qualitatively differ from the ones in the traditional IP-regime. In particular, excitations in the OOP-regime do not have a resonance character and exist in a wide range of driving frequencies. Also, the output DC voltage of an STMD in the OOP regime is nearly independent of the input microwave power, provided it exceeds a certain threshold value. We believe that these properties of an STMD in the OOP-regime will be useful for the development of nano-sized threshold detectors with a large output DC voltage, spectrum analyzers, and also for applications in microwave energy harvesting in the low-frequency region of the microwave band.

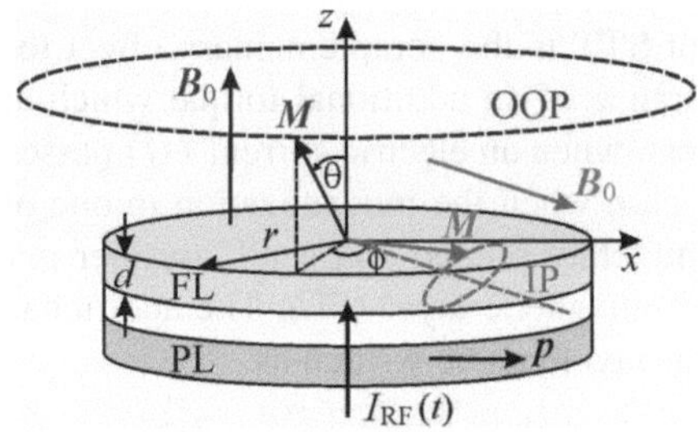

Fig. 11.1 The model of an STMD: circular nano-pillar of radius r consists of the "free" magnetic layer (FL) of thickness d and the "pinned" magnetic layer (PL). Under the action of microwave current $I_{RF}(t)$, the magnetization vector $\mathbf{M}$ (shown by a *blue or red arrow*) is precessing along small-angle in-plane (*red dashed curve*, IP-regime) or large-angle out-of-plane (*blue dashed curve*, OOP-regime) trajectory about the direction of the bias magnetic field $\mathbf{B}_0$ (shown by a *red or blue arrow*), $\mathbf{p} = \mathbf{x}$ is the unit vector in the direction of the magnetization of the PL, $\mathbf{x}$ is the unit vectors of the x-axis (Color figure online)

11.2 Basic Physics of STT and TMR

The operation of a STMD based on the MTJ is based on two related phenomena: tunneling magnetoresistance (TMR) and the above mentioned STT.

The phenomenon of TMR exists because of the dependence of the resistance $R(\beta)$ of a MTJ on the angle β between magnetizations of the two ("free" and "pinned") magnetic layers of the junction. Here we shall consider only symmetric unbiased junctions, for which $R(\beta)$ can be written as [18]

$$R(\beta) = \frac{R_\perp}{1 + P^2 \cos\beta}, \tag{11.1}$$

where $R_\perp$ is the junction's resistance in the perpendicular ($\beta = \pi/2$) magnetic state and P is the dimensionless spin-polarization efficiency. Conventional parallel (R_P) and antiparallel (R_{AP}) resistances are connected with $R_\perp$ and P by

$$R_P = \frac{R_\perp}{1 + P^2}, \qquad R_{AP} = \frac{R_\perp}{1 - P^2}, \tag{11.2}$$

whereas conventional TMR ratio can be calculated as

$$\mathrm{TMR} = \frac{R_{AP} - R_P}{R_P} = \frac{2P^2}{1 - P^2}. \tag{11.3}$$

Typical values of P are about 0.5–0.7, whereas $R_\perp$ can be estimated as

$$R_\perp = \frac{RA}{S}, \tag{11.4}$$

where $RA \sim 1$–$10\ \Omega\,\mu\mathrm{m}^2$ is the resistance–area product of the MTJ and S is its area. For typical sub-micron sizes of the junction, $R_\perp$ is much higher than the characteristic impedance $Z_{TL} = 50\ \Omega$ of standard microwave transmission lines.

The phenomenon of STT is the complementary effect to TMR. The STT phenomenon is the appearance of an additional torque which acts on the magnetizations of junctions' layers, when an electric current $I(t)$ passes through the junction. We shall consider the case when the magnetization in one of the layers ("pinned" magnetic layer) is completely pinned and shall consider magnetization dynamics only in another ("free") magnetic layer (FL). The additional STT that acts on the magnetization $\mathbf{M}$ of this layer can be written as

$$\left(\frac{d\mathbf{M}}{dt}\right)_{\mathrm{STT}} = \frac{\sigma(\beta)I(t)}{M_s}\big[\mathbf{M}\times[\mathbf{M}\times\mathbf{p}]\big], \tag{11.5}$$

where $M_s = |\mathbf{M}|$ is the saturation magnetization of the FL, $\mathbf{p}$ is the unit vector in the direction of the magnetization of the PL, and $\sigma(\beta)$ is the current–torque proportionality coefficient that is equal to

$$\sigma(\beta) = \frac{\sigma_\perp}{1+P^2\cos\beta}, \qquad \sigma_\perp = \frac{\gamma\hbar}{2e}\frac{P}{M_s V}. \tag{11.6}$$

Here $\gamma \approx 2\pi \cdot 28$ GHz/T is the modulus of the gyromagnetic ratio, $\hbar$ is the reduced Planck constant, e is the modulus of the electron charge, and $V = Sd$ is the volume of the FL (d is its thickness).

11.3 Small-Angle In-Plane Dynamical Regime of STMD Operation

In the traditional regime of operation of an STMD (based on the above mentioned spin-torque diode effect), the STT excites a small-angle in-plane magnetization precession about the equilibrium direction of the magnetization in the "free" magnetic layer (FL) of a junction (the trajectory of magnetization precession is shown by a red dashed curve in Fig. 11.1). The precessional motion of magnetization in the IP-regime determines the following typical properties of a traditional STMD [13, 14, 18]:

(a) The STMD operates as a frequency-selective microwave detector with a resonance frequency that is close to the frequency of the ferromagnetic resonance (FMR) $\omega_0 = 2\pi f_0$ of the FL;
(b) The frequency operation range of the detector has an order of the FMR linewidth Γ;
(c) The output DC voltage U_{DC} of the STMD is proportional to the input microwave power $P_{\mathrm{RF}} = I_{\mathrm{RF}}^2 R_0/2$ (R_0 is the equilibrium MTJ resistance), so the spin-torque diode operates as a resonance-type, quadratic microwave detector:

$$U_{\mathrm{DC}} = \varepsilon_{\mathrm{res}} P_{\mathrm{RF}} \frac{\Gamma^2}{\Gamma^2 + (\omega_s - \omega_0)^2}, \tag{11.7}$$

where

$$\varepsilon_{\text{res}} = \frac{\gamma \hbar}{4e} \frac{P^3}{M_s V \Gamma} Q(\beta_0) \tag{11.8}$$

is the resonance (at $\omega_s = \omega_0$) diode volt–watt sensitivity (see [18]). Here $Q(\beta_0)$ is the geometrical factor that depends on the angle β_0 between the directions of the equilibrium magnetization in FL and PL of the MTJ. For an in-plane magnetized MTJ:

$$Q(\beta_0) = \frac{\sin^2(\beta_0)}{(1 + P^2 \cos \beta_0)^2}; \tag{11.9}$$

(d) The diode resonance sensitivity ε_{res} strongly depends on the angle β between magnetization directions of the FL and PL – see (11.9).

The resonance sensitivity of traditional STMD $\varepsilon_{\text{res}} = U_{\text{DC}}/P_{\text{RF}}$ is predicted to be approximately I_{RF} (see [18]), while the best achieved experimental result to date is $\varepsilon_{\text{res}} \approx 300$ V/W [14].

11.3.1 Analytical Theory of Noise Properties of a STMD in IP-Regime

It is known that the operation and the minimum detectable power of all types of microwave detectors are limited by noise (in particular, by the low-frequency Johnson–Nyquist noise in the case of unbiased Schottky diodes [19]), and, therefore, it is important to understand the noise-handling properties of the novel STMD based on MTJ – this is the main aim of this part of the paper.

Below we present a theoretical analysis of the noise properties of a passive resonance-type STMD (no DC bias current) using the STMD model developed in [18] with additional terms describing the influence of thermal fluctuations. The results presented below are expanded and are complementary to the results presented in [20].

During the theoretical analysis we use the following assumptions:

(a) We assume that the lateral sizes of STMD FL are sufficiently small, so the internal magnetization dynamics can be described in the macrospin approximation;
(b) The PL is assumed to be truly "pinned", i.e., we neglect any possible fluctuations or oscillations of magnetization in this layer;
(c) We consider unbiased STMD, i.e., the current $I(t)$ flowing through STMD has only microwave component $I_{\text{RF}}(t)$, $I(t) = I_{\text{RF}}(t)$;
(d) The amplitude of the detected current I_{RF}, as well as the temperature of the system T, is assumed to be sufficiently small to provide a linear response of the magnetization to the current and thermal fluctuations. This also means that in all expressions we shall keep only the lowest-order (of the current amplitude and temperature) non-zero terms;

(e) We shall consider only the case of a low-damping STMD $f_0 \gg \Gamma/2\pi$, in which case one can drop non-resonant terms in the detected output voltage U_{DC};
(f) We also assume that the frequency interval of the detection Δf (inversely proportional to the measurement time) is much smaller than the FMR linewidth $\Gamma/2\pi$ in the FL, $\Delta f \ll \Gamma/2\pi$. This means that we are interested only in the zero-frequency limit of the noise spectral density of the output voltage;
(g) Lastly, we assume that the noise temperature T for all considered below noise sources is the same.

There are several possible sources of noise in STMD. Moreover, since the operation of the STMD is necessary nonlinear, the noise is, in general, not additive and can depend on the amplitude and frequency of the detected signal. Among all possible noise sources we shall consider only the following:

(a) *Low-frequency Johnson–Nyquist* (JN) *noise*, i.e., low-frequency voltage fluctuations $U_N(t)$ associated with the equilibrium electrical resistance of the STMD R_0. This type of the noise is additive and is independent of the magnetization dynamics.
(b) *High-frequency Johnson–Nyquist noise*. In contrast with the low-frequency components, microwave part of the JN noise does not contribute to the output detected signal directly. A microwave noise current $I_N(t)$, however, can induce additional fluctuations of the magnetization direction in the FL and, respectively, fluctuations of the electrical resistance of the STMD. Being mixed with the detected current $I_{\mathrm{RF}}(t)$, these resistance fluctuations will create non-additive low-frequency noise in the output voltage U_{DC}.
(c) *Magnetic noise* (MN), which is caused by the thermal fluctuations of the magnetization direction in the MTJ FL, is modeled by the action of a random white Gaussian magnetic field $\mathbf{B}_N(t)$, and leads to the fluctuations of the electric resistance of the STMD. It is transformed into low-frequency noise after mixing with the driving current $\tilde{q}$.

The other noise sources, such as shot noise and flicker noise, might be important for STMDs biased by a DC current.

To calculate the noise spectrum in a STMD we shall, first, calculate the linear response of the magnetization of the FL to arbitrary microwave currents and magnetic fields. Then we shall calculate corresponding variations of the electrical resistance of the STMD, i.e., calculate the electrical response of STMD to currents and fields. After that, using a simplified electrical scheme of STMD, we shall calculate the STMD noise spectrum.

Dynamics of the magnetization $\mathbf{M}$ of the FL under the action of microwave current $I(t)$ and magnetic field $\mathbf{B}(t)$ are described by the Landau–Lifshits–Gilbert–Slonczewski equation:

$$\frac{d\mathbf{M}}{dt} = \gamma\big[\mathbf{B}_{\mathrm{eff}}(\mathbf{M}) \times \mathbf{M}\big] + \frac{\alpha}{M_s}\left[\mathbf{M} \times \frac{d\mathbf{M}}{dt}\right] + \frac{\sigma(\beta)I(t)}{M_s}\big[\mathbf{M} \times [\mathbf{M} \times \mathbf{p}]\big] + \gamma\big[\mathbf{B}(t) \times \mathbf{M}\big]. \tag{11.10}$$

Here $\mathbf{B}_{\mathrm{eff}}(\mathbf{M})$ is the effective magnetic field which includes the external bias magnetic field $\mathbf{B}_0$ and the demagnetization field, α is the Gilbert damping constant. The first term in the right-hand side of (11.10) describes conservative magnetization precession, the second term describes magnetic dissipation, and the last two terms describe excitation of the magnetization precession by the microwave current $I(t)$ and magnetic field $\mathbf{B}(t)$, respectively.

To find the linear response of the magnetization, we represent $\mathbf{M}$ in the form

$$\mathbf{M}(t) = \big(\mathbf{m}_0 + \mathbf{m}(t)\big)M_s, \tag{11.11}$$

where $\mathbf{m}_0$ is the unit vector in the direction of the equilibrium orientation of the magnetization vector and $\mathbf{m}(t)$ is the dimensionless amplitude of small magnetization precession. In the considered linear regime, the vectors $\mathbf{m}_0$ and $\mathbf{m}(t)$ are orthogonal to each other. Substituting (11.11) into (11.10) and keeping only linear terms in $\mathbf{m}$, $I(t)$, and $\mathbf{B}(t)$ yields the linear equation for $\mathbf{m}(t)$:

$$\frac{d\mathbf{m}}{dt} = -i\hat{\mathbf{L}}\cdot\mathbf{m} + \alpha\left[\mathbf{m}_0 \times \frac{d\mathbf{m}}{dt}\right] + \sigma_0 I(t)\big[\mathbf{m}_0 \times [\mathbf{m}_0 \times \mathbf{p}]\big] + \gamma\big[\mathbf{B}(t) \times \mathbf{m}_0\big] \tag{11.12}$$

(operator above $\mathbf{L}$ is not evident in WORD), where $\sigma_0 = \sigma(\beta_0)$ and the linear operator $\hat{\mathbf{L}}$ is determined by the linearization of the first conservative term in the right hand side of (11.10). It is clear (and can be easily proven rigorously) that the vector $\mathbf{m}_0$ is the eigenvector of the operator $\hat{\mathbf{L}}$ corresponding to the eigenvalue 0. The two other eigenvectors are complex conjugates (i.e., $\boldsymbol{\mu}$ and $\boldsymbol{\mu}^*$) and correspond to opposite eigenvalues (i.e., $\omega_0 > 0$ and $-\omega_0 < 0$). These eigenvectors are orthogonal to $\mathbf{m}_0$ and, respectively, span the vector space of small magnetization oscillations $\mathbf{m}(t)$. Taking into account that $\mathbf{m}(t)$ is a real-valued vector, we can represent it as

$$\mathbf{m}(t) = \boldsymbol{\mu}c(t) + \text{c.c.}, \tag{11.13}$$

where $c(t)$ is the complex amplitude of precession and c.c. denotes complex-conjugated part.

Substituting (11.13) into (11.12) and taking scalar product of both sides of this equation with $[\mathbf{m}_0 \times \boldsymbol{\mu}^*]$ gives a simple equation for the rate of change of $c(t)$:

$$\frac{dc}{dt} = -i\omega_0 c - \Gamma c + \sigma_0 \tilde{q} I(t) + \gamma\tilde{\mathbf{u}}\cdot\mathbf{B}(t). \tag{11.14}$$

Here Γ is the damping rate of magnetization precession,

$$\Gamma = i\alpha\omega_0 \frac{|\boldsymbol{\mu}|^2}{(\boldsymbol{\mu}, \mathbf{m}_0, \boldsymbol{\mu}^*)}, \tag{11.15}$$

and dimensionless current $\tilde{q}$ and field $\tilde{\mathbf{u}}$ "susceptibilities" are equal to

$$\tilde{q} = -\frac{(\mathbf{p}, \mathbf{m}_0, \boldsymbol{\mu}^*)}{(\boldsymbol{\mu}, \mathbf{m}_0, \boldsymbol{\mu}^*)}, \qquad \tilde{\mathbf{u}} = -\frac{\boldsymbol{\mu}^*}{(\boldsymbol{\mu}, \mathbf{m}_0, \boldsymbol{\mu}^*)}. \tag{11.16}$$

Here the notation $(\mathbf{v}_1, \mathbf{v}_2, \mathbf{v}_3)$ denotes scalar triple product, $(\mathbf{v}_1, \mathbf{v}_2, \mathbf{v}_3) = \mathbf{v}_1 \cdot [\mathbf{v}_2 \times \mathbf{v}_3]$. Note that the "tilded" susceptibilities $\tilde{q}$ and $\tilde{\mathbf{m}}$ depend on the normalization of the eigenvector $\boldsymbol{\mu}$ and, respectively, do not have a physical meaning.

Equation (11.14) can be easily solved in the frequency representation:

$$c_\omega = \frac{\sigma_0 \tilde{q}}{\Gamma_\omega} I_\omega + \frac{\gamma \tilde{\mathbf{u}}}{\Gamma_\omega} \cdot \mathbf{B}_\omega, \tag{11.17}$$

where

$$\Gamma_\omega = \Gamma - i(\omega - \omega_0). \tag{11.18}$$

Finally, the magnetization response $\mathbf{m}$ in the frequency representation is given by

$$\mathbf{m}_\omega = \sigma_0 \left(\frac{\tilde{q}\boldsymbol{\mu}}{\Gamma_\omega} + \frac{\tilde{q}^* \boldsymbol{\mu}^*}{\Gamma^*_{-\omega}} \right) I_\omega + \gamma \left(\frac{\boldsymbol{\mu}\tilde{\mathbf{u}}}{\Gamma_\omega} + \frac{\boldsymbol{\mu}^* \tilde{\mathbf{u}}^*}{\Gamma^*_{-\omega}} \right) \cdot \mathbf{B}_\omega, \tag{11.19}$$

where we took into account that both $I(t)$ and $\mathbf{B}(t)$ are real-valued and, respectively, $I_\omega = I^*_{-\omega}$ and $\mathbf{B}_\omega = \mathbf{B}^*_{-\omega}$.

The expressions derived above give the complete information about the magnetization response on an arbitrary microwave current $I(t)$ and magnetic field $\mathbf{B}(t)$ provided that equilibrium magnetization direction $\mathbf{m}_0$, vector structure of the precession mode $\boldsymbol{\mu}$, PL magnetization orientation $\mathbf{p}$, and FMR frequency ω_0 are known. Calculation of all these parameters can be easily done in each particular case.

From the "electrical" point of view, STMD is a resistance $R(\beta)$ that depends on the angle $\beta = \arccos(\mathbf{M} \cdot \mathbf{p}/M_s)$ between magnetizations of the "free" and "pinned" layers. Since the magnetization of the FL oscillates, the resistance $R(\beta)$ changes in time and can be written as

$$R = R_0 + \delta R(t), \qquad \delta R(t) = R_0 \rho \delta\beta(t), \tag{11.20}$$

where $R_0 = R(\beta_0)$ is the equilibrium STMD resistance, $\beta_0 = \arccos(\mathbf{m}_0 \cdot \mathbf{p})$ is the equilibrium angle, and

$$\rho = \frac{1}{R_0} \left(\frac{dR}{d\beta} \right)_{\beta=\beta_0} = \frac{P^2 \sin\beta_0}{1 + P^2 \cos\beta_0} \tag{11.21}$$

is the dimensionless slope of the angular dependence of the resistance $R(\beta)$ at the equilibrium and angular variations $\delta\beta$ can be written as

$$\delta\beta(t) = -\frac{\mathbf{m}(t) \cdot \mathbf{p}}{\sin\beta_0}. \tag{11.22}$$

The output signal of a STMD consists of the low-frequency components of the voltage drop across the STMD. We assume that the high-frequency components are completely filtered out by an external circuit and that there is no low-frequency current (electrical circuit of STMD is open at low frequencies). Then, the output voltage $U_{\text{out}}(t)$ can be written as

$$U_{\text{out}}(t) = \left[U_{\text{JN}}(t) + \delta R(t) I(t)\right]_{\omega \to 0} \tag{11.23}$$

where $U_{\text{JN}}(t)$ is the JN voltage noise, $I(t)$ is the microwave current through the STMD, and notation $[\ldots]_{\omega\to 0}$ symbolically denotes filtering of low-frequency components. The term $R_0 I(t)$ does not give any contribution to the low-frequency signal and has been omitted.

The current $I(t)$ consists of the input signal current $I_{\text{RF}}(t) = I_{\text{RF}} \sin(\omega_s t)$, where ω_s is the signal frequency, and noise current $I_N(t) = U_N(t)/R_0$ created by the microwave components of the JN voltage noise. The resistance fluctuations $\delta R(t)$ are created by the current $I(t)$ and thermal noise magnetic field $\mathbf{B}_N(t)$.

The detected voltage U_{DC} is obtained from (11.23) by taking into account only the signal components:

$$U_{\text{DC}} = \text{Re}\left(\frac{\rho q}{2} \frac{R_0 \sigma_0}{\Gamma_{\omega_s}}\right) I_{\text{RF}}^2, \tag{11.24}$$

where we have dropped the non-resonant term (proportional to $1/\Gamma_{-\omega_s}$), which can be done for low-damping STMD ($\omega_0 \gg \Gamma$); $q = -(\boldsymbol{\mu}\cdot\mathbf{p})\tilde{q}/\sin\beta_0$.

To calculate the spectrum of the low-frequency noise, one has to take into account that the JN voltage $U_{\text{JN}}(t)$ and thermal magnetic field $\mathbf{B}_N(t)$ are independent random white Gaussian processes with constant spectral densities (the spectral density $S_f(\omega)$ of a stationary random process $f(t)$ is defined by $\langle f_\omega f^*_{\omega'}\rangle = 2\pi\delta(\omega-\omega')S_f(\omega)$):

$$S(U_N) = S(I_N)R_0^2 = 2k_{\text{B}} T R_0, \qquad S(B_N) = \frac{2\alpha k_{\text{B}} T}{\gamma M_s V}, \tag{11.25}$$

where k_{B} is the Boltzmann constant. The spectral density $S(B_N)$ is the correlator of the same vector components of $\mathbf{B}_N(t)$; different components of $\mathbf{B}_N$ are independent.

Calculation of the output voltage noise spectrum $S_{\text{out}}(\Omega)$, where Ω is the frequency of the output voltage fluctuations, from (11.23) gives

$$S_{\text{out}}(\Omega) = S_{\text{out}}^{(1)}(\Omega) + S_{\text{out}}^{(2)}(\Omega) + S_{\text{out}}^{(3)}(\Omega), \tag{11.26}$$

where

$$S_{\text{out}}^{(1)}(\Omega) = S(U_N), \tag{11.27}$$

$$S_{\text{out}}^{(2)}(\Omega) = \frac{1}{4} S(U_N)\sigma_0^2\rho^2 I_{\text{RF}}^2 \left[\left|\frac{q}{\Gamma_{\omega_s}} + \frac{q^*}{\Gamma^*_{\omega_s-\Omega}}\right|^2 + \left|\frac{q}{\Gamma_{\omega_s}} + \frac{q^*}{\Gamma^*_{\omega_s+\Omega}}\right|^2\right], \tag{11.28}$$

$$S_{\text{out}}^{(3)}(\Omega) = \frac{1}{4} S(B_N) R_0^2 \gamma^2 \rho^2 u^2 I_{\text{RF}}^2 \left[|\Gamma_{\omega_s-\Omega}|^{-2} + |\Gamma_{\omega_s+\Omega}|^{-2}\right]. \tag{11.29}$$

The three terms $S_{\rm out}^{(1,2,3)}(\Omega)$ are caused, respectively, by (i) low-frequency components of the JN noise, (ii) high-frequency components of the JN noise and related fluctuations of the MTJ resistance, and (iii) fluctuations of the MTJ resistance due to thermal magnetic field.

As one can see from (11.26)–(11.29), the frequency dependence of the detector noise is completely determined by the frequency dependence of the damping factor Γ_ω (see (11.18)). In the following we assume, for simplicity, that the signal frequency ω_s exactly coincides with the FMR frequency ω_0 and that the frequency interval of detection $\Delta f \approx \Omega/2\pi$ (inversely proportional to the measurement time) is much smaller than the FMR linewidth $\Gamma/2\pi$.

In such a resonant case (11.24) becomes

$$U_{\rm DC} = \left(\frac{\rho q_r}{2}\frac{R_0\sigma_0}{\Gamma}\right) I_{\rm RF}^2 = \rho q_r \frac{\sigma_0}{\Gamma} P_{\rm RF}, \tag{11.30}$$

where $q_r = {\rm Re}(q)$.

The noise spectrum, $S_{\rm out}$, in this case can be written in a very simple form

$$S_{\rm out} = \left(1 + \frac{U_{\rm DC}}{U_{\rm IM}} + \frac{U_{\rm DC}}{U_{\rm MN}}\right) S(U_N), \tag{11.31}$$

where

$$U_{\rm IM} = \frac{R_0 \Gamma}{4\rho q_r \sigma_0}, \qquad U_{\rm MN} = \frac{q_r \sigma_0 M_s V \Gamma}{\alpha \rho u^2 \gamma}. \tag{11.32}$$

Finally, root mean square fluctuations of the output voltage $\Delta U_{\rm DC}$ can be determined as

$$\Delta U_{\rm DC} = \sqrt{2 S_{\rm out} \Delta f} = \Delta U_{\rm JN} \sqrt{1 + \frac{U_{\rm DC}}{U_{\rm IM}} + \frac{U_{\rm DC}}{U_{\rm MN}}}, \tag{11.33}$$

where Δf is the frequency bandwidth of measurements and

$$\Delta U_{\rm JN} = \sqrt{2 S(U_N) \Delta f} \tag{11.34}$$

is the voltage fluctuation caused by the low-frequency JN noise only. The minimum detected voltage can be estimated from the condition $\Delta U_{\rm DC} = U_{\rm DC}$.

The signal-to-noise ratio (SNR) of a STMD can be calculated as

$$SNR = \frac{U_{\rm DC}}{\Delta U_{\rm DC}} = \frac{P_{\rm RF}}{P_{\rm JN}} \sqrt{\frac{P_{\rm MN}}{P_{\rm MN} + P_{\rm RF}}}, \tag{11.35}$$

where we introduced the noise powers $P_{\rm JN} = U_{\rm JN}/\varepsilon_{\rm res}$, $P_{\rm MN} = U_{\rm MN}/\varepsilon_{\rm res}$ and took into account that $U_{\rm IM} \gg U_{\rm JN}, U_{\rm MN}$ for the typical STMD [20].

11.3.2 The Performance of a STMD in the Presence of Thermal Noise

The simple analysis of (11.33), (11.35) demonstrates that there are two distinct regimes of operation of the resonance STMD in the presence of thermal noise. We shall classify them by the type of noise that limits the minimum detectable power of STMD P_{min} (power corresponding to $SNR = 1$).

The first regime corresponds to the case of relatively high frequencies of the input microwave signal, when $P_{\text{MN}} \gg P_{\text{RF}}$ (for $P_{\text{RF}} \sim P_{\text{min}}$). In this regime, similar to the conventional semiconductor diodes, the minimum detectable power is limited by the low-frequency JN noise, $P_{\text{min}} = P_{\text{JN}}$, and the SNR of STMD is linearly proportional to the input microwave power P_{RF} ($SNR \simeq P_{\text{RF}}/P_{\text{JN}}$).

The second regime takes place in the opposite limiting case of relatively low input frequencies, when $P_{\text{MN}} \ll P_{\text{RF}}$. In this case, the SNR of the STMD increases with P_{RF} much slower than in conventional diodes, and is proportional to the square root of the input microwave power: $SNR \simeq \sqrt{P_{\text{RF}}/P_{\text{min}}}$. The minimum detectable power $P_{\text{min}} = P_{\text{JN}}^2/P_{\text{MN}}$ in this regime is limited by the magnetic noise in the FL of the MTJ.

The existence of two distinct regimes of STMD operation is illustrated in Fig. 11.2, where two curves calculated from (11.35) show the STMD SNR as functions of the input power for signal frequencies $f_1 = 1$ GHz (dashed blue line) and $f_2 = 25$ GHz (red solid curve). It can be seen, that both curves (presented in logarithmic coordinates) demonstrate the clear change of slope from 1 to $1/2$ in the region, where the input power P_{RF} is close to the characteristic power of the magnetic noise P_{MN} (which increases with the increase of the input signal frequency). The minimum detectable power P_{min} (corresponding to $SNR = 1$) in the high-frequency case is smaller than P_{MN} and lies in the region of the linear dependence of SNR on P_{RF} (solid red line in Fig. 11.2). The situation is opposite in the low frequency case (blue dashed curve in Fig. 11.2), when $P_{\text{min}} > P_{\text{MN}}$ and lies in the region, where the slope of the SNR curve is equal to $1/2$.

The evolution of the characteristic powers P_{MN} and P_{min} with an increase of frequency of the input microwave signal is shown in Fig. 11.3. The curve $P_{\text{MN}}(f)$ separates the plane into the region, where magnetic noise is dominant (above the curve), and the region, where the STMD operation is limited by the JN noise (below the curve). It is clear that the smallest detectable power is achieved near the border of these two regimes.

When an STMD based on an MTJ nanopillar is used as a sensor of microwave radiation, it is typically connected to a standard transmission line with an impedance of $Z_{\text{TL}} = 50\ \Omega$. The minimum detectable microwave power delivered to a 50-Ω transmission line can be written as [21]

$$P_{\text{min}}[50\ \Omega] = \frac{1}{4}\frac{(R_0 + Z_{\text{TL}})^2}{Z_{\text{TL}} R_0} P_{\text{min}}. \tag{11.36}$$

Using this expression and taking into account the size dependence of the STMD resistance ($R_0 \propto 1/r^2$), it is possible to show that $P_{\text{min}}[50\ \Omega]$ has a clear minimum

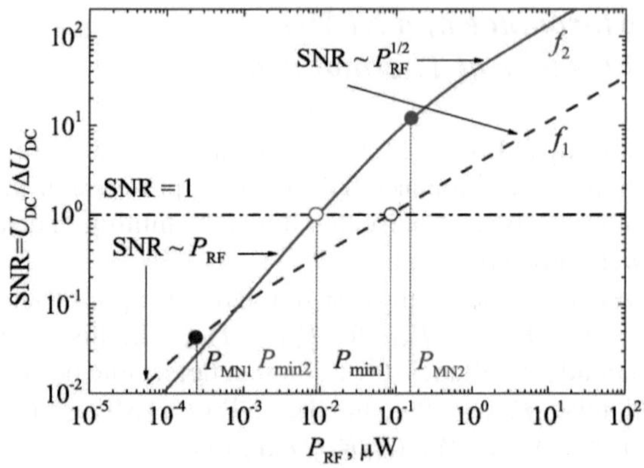

Fig. 11.2 Typical dependence of the SNR of STMD on the input microwave power P_{RF} calculated from (11.35) for two different frequencies of the input microwave signal: $f_1 = 1$ GHz (*dashed blue line*) and $f_2 = 25$ GHz (*solid red line*) [20]. P_{min} is the minimum detectable power of STMD (at $SNR = 1$) and P_{MN} is the frequency-dependent characteristic power of magnetic noise (Color figure online)

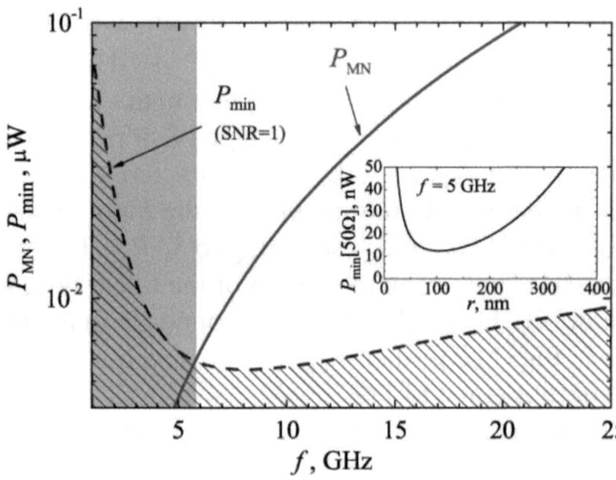

Fig. 11.3 The characteristic power of magnetic noise P_{MN} (*solid red line*) and minimum detectable power P_{min} of STMD (*dashed blue line*) as functions of the input microwave frequency f. The *blue dashed area* corresponds to undetectable signals ($P_{RF} < P_{min}$) and *gray shaded area* shows the low-frequency STMD regime, where the magnetic noise is dominant in the whole practical region $P_{RF} > P_{min}$. *Inset*: minimum detectible microwave power delivered to a 50-Ω transmission line $P_{min}[50\ \Omega]$ for $f = 5$ GHz as a function of the radius r of the MTJ nanopillar (Color figure online)

as a function of the nanopillar radius r. For instance, the optimum value of the nanopillar radius is $r_{opt} \approx 100$ nm for the input frequency $f = 5$ GHz (see the inset in Fig. 11.3).

As one can see from Figs. 11.2 and 11.3, the STMD operating in IP-regime has a performance comparable to the performance of conventional Schottky diode. Taking this into account, we believe that such STMDs might be useful for developing practical devices involving microwave detectors, sensor systems, frequency analyzers, etc.

11.4 Large-Angle Out-of-Plane Dynamical Regime of STMD Operation

In contrast to the well-known IP-regime of STMD operation, here we consider a different regime of operation of an STMD, based on the excitation of large-angle out-of-plane (OOP) magnetization precession under the action of an input microwave current $I_{\mathrm{RF}}(t)$. We show that this regime of STMD operation is realized, when the STMD is biased by the perpendicular magnetic field $\mathbf{B}_0 = \mathbf{z}B_0$, which is smaller than the saturation magnetic field of the FL, i.e., $B_0 < \boldsymbol{\mu}_0 M_s$ ($\boldsymbol{\mu}_0$ is the vacuum permeability).

11.4.1 Analytical Description of OOP-Regime

We consider a simple model of an STMD, formed by a circular MTJ nano-pillar (see Fig. 11.1). The magnetization of the pinned layer (PL) of the MTJ is assumed to be completely fixed and lie in the plane of the layer. The direction of the PL magnetization $\mathbf{p} = \mathbf{x}$ determines the spin-polarization axis. The radius r of the MTJ nano-pillar is assumed to be sufficiently small, so that the magnetization of the free layer (FL) $\mathbf{M} \equiv \mathbf{M}(t)$ is spatially-uniform and can be treated in the macrospin approximation. For simplicity, we neglect any in-plane anisotropy of the FL.

The dynamics of the unit magnetization vector $\mathbf{m}(t) = \mathbf{M}(t)/M_s$ in the FL under the action of a microwave current $I_{\mathrm{RF}}(t) = I_{\mathrm{RF}} \sin(\omega_s t)$ is governed by the Landau–Lifshits–Gilbert–Slonczewski (LLGS) equation:

$$\frac{d\mathbf{m}}{dt} = \gamma[\mathbf{B}_{\mathrm{eff}} \times \mathbf{m}] + \alpha\left[\mathbf{m} \times \frac{d\mathbf{m}}{dt}\right] + \sigma I_{\mathrm{RF}}(t)\big[\mathbf{m} \times [\mathbf{m} \times \mathbf{p}]\big], \tag{11.37}$$

$\mathbf{B}_{\mathrm{eff}} = (B_0 - \mu_0 M_z)\mathbf{z}$ is the effective magnetic field, M_z is the z-component of vector $\mathbf{M}$.

Using the spherical coordinate system for the magnetization vector $\mathbf{m} = \mathbf{x}\sin\theta\cos\varphi + \mathbf{y}\sin\theta\sin\varphi + \mathbf{z}\cos\theta$, one can obtain the equations for the polar θ and azimuthal φ angles:

$$\frac{d\theta}{dt} = -\alpha\omega_P \sin\theta - \sigma I_{\mathrm{RF}} \sin(\omega_s t)\cos\theta\cos\varphi, \tag{11.38a}$$

$$\frac{d\varphi}{dt} = \omega_P + \sigma I_{\mathrm{RF}} \sin(\omega_s t)\cos\theta\sin\varphi. \tag{11.38b}$$

Here $\omega_P \equiv \omega_P(\theta) = \omega_H - \omega_M \cos\theta$ is the frequency of magnetization precession in the OOP-regime, $\omega_H = \gamma B_0$, and $\omega_M = \gamma\mu_0 M_s$. For simplicity, we neglected in (11.38a), (11.38b), second-order non-conservative terms ($\propto \alpha^2$ and $\propto \alpha I_{\mathrm{RF}}$), which have a negligible effect on the magnetization dynamics.

In the OOP precessional regime, the magnetization precesses around the **z**-axis along an approximately circular orbit, $\theta \approx$ const, $\varphi \approx \omega_s t + \psi$, where ψ is the phase shift between the magnetization precession and the driving current. To analyze the conditions, under which the OOP regime is possible, one can average (11.38a), (11.38b) over the period of precession $2\pi/\omega_s$ and obtain the following equations for the slow variables θ and ψ:

$$\left\langle \frac{d\theta}{dt} \right\rangle = -\alpha\omega_P \sin\theta + v(a)\frac{\sigma_\perp I_{\mathrm{RF}}}{2}\cos\theta\sin\psi, \tag{11.39a}$$

$$\left\langle \frac{d\psi}{dt} \right\rangle = \omega_P - \omega_s + u(a)\frac{\sigma_\perp I_{\mathrm{RF}}}{2}\frac{1}{\sin\theta}\cos\psi. \tag{11.39b}$$

Here $a = P^2 \sin\theta$ and

$$v(a) = \frac{1}{\sqrt{1-a^2}}\left[1 + \left(\frac{\sqrt{1-a^2}-1}{a}\right)^2\right], \tag{11.40}$$

$$u(a) = \frac{1}{\sqrt{1-a^2}}\left[1 - \left(\frac{\sqrt{1-a^2}-1}{a}\right)^2\right]. \tag{11.41}$$

Note that for typical spin-polarization values $P \leq 0.7$, both dimensionless functions $u(a)$ and $v(a)$ are very close to 1 for all angles θ.

The OOP regime of magnetization precession corresponds to a stationary solution of (11.39a), (11.39b) $\theta = \theta_s =$ const, $\psi = \psi_s =$ const. Solving (11.39a), (11.39b) in this case, one can find the stationary value of the phase shift ψ_s:

$$\sin\psi_s = 2\frac{\alpha}{v}\frac{\omega_P}{\sigma_\perp I_{\mathrm{RF}}}\tan\theta_s, \qquad \cos\psi_s = 2\frac{1}{u}\frac{\omega_s - \omega_P}{\sigma_\perp I_{\mathrm{RF}}}\sin\theta_s. \tag{11.42}$$

Eliminating ψ_s from the above equations, one obtains the characteristic equation for θ_s:

$$(\omega_s - \omega_P)^2\sin^2\theta_s + \frac{\alpha^2 u^2}{v^2}\omega_P^2\tan^2\theta_s = \frac{u^2}{4}\sigma_\perp^2 I_{\mathrm{RF}}^2. \tag{11.43}$$

This equation for θ_s is a nonlinear equation which, in general, can be only solved numerically.

One can see that (11.43) has solutions only for RF currents I_{RF} larger than a certain critical current I_{th}. At the threshold, $\omega_P(\theta_s) \approx \omega_s$, one can obtain an approximate expression for the threshold precession angle

$$\theta_{\mathrm{th}} \approx \frac{\pi}{2} - \frac{\omega_H - \omega_s}{\omega_M} \tag{11.44}$$

and determine the threshold microwave current $I_{\text{th}}(\omega_s)$ needed for the excitation of the OOP precession:

$$I_{\text{th}}(\omega_s) = 2\frac{\alpha}{v}\frac{\omega_M}{\sigma_\perp}\frac{\omega_s}{\omega_H - \omega_s}. \tag{11.45}$$

In the last expression, we used the approximation $\sin(\theta_{\text{th}}) \approx 1$, which is valid for moderate magnetic fields and frequencies $\omega_H, \omega_s \ll \omega_M$.

One can show from (11.39a), (11.39b) that the OOP-regime will be stable if the following approximate conditions are fulfilled:

$$0 < \cos\theta_s < \frac{\omega_H}{\omega_M}, \qquad \omega_s < \omega_H. \tag{11.46}$$

The first condition given by (11.46) means that the precession angle θ_s must be sufficiently large because $\omega_H \ll \omega_M$ (we consider the case of a weak DC magnetic field) and $\cos\theta_s \ll 1$. The condition of (11.46) restricts the region of existence of the OOP-regime to sufficiently low driving frequencies, determined by the bias magnetic field B_0.

The output STMD voltage in the OOP-regime can be computed as

$$U_{\text{DC}} = I_{\text{RF}}\langle R(\beta)\sin(\omega_s t)\rangle = w(a) I_{\text{RF}} R_\perp \sin\psi_s, \tag{11.47}$$

where

$$w(a) = \frac{1}{\sqrt{1-a^2}}\left(\frac{1-\sqrt{1-a^2}}{a}\right). \tag{11.48}$$

Using (11.42) for $\sin\psi_s$ allows one to get the DC voltage as a function of θ_s only:

$$U_{\text{DC}} = 2\alpha\frac{w}{v}\frac{\omega_P}{\sigma_\perp} R_\perp \tan\theta_s. \tag{11.49}$$

This equation for U_{DC} can be simplified further by using the approximation $\sin\theta_s \approx 1$, $\cos\theta_s \approx \cos\theta_{\text{th}} \approx (\omega_H - \omega_s)/\omega_M$, valid for not very large driving currents:

$$U_{\text{DC}} \approx 2\alpha\frac{w}{v}\frac{\omega_M}{\sigma_\perp} R_\perp \frac{\omega_s}{\omega_H - \omega_s}. \tag{11.50}$$

Comparing (11.50) with (11.45) for threshold current $I_{\text{th}}(\omega_s)$, one can rewrite the output DC voltage of the STMD as the function of threshold current $I_{\text{th}}(\omega_s)$:

$$U_{\text{DC}} \approx w I_{\text{th}}(\omega_s) R_\perp. \tag{11.51}$$

It follows from (11.50), (11.51) that the output DC voltage of the STMD practically does not depend on the amplitude of RF current, provided that it is larger than the threshold current I_{th}.

11.4.2 Performance of a STMD in OOP-Regime

Below we shall analyze an analytical solution for the STMD in the OOP-regime of operation, compare this solution with the results of numerical calculations, and then also compare the performance of the STMD in IP- and OOP-regimes.

We shall consider the case of the STMD with the following typical parameters (see, e.g., [14, 17]): radius of the STMD FL $r = 50$ nm, thickness of the STMD FL $d = 1$ nm, spin-polarization efficiency of current $P = 0.7$, resistance of STMD in perpendicular magnetic state ($\beta = \pi/2$) $R_{\perp} = RA/(\pi r^2) = 1\ \mathrm{k\Omega}$ (giving resistance–area product of MTJ $RA = 7.854\ \Omega\,\mu\mathrm{m}^2$), Gilbert damping constant $\alpha = 0.01$, saturation magnetization of the FL $\mu_0 M_s = 800$ mT.

We choose the magnitude of the external out-of-plane DC magnetic field as $B_0 =$ 200 mT for the STMD in OOP-regime, which corresponds to the maximum OOP frequency $\omega_H = 2\pi \times 5.6$ GHz.

In the IP-regime of operation, the STMD will be characterized by the equilibrium angle $\beta_0 = \pi/2$ between the equilibrium magnetization of the FL and the magnetization of the PL. Hence the equilibrium resistance of the STMD in the IP-regime is $R_0 = R_{\perp} = 1\ \mathrm{k\Omega}$. We choose the magnitude of the external DC in-plane magnetic field as $B_0 = 14.1$ mT for the STMD in the IP-regime that, in accordance with the expression for FMR frequency $f_0 = \omega_0/2\pi = (\gamma/2\pi)\sqrt{B_0(B_0 + \mu_0 M_s)}$, gives $f_0 = 3$ GHz. The resonance STMD sensitivity in the passive regime for such parameters is $\varepsilon_{\mathrm{res}} \approx 2700$ V/W (see, e.g., [18]), which is greater or comparable to the sensitivity of a typical unbiased Schottky diode [14].

We use (11.7) for the calculation of the output DC voltage of an STMD in the IP-regime as a function of the input microwave current magnitude I_{RF} and frequency $f = f_s = \omega_s/2\pi$. These curves are indicated below in Figs. 11.4 and 11.5 by red dashed lines.

In order to verify the conclusions of the analytical theory of an STMD in the OOP-regime we numerically solved the LLGS (11.37) and then numerically calculated the output DC voltage of the detector as $U_{\mathrm{DC}} = \langle I_{\mathrm{RF}}(t) R(\beta) \rangle$. The results of our calculations are presented in Figs. 11.4 and 11.5. Here solid blue lines and red dashed lines present the analytical dependencies of U_{DC} in the OOP- and IP-regimes (see (11.51) and (11.7)), respectively. Dots are the results of our numerical calculations. Black crosses and green circles correspond to the cases of increasing and decreasing of the parameter (frequency ω_s or magnitude I_{RF} of the RF current), respectively. As one can see, the results of analytical theory are in reasonable agreement with the results of our numerical calculations.

As one can see from Fig. 11.4, in the OOP-regime the STMD works as a broadband low-frequency, non-resonant microwave detector in contrast to the traditional resonance IP-regime. The response of the STMD to an input microwave current with magnitude I_{RF} is also substantially different in the cases of OOP- and IP-regimes of operation of an STMD (see Fig. 11.5). In the IP-regime, the output DC voltage U_{DC} of the detector is proportional to the input microwave power $P_{\mathrm{RF}} = (1/2) I_{\mathrm{RF}}^2 R_0$ (red dashed curve in Fig. 11.5, see also (11.7)). In contrast, the output DC voltage U_{DC} of the detector in the OOP-regime has a step-like dependence (blue solid curve and

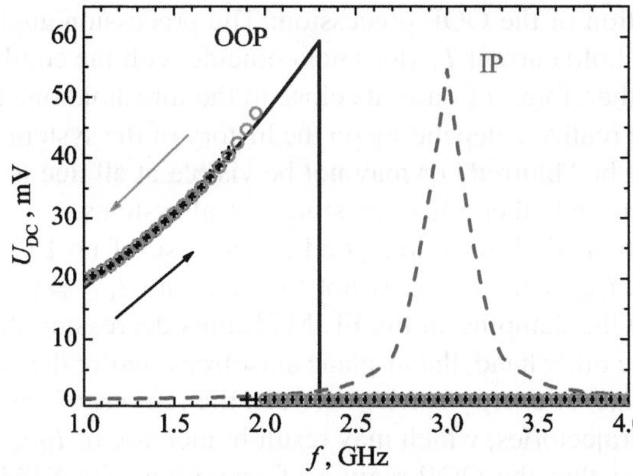

Fig. 11.4 Typical dependence of the output DC voltage U_{DC} of a STMD on frequency of input RF signal f in OOP- (*solid line and points*) and IP-regime (*dashed line*), respectively. *Blue solid line* is the analytical dependence given by (11.51), *red dashed line* is the analytical dependence given by (11.7). Points are the results of numerical simulations. *Black crosses* and *green circles* corresponded to the case when frequency is increased and decreased, respectively. $I_{RF} = 0.2$ mA (Color figure online)

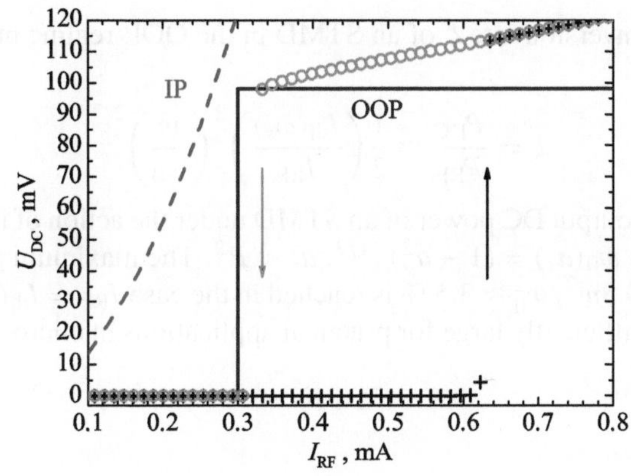

Fig. 11.5 Typical dependence of the output DC voltage U_{DC} of a STMD on input microwave current I_{RF} in OOP- (*solid line and points*) and IP-regime (*dashed line*), respectively. *Blue solid line* is the analytical dependence given by (11.51), *red dashed line* is the analytical dependence given by (11.7). Points are the results of numerical simulations. *Black crosses* and *green circles* corresponded to the case when the current is increased and decreased, respectively. $f = 3$ GHz (Color figure online)

points in Fig. 11.5): $U_{DC} \approx 0$ for $I_{RF} < I_{th}(\omega_s)$ and $U_{DC} \approx \text{const}$ for $I_{RF} > I_{th}(\omega_s)$. Thus, in the OOP-regime, the STMD operates as a non-resonant broadband threshold microwave detector of low frequency RF signals.

It is important to note that the results of our numerical simulations show the existence of a hysteresis in the curves $U_{DC}(f_s)$ and $U_{DC}(I_{RF})$ in the OOP-regime (see Figs. 11.3 and 11.4). The origin of this hysteresis lies in the "hard", or subcritical, scenario of excitation of the OOP precession: The precession angle θ_{th} that corresponds to the threshold current I_{th} does not coincide with the equilibrium magnetization angle and, therefore, for currents close to the threshold one the OOP regime may or may not be realized, depending on the history of the system. In experiments, the hysteresis may be "blurred" or may not be visible at all due to the influence of thermal fluctuations and other noises existing in real systems.

The results presented above correspond to the case of no DC bias current applied to the MTJ ($I_{DC} = 0$). If this is not the case and $I_{DC} \neq 0$, this current will partly compensate the damping in the FL MTJ, thus decreasing the threshold current $I_{th}(\omega_s)$. On the other hand, the in-plane anisotropy and/or the in-plane bias field in the FL may create an energy barrier between the regions of small-angle IP- and large-angle OOP-trajectories, which may result in increase of I_{RF}.

We also suggest that the OOP-regime of operation of a STMD might be responsible for an extremely large diode volt-watt sensitivity $\varepsilon \sim 10^5$ V/W observed in recent experiments with thermally-activated "non-adiabatic stochastic resonance" [22].

The STMD in the OOP-regime could be used as a base element for new energy harvesting devices, inasmuch as it has no resonance frequency, and, therefore, could accumulate energy from all the low-frequency region ($\omega_s < \omega_H$) of the microwave spectrum.

The energy conversion rate ζ of an STMD in the OOP-regime may be estimated as

$$\zeta = \frac{P_{DC}}{P_{RF}} \approx \frac{1}{2}\left(\frac{I_{th}(\omega_s)}{I_{RF}}\right)^2 \left(\frac{w}{w_0}\right)^2, \tag{11.52}$$

where P_{DC} is the output DC power of an STMD under the action of input microwave power P_{RF}, $w_0 \equiv w_0(a_s) = (1 - a_s^2)^{-1/2}$, $a_s \approx P^2$. The maximum possible conversion rate $\zeta_{max} \approx 0.5w^2/w_0^2 \approx 3.5$ % is reached in the case $I_{RF} = I_{th}(\omega_s)$. We believe that this ratio is sufficiently large for practical applications in microwave energy harvesting.

11.5 Summary

In conclusion, we have demonstrated that there are two distinct regimes of STMD operation. One of them is characterized by a small-angle in-plane magnetization precession (IP-regime), while the other is characterized by a large-angle out-of-plane magnetization precession (OOP-regime). The performance of a STMD in these regimes is substantially different. In IP-regime, STMD in the presence of noise can operate in two distinct sub-regimes, one of which is limited by magnetic noise and is different from the regime of operation of traditional semiconductor detectors. The developed formalism for STMD operating in IP-regime can be used for the optimization of noise-handling parameters of a STMD. We also has been demonstrated that there is a novel regime of operation of an STMD, based on the excitation of large-angle out-of-plane (OOP) magnetization precession. In this regime,

STMD operates as threshold detector of low-frequency RF signals. We believe that the OOP regime of STMD operation can be used for the development of novel types of threshold microwave detectors and might be responsible for extremely large diode volt-watt sensitivity observed in recent experiments with "non-adiabatic stochastic resonance" [22]. This regime of operation of an STMD might also be useful for the creation of energy harvesting devices based on STMD.

Acknowledgements This work was supported in part by the Contract from the U.S. Army TARDEC, RDECOM, by the grants ECCS-1001815 and DMR-1015175 from the National Science Foundation of the USA, by the grant from DARPA, by the Grant No. M/212-2011 from the State Agency of Science, Innovations and Informatization of Ukraine, and by the Grant No. UU34/008 from the State Fund for Fundamental Research of Ukraine.

References

1. J.C. Slonczewski, J. Magn. Magn. Mater. **159**, L1 (1996)
2. J.C. Slonczewski, J. Magn. Magn. Mater. **195**, L261 (1999)
3. L. Berger, Phys. Rev. B **54**, 9353 (1996)
4. J.A. Katine, F.J. Albert, R.A. Buhrman, E.B. Myers, D.C. Ralph, Phys. Rev. Lett. **84**, 3149 (2000)
5. S. Urazhdin, N.O. Birge, W.P. Pratt Jr., J. Bass, Phys. Rev. Lett. **91**, 146803 (2003)
6. S.I. Kiselev, J.C. Sankey, I.N. Krivorotov, N.C. Emley, R.J. Schoelkopf, R.A. Buhrman, D.C. Ralph, Nature (London) **425**, 308 (2003)
7. S. Kaka, M.R. Pufall, W.H. Rippard, T.J. Silva, S.E. Russek, J.A. Katine, Nature (London) **437**, 389 (2005)
8. K.J. Lee, A. Deac, O. Redon, J.-P. Nozières, B. Dieny, Nat. Mater. **3**, 877 (2004)
9. I.N. Krivorotov, N.C. Emley, J.C. Sankey, S.I. Kiselev, D.C. Ralph, R.A. Buhrman, Science **307**, 228 (2005)
10. F.B. Mancoff, N.D. Rizzo, B.N. Engel, S. Tehrani, Nature **437**, 393 (2005)
11. D. Houssameddine, U. Ebels, B. Delaët, B. Rodmacq, I. Firastrau, F. Ponthenier, M. Brunet, C. Thirion, J.-P. Michel, L. Prejbeanu-Buda, M.-C. Cyrille, O. Redon, B. Dieny, Nat. Mater. **6**, 447 (2007)
12. A. Ruotolo, V. Cros, B. Georges, A. Dussaux, J. Grollier, C. Deranlot, R. Guillemet, K. Bouzehouane, S. Fusil, A. Fert, Nat. Nanotechnol. **4**, 528 (2009)
13. A.A. Tulapurkar, Y. Suzuki, A. Fukushima, H. Kubota, H. Maehara, K. Tsunekawa, D.D. Djayaprawira, N. Watanabe, S. Yuasa, Nature **438**, 339 (2005)
14. S. Ishibashi, T. Seki, T. Nozaki, H. Kubota, S. Yakata, A. Fukushima, S. Yuasa, H. Maehara, K. Tsunekawa, D.D. Djayaprawira, Y. Suzuki, Appl. Phys. Express **3**, 073001 (2010)
15. A. Slavin, V. Tiberkevich, IEEE Trans. Magn. **45**, 1875 (2009)
16. J.C. Sankey, Y. Cui, J.Z. Sun, J.C. Slonczewski, R.A. Buhrman, D.C. Ralph, Nat. Phys. **4**, 67 (2008)
17. C. Wang, Y.-T. Cui, J.Z. Sun, J.A. Katine, R.A. Buhrman, D.C. Ralph, Phys. Rev. B **79**, 224416 (2009)
18. C. Wang, Y.-T. Cui, J.Z. Sun, J.A. Katine, R.A. Buhrman, D.C. Ralph, J. Appl. Phys. **106**, 053905 (2009)
19. N.B. Lukyanchikova, *Noise Research in Semiconductor Physics* (CRC Press, Amsterdam, 1996)
20. O. Prokopenko, G. Melkov, E. Bankowski, T. Meitzler, V. Tiberkevich, A. Slavin, Appl. Phys. Lett. **99**, 032507 (2011)
21. D.M. Pozar, *Microwave Engineering*, 3rd edn. (Wiley, New York, 2005)
22. X. Cheng, C.T. Boone, J. Zhu, I.N. Krivorotov, Phys. Rev. Lett. **105**, 047202 (2010)

Chapter 12
Spin-Wave Emission from Spin-Torque Nano-Oscillators and Its Control by Microwave Pumping

Vladislav E. Demidov, Sergei Urazhdin, and Sergej O. Demokritov

Abstract Spin transfer torque is a unique mechanism enabling excitation of microwave-frequency oscillations in magnetic nanodevices by dc current, making them attractive for applications in microwave electronics as nanoscale microwave oscillators. Here we review our recent investigations of spin-wave emission by spin torque oscillators and show that these devices can be also used as emitters of high-frequency spin waves for the emerging field of magnonics, where spin waves are utilized as a medium for high-speed transmission and processing of electrical signals on the nanometer scale. We demonstrate that, due to the anisotropy of the spin-wave spectrum, the emission is strongly directional, and the direction of the spin-wave propagation is steerable by the magnetic field. Moreover, we show that the characteristics of the emitted spin waves can be controlled by using nonlinear dynamic magnetic processes.

12.1 Introduction

It is now well established that spin-polarized electric current injected into a ferromagnetic layer through a nanocontact exerts spin-transfer torque on the magnetization, resulting in microwave-frequency precession [1–9]. Since the frequency of the excited precession can be varied by the static magnetic field in a wide range from hundreds of megahertz to hundreds of gigahertz, devices based on this phenomenon have been recently attracting significant attention due to their potential for implementation of highly-flexible on-chip microwave oscillators for integrated electronics, the so-called spin-torque nano-oscillators (STNOs) [10–12].

In STNOs formed by a point contact on an extended magnetic film, magnetization oscillations excited in the active device area can propagate through the film as spin waves [13–15]. Such emission has a number of important applications. It is

V.E. Demidov (✉) · S.O. Demokritov
Institute for Applied Physics, University of Muenster, Corrensstr. 2-4, 48149 Muenster, Germany
e-mail: demidov@uni-muenster.de

S. Urazhdin
Department of Physics, Emory University, Atlanta, GA 30322, USA

S.O. Demokritov, A.N. Slavin (eds.), *Magnonics*, Topics in Applied Physics 125,
DOI 10.1007/978-3-642-30247-3_12,

believed, for example, to provide a possibility for mutual synchronization in groups of STNOs via the exchange of spin waves [16–19], increasing the generated microwave power to the level useful for technical applications.

With the recent developments in the field of spin-wave-based electronics (magnonics) [20–23], efficient excitation of spin waves in magnetic nanostructures became important for applied nano-magnetism [24–27]. Usually this excitation is realized using dynamic magnetic fields created by conductors carrying microwave currents. Such inductive excitation was widely used in spin-wave devices with macroscopic dimensions based on dielectric garnet films. However, this mechanism has a number of disadvantages, which become especially detrimental to micro- and nano-structured magnonic devices. Specifically, in such devices, the matching of microwave-frequency transmission lines and elimination of the parasitic coupling between them is difficult because of restrictions imposed by their small spatial dimensions.

Spin-torque nano-devices capable of on-chip conversion of the energy of a dc current into microwave-frequency magnetization oscillations are free from these disadvantages and, therefore, have large technological potential for implementation of nano-scale spin-wave emitters for magnonic applications. In spite of this, emission of spin waves by STNOs remains largely experimentally unexplored, since such experiments require advanced experimental techniques enabling direct access to spin waves on the nano-scale and providing the possibility to map spin-wave propagation with high spatial resolution.

Here we review our recent experimental work on spin-wave emission by STNOs and control of their emission characteristics by microwave pumping. The experiments were performed using micro-focus Brillouin light scattering (BLS) spectroscopy technique characterized by the sub-micrometer spatial resolution and ultra-high sensitivity [28, 29]. We show that STNOs in practically useful in-plane magnetization configuration can emit spin waves into the surrounding magnetic film. Because of the anisotropy imposed by the in-plane static magnetic field, the emission is strongly directional and its direction can be steered by varying the direction of the field. We also show that the characteristics of emitted spin waves can be controllably manipulated by using the phenomenon of nonlinear frequency conversion.

12.2 Studied Samples and Their Electronic Characterization

The layout of the experiment is shown in Fig. 12.1. The samples were fabricated by a multistep e-beam lithography process, as follows. A multilayer Cu(40)Py(6)Cu(3)$Co_{70}Fe_{30}$(9)Au(5), where Py = $Ni_{80}Fe_{20}$ and thicknesses are given in nanometers, was deposited on a sapphire substrate with electrical leads patterned into coplanar microstrip lines. The top part of the multilayer down to the Py(6) layer was patterned by Ar ion milling through an evaporated Al(70) mask shaped as ellipse with dimensions of 100 by 50 nm, while the bottom electrode and the Py(6) layer were left extended. A SiO_2(40) was subsequently sputtered to

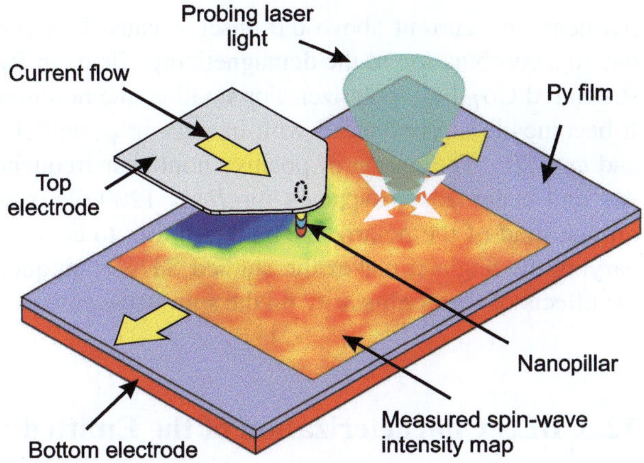

Fig. 12.1 Schematic of the experiment

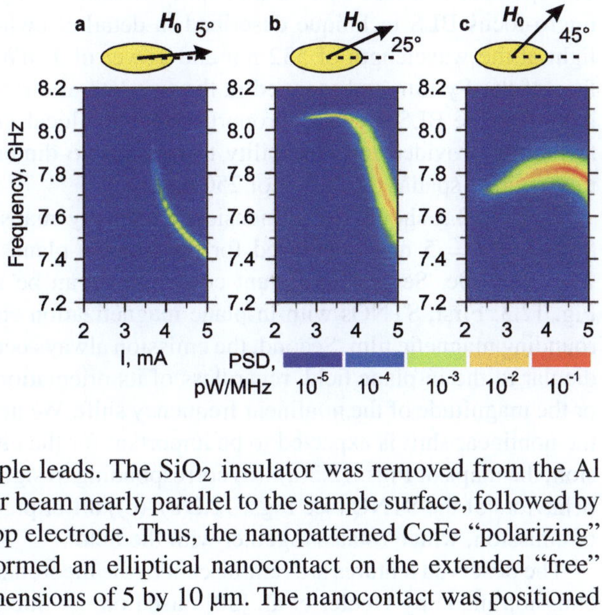

Fig. 12.2 Pseudo-color logarithmic maps of the power spectral density (PSD) of the signal generated by the device due to the magnetoresistance effect at different angles φ between the in-plane bias magnetic field and the major axis of the nanopillar, as labeled

electrically isolate the sample leads. The SiO_2 insulator was removed from the Al mask by ion milling with Ar beam nearly parallel to the sample surface, followed by sputtering of an Au(200) top electrode. Thus, the nanopatterned CoFe "polarizing" layer and the Cu spacer formed an elliptical nanocontact on the extended "free" Py(6) layer with lateral dimensions of 5 by 10 μm. The nanocontact was positioned at a distance of 200 nm from the edge of the top electrode, and its major axis was oriented at 45° with respect to the axis of the electrode. The samples were magnetized by a static magnetic field $H_0 = 300$–1500 Oe applied in the plane of the Py film at an angle φ with respect to the major axis of the nanostructured polarizer.

The oscillation characteristics of STNOs were determined from the measurements of the microwave signals generated due to the magnetoresistance effect [6], as shown in Figs. 12.2(a)–(c). The plots of the power spectral density (PSD) illustrate the dependence of the oscillation frequency on the bias current I for three different angles φ between the in-plane bias magnetic field $H_0 = 900$ Oe and the major axis of the nanostructured polarizer. The microwave generation starts at an onset current $I = 2.5$–3.5 mA that depends on φ. The dependence of the generation frequency on current above the onset is caused by the nonlinear frequency shift, due to a combination of the demagnetizing effects in Py and the dipolar field of the structured $Co_{70}Fe_{30}$ polarizer. For small φ, the nonlinear shift is strongly negative. It becomes less pronounced with increasing φ, and changes to positive at small I and $\varphi > 20°$. The region of positive nonlinear frequency shift is reduced at larger H_0, and eventually disappears for $H_0 > 1200$ Oe, suggesting its origin from the dipolar field of the polarizer. The possibility to control the nonlinear behaviors by varying the angle φ makes the studied STNOs uniquely suited for the analysis of the effects of the nonlinearity on the spin-wave emission.

12.3 BLS Characterization of the Emitted Spin Waves

The spatially resolved detection of spin waves in the Py film was performed by the micro-focus BLS technique described in detail elsewhere [29]. The probing laser light at the wavelength of 532 nm and power of 1 mW was focused onto the surface of the Py film and scanned in the two lateral directions (see Fig. 12.1), while recording the BLS intensity proportional to the local intensity of spin waves. This technique provided the possibility to record two-dimensional spin-wave intensity maps with a spatial resolution of 250 nm.

Figure 12.3 shows two-dimensional intensity maps of spin waves emitted by STNO at $I = 5$ mA, measured for different in-plane directions of the bias field $H_0 = 900$ Oe. Several important conclusions can be made based on the data of Fig. 12.3. First, STNOs with in-plane magnetization emit spin waves into the surrounding magnetic film. Second, the emission always occurs in the direction perpendicular to the in-plane field, regardless of its orientation, the generation frequency, or the magnitude of the nonlinear frequency shift. We note that, although the sign of the nonlinear shift is expected to be important for the efficiency of spin-wave emission, the maps of Figs. 12.3(a)–(d) corresponding to significantly different nonlinear behaviors of the STNO (see Figs. 12.2(a)–(c)) differ predominantly by the direction of emission, which rotates together with the field.

The observed features are reminiscent of the theoretical predictions of anisotropic emission made by Hoefer et al. [30] based on the interplay between the effects of the bias field and the Oersted field of the current. Indeed, we find evidence for the effects of Oersted field in the noticeable asymmetry of the two oppositely directed emission lobes in Fig. 12.3(d). The asymmetry inverted when H_0 was reversed, and decreased at larger H_0, suggesting its origin from the effect of the Oersted field of the current, which is aligned with H_0 in one direction of emission, and is opposed to it in the other direction. Nevertheless, the Oersted field cannot account for the observed directional emission perpendicular to the magnetic field, since in our measurements the directionality of emission did not depend on the magnitudes of the current or the magnetic bias field.

Figure 12.4 illustrates the spin-wave characteristics determined at the location of the maximum spin wave intensity. Figures 12.4(a) and (b) show the BLS spectra

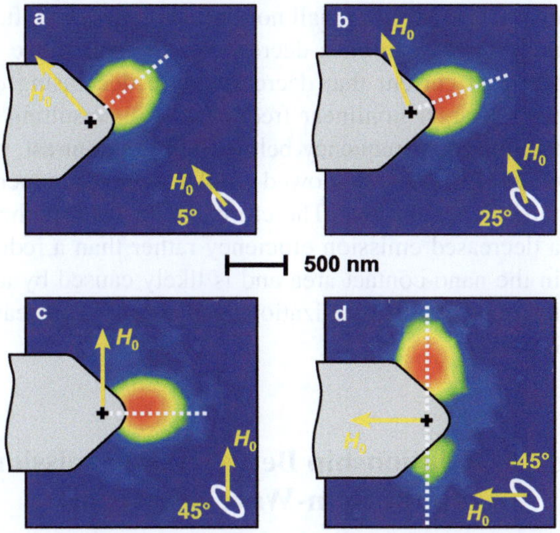

Fig. 12.3 Spatial maps of the intensity of spin waves emitted by the STNO for different angles φ, as labeled. The schematic of the top electrode is superimposed on each map, with a cross indicating the location of the nanocontact. *Dashed lines* indicate the direction of the spin-wave emission

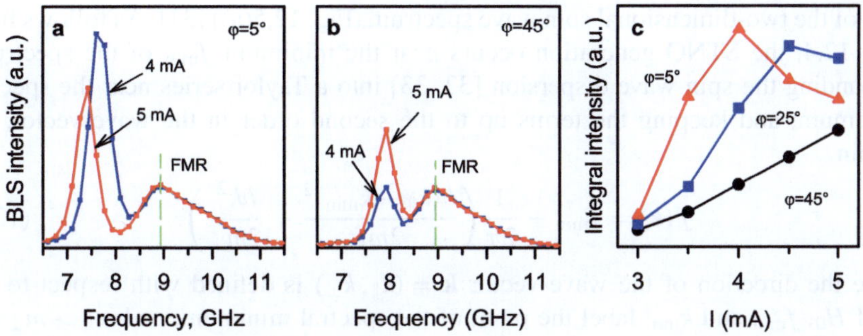

Fig. 12.4 (**a**) and (**b**) Dependences of BLS spectra on current for $\varphi = 5$ and $45°$, respectively. *Dashed vertical lines* mark the frequency of the ferromagnetic resonance (FMR). (**c**) Dependences of the integrated intensity of emitted spin waves on current for $\varphi = 5°$ (*triangles*), $25°$ (*squares*), and $45°$ (*dots*)

of spin waves in the Py film for $I = 4$ and 5 mA, for different angles φ, as indicated. The narrow peak in the spectra corresponds to spin waves emitted by the STNO, whereas the wide peak corresponds to spin waves always existing in the film due to thermal fluctuations. The latter peak provides information about the frequency of the ferromagnetic resonance (FMR) in the Py film, corresponding to its maximum. As seen from Fig. 12.4, irrespectively of the angle φ, the frequency of emitted spin waves determined by the oscillation frequency of the STNO is situated below FMR. At $\varphi = 45°$, the spectra exhibit a small nonlinear frequency shift which increases as φ is reduced, in agreement with the electrical measurements shown in Figs. 12.2(a)–(c). Figure 12.4(c) summarizes the dependences of the frequency-integrated spin-wave intensity on the current I. As seen from these data, the intensity of the emitted spin waves increases linearly with current for the angle $\varphi = 45°$

characterized by a small nonlinear frequency shift. In contrast, the data for $\varphi = 25°$ and $\varphi = 5°$ exhibit a decrease of the spin-wave intensity starting from a certain value of current that decreases with decreasing φ. These findings are correlated with a larger nonlinear frequency shift, resulting in more significant reduction of the emission frequency below FMR. In contrast, magnetoresistance measurements (Figs. 12.2(a)–(c)) showed similar monotonic increases of generated power for all three configurations. Therefore, the decrease in the BLS intensity is associated with a decreased emission efficiency rather than a reduced amplitude of the oscillation in the nano-contact area and is likely caused by a partial nonlinear localization of the generated magnetization oscillations underneath the nano-contact, as discussed below.

12.4 Relationship Between the Emission Characteristics and the Spin-Wave Spectrum

We base our interpretation of the observed directional emission patterns on the analysis of the two-dimensional spin wave spectrum (Fig. 12.5(a)) [31]. As follows from Fig. 12.4, the STNO generation occurs near the minimum $f_{\min}$ of the spectrum. Expanding the spin-wave dispersion [32, 33] into a Taylor series near the spectral minimum, and keeping the terms up to the second order in the wave-vector, we obtain

$$f(k) = f_{\min} + \frac{1}{2\pi}\left(\frac{\hbar(k_{\parallel} - k_{\min})^2}{2m_{\parallel}} + \frac{\hbar k_{\perp}^2}{2m_{\perp}}\right). \qquad (12.1)$$

Here the direction of the wave-vector $\mathbf{k} = (k_{\parallel}, k_{\perp})$ is defined with respect to the field H_0, $f_{\min}$ and $k_{\min}$ label the point of the spectral minimum, and $m_{\parallel} > m_{\perp}$ are the effective masses defined by the curvature of the dispersion at the minimum, $m_{\parallel,\perp} = \frac{\hbar}{2\pi}(\frac{\partial^2 f}{\partial k_{\parallel,\perp}^2})^{-1}$. This definition is similar to the usual definition of effective masses of electrons in semiconductors.

We now demonstrate that the observed directionality of spin-wave emission is caused by the anisotropy of the effective masses. First of all, we emphasize that the direction of the energy flow due to spin-wave propagation and, as a consequence, the geometry of the spin-wave pattern is determined by the direction of the group velocity, $\mathbf{V}_g = 2\pi\nabla f(k_{\parallel}, k_{\perp})$, which is not necessarily collinear with the wave-vector. Let us assume that the STNO can be approximated as a point source generating spin waves at the frequency f_0 of the oscillation. In this case, all spin waves satisfying $f(\mathbf{k}) = f_0$ (an ellipse in the momentum space, Fig. 12.5(b)) are excited with similar efficiency for all $k_{\parallel} < \pi/a_{\parallel}$ and $k_{\perp} < \pi/a_{\perp}$, where $a_{\parallel}$ and $a_{\perp}$ are the lateral dimensions of the STNO parallel and perpendicular to the field, respectively. Consequently, the intensity $\delta J_0(\theta)$ of spin waves emitted into an in-plane angle interval $\delta\theta$ is given simply by the phase volume $\delta\Gamma = (d\Gamma/d\theta)\delta\theta$, i.e., the length of the segment of the $f(\mathbf{k}) = f_0$ contour corresponding to the direction of the group velocity $\mathbf{V}_g$ within this angular interval. As illustrated in Fig. 12.5(b), the function $d\Gamma/d\theta$ is

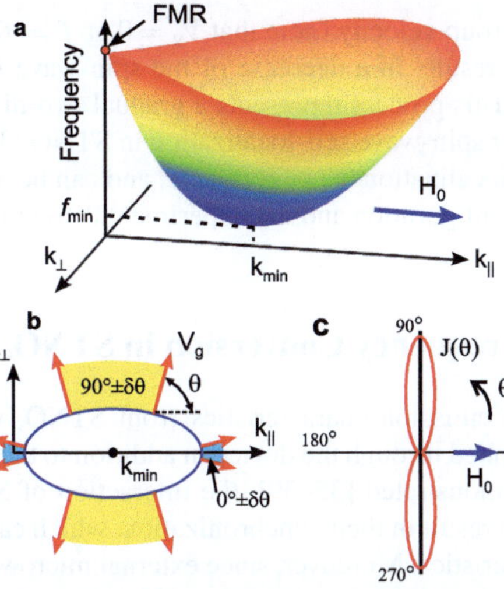

Fig. 12.5 (**a**) Dependence of the spin-wave frequency f on the wave-vector $\mathbf{k} = (k_\parallel, k_\perp)$ oriented at arbitrary direction with respect to the bias field. (**b**) Projection of the contour of constant frequency onto the $(k_\parallel, k_\perp)$ plane. *Arrows* show the directions of the spin-wave group velocity at different locations on the contour. *Arcs* indicate the phase volumes corresponding to the emission of spin waves parallel and perpendicular to H_0, for the same angular half-intervals $\delta\theta = 15°$. (**c**) Angular dependence of spin wave intensity calculated for $\xi = 500$ nm, $\lambda_r = 400$ nm, and the ratio of the effective masses $m_\parallel/m_\perp = 4$ obtained from the analysis of the spin-wave spectra

strongly anisotropic: for a given $\delta\theta$ the segment is significantly larger for $\theta = 90°$ than for $\theta = 0°$, resulting in a higher intensity of spin waves emitted perpendicular to H_0.

In addition to the anisotropic emission, anisotropic spatial decay also contributes to the angular dependence of spin wave intensity $J(\theta)$ at a given distance from the STNO. In polar coordinates (θ, ξ), the effect of damping can be accounted by $J(\xi, \theta) = J_0(\theta)\exp(-\xi/\lambda_r) = J_0(\theta)\exp(-2\xi\omega_r/|V_g(\theta)|)$, where $\lambda_r = V_g/2\omega_r$ is the decay length determined by the spin-wave relaxation rate ω_r, which is considered to be independent of θ, and the group velocity $V_g(\theta)$. According to (12.1), the group velocity reaches a maximum at $\theta = 90°$ and a minimum at $\theta = 0°$. As a consequence, the ratio of the intensities for these orthogonal directions increases exponentially with the distance from STNO.

A calculation based on the above analysis of the angular dependence of intensity $J(\theta)$ for a fixed distance $\xi = 500$ nm and the measured value $\lambda_r(90°) = 400$ nm shows a strong anisotropy of the spin-wave emission pattern characterized by two maxima at $\theta = \pm 90°$ and minima at $\theta = 0°$ and 180° (Fig. 12.5(c)), in excellent agreement with the experiment.

Analysis of the two-dimensional spectrum in Fig. 12.5(a) also provides a qualitative explanation for the decrease of the emitted spin-wave intensity at large currents for $\varphi < 45°$ (Fig. 12.4). Indeed, the strong nonlinear frequency shift at small φ leads to lowering of the generation frequency towards f_{min} and, as a consequence, to a reduction of the group velocity (note that $V_g = 0$ at $f = f_{min}$) and of the decay length λ_r. Smaller λ_r results in a decrease of the spin-wave intensity at the same point of observation. This process represents a gradual two-dimensional version of the predicted nonlinear spin-wave self-localization in STNO [34]. Nevertheless, our results show that this localization is not complete, and can be avoided by the proper choice of the STNO configuration and the experimental conditions.

12.5 Nonlinear Frequency Conversion in STNO

To attain controllable emission characteristics from STNO, we used microwave-frequency currents applied through the device in addition to the driving dc current I. As was previously demonstrated [35–39], the interaction of STNOs with external microwave signals can result in their synchronization, which can be used to improve the oscillation characteristics. Moreover, since external microwave signals affect the oscillation of STNOs, one can expect that they also affect the spin-wave emission by STNOs into the surrounding magnetic film.

In our experiments, we applied a "pumping" microwave current at frequency f_P significantly larger than the STNO oscillation frequency f_0. In all the measurements described below, the applied microwave power was $P = 1$ mW and the static magnetic field $H_0 = 700$ Oe was applied at an angle $\varphi = -45°$. The STNO devices used in these experiments were similar to those discussed in Sect. 12.2, except for slightly different lateral dimensions of the nano-contact (150 nm by 30 nm).

Figure 12.6(a) shows the BLS spectra of spin waves in the Py film acquired with the probing laser spot positioned close to the edge of the top electrode of the studied device. The shaded area shows the spectrum acquired at $I = 0$, which characterizes the spectral amplitudes of thermally excited spin waves. At $I > 2.7$ mA, the BLS spectra exhibit an additional narrow peak caused by the spin waves emitted by the STNO. The intensity of the peak increases monotonically with the dc bias current and its frequency stays nearly constant, in agreement with the data of Figs. 12.4(b) and (c).

As the microwave pumping was applied, the measurements revealed two qualitatively different regimes depending on the value of f_P, as illustrated in Fig. 12.6(b). At $f_P = 2f_0 = 14.3$ GHz, the microwave pumping leads to an approximately two-fold increase of the intensity of the auto-oscillation peak, which is indicative of the synchronization of STNO to the external microwave signal with synchronization index $f_P/f_0 = 2$. In contrast, at a larger pumping frequency $f_P = 15.0$ GHz, the intensity of the auto-oscillation peak decreases and its frequency is slightly reduced. Moreover, a second peak appears in the spectrum at a frequency different from both f_0 and $f_P/2$.

To quantify this phenomenon, we measured the BLS spectra for f_P varying from 13.0 to 17.0 GHz with a step size of 0.1 GHz. These spectra were analyzed to extract the frequency f_1 of the auto-oscillation peak and the frequency f_2 of the secondary peak induced by the microwave pumping (Fig. 12.7(a)), as well as their intensities J_1 and J_2 (Fig. 12.7(b)). We identify three regimes in these data, characterized

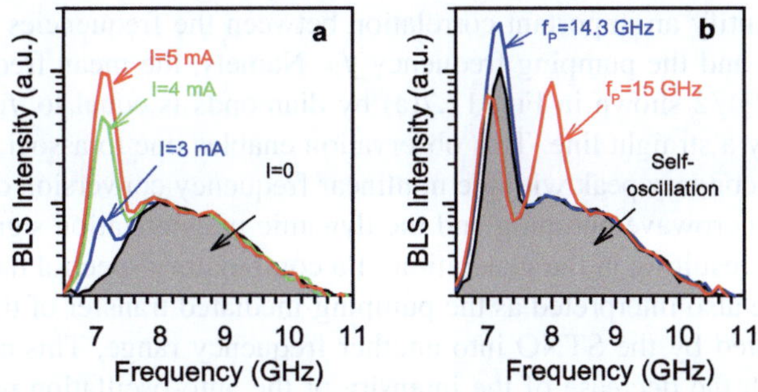

Fig. 12.6 (**a**) BLS spectra of spin waves in the Py film in the absence of the external microwave signal, acquired at different values of I, as labeled. The *shaded region* is the spectrum of thermally excited spin waves acquired at $I = 0$. (**b**) BLS spectra acquired at $I = 5$ mA in the presence of an additional microwave current with power $P = 1$ mW, applied at the labeled values of frequency f_P. The *shaded region* shows the reference spectrum acquired in the free-running auto-oscillation regime, at $I = 5$ mA

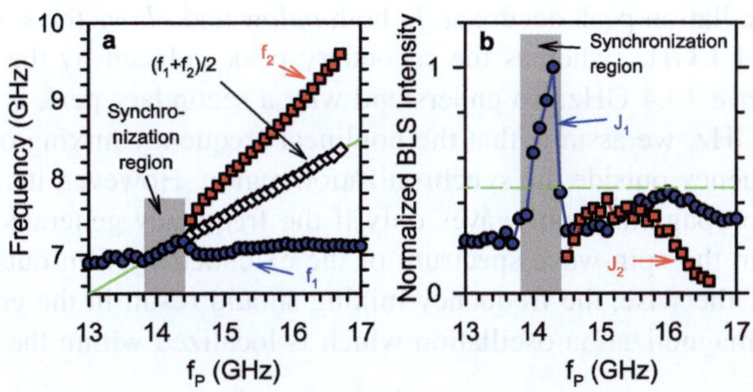

Fig. 12.7 Dependencies of the frequencies (**a**) and intensities (**b**) of the oscillation peaks on the pumping frequency. *Circles* show the data for the auto-oscillation peak and squares – for the pumping-induced secondary peak. In (**a**), *diamonds* show the mean frequency $(f_1 + f_2)/2$, and the *solid straight line* shows $f_P/2$. *Dashed horizontal line* in (**b**) is the intensity of the auto-oscillation peak in the absence of the external microwave signal

by different behaviors of the frequencies and the intensities of the peaks. In the first regime at $f_P < 13.8$ GHz, there is only one peak in the spectrum whose frequency $f_1 \approx 6.9$ GHz is nearly independent of f_P. In the second regime, in the range 13.8 GHz $< f_P <$ 14.4 GHz, there is still only one peak, but its frequency follows an exact linear relationship with the pumping frequency, $f_1 = f_P/2$. This behavior is consistent with the synchronization of the STNO to the pumping signal characterized by the synchronization index $f_P/f_1 = 2$. In the third regime, at f_P above the upper boundary of the synchronized regime at 14.4 GHz, a second peak with frequency f_2 appears in the spectrum, while the frequency f_1 of the first peak abruptly decreases. The frequency f_2 of the secondary peak increases with increasing f_P, whereas f_1 remains unchanged.

We can identify an important correlation between the frequencies f_1 and f_2 of the two peaks and the pumping frequency f_P. Namely, the mean frequency of the peaks $(f_1 + f_2)/2$ shown in Fig. 12.7(a) by diamonds is equal to $f_P/2$ shown in Fig. 12.7(a) by a straight line. This observation enables one to associate the generation of the secondary peak with the nonlinear frequency conversion due to mixing between the microwave pumping and the dynamic magnetization signal generated by the STNO, resulting in the generation of a combinatory spectral harmonic. This process can be also interpreted as the pumping-mediated transfer of the microwave energy generated by the STNO into another frequency range. This explanation is consistent with the decrease of the intensity of the auto-oscillation peak observed simultaneously with the appearance of the secondary peak (Fig. 12.6(b)). The suppression of the primary peak is also clearly seen in the dependencies of the BLS intensities of the peaks on f_P (Fig. 12.7(b)). In comparison to the free-running auto-oscillation intensity shown in Fig. 12.7(b) by a dashed horizontal line, the intensity J_1 of the primary oscillation peak is noticeably suppressed at all pumping frequencies, except for the enhancement in the synchronization region and a less pronounced suppression around $f_P = 16$ GHz. We also note that the suppression of the auto-oscillation peak occurs at f_P both *below* and *above* the synchronization region 13.8–14.4 GHz, whereas the secondary peak induced by the pumping appears *only above* 14.4 GHz. To understand why a secondary peak is not observed at $f_P < 13.8$ GHz, we assume that the nonlinear frequency mixing occurs for any pumping frequency outside the synchronization region. However, it can lead to the excitation of propagating spin waves only if the frequency generated by the mixing falls within the spin-wave spectrum of the extended Py film outside the point contact area. Otherwise, the frequency mixing should result in the generation of a non-resonant magnetization oscillation which is localized within the point contact area.

12.6 Effect of the Microwave Pumping on the Spin-Wave Emission Characteristics

We now address the spatial characteristics of the emitted spin waves in different regimes. Figure 12.8(a) shows a two-dimensional map of the spin-wave intensity recorded at the auto-oscillation frequency in the absence of pumping. This map is similar to those shown in Fig. 12.3: The spin-wave emission is strongly directional and occurs in the two opposite directions perpendicular to the static field H_0. While the map shown in Fig. 12.8(a) was recorded in the absence of pumping, very similar maps were also observed in the presence of pumping for spin waves corresponding to the auto-oscillation peak at f_1. Within the entire frequency range $f_p = 13$–17 GHz, the spatial decay of the emitted waves remained unchanged.

A different behavior was found for spin waves corresponding to the pumping-induced combinatory frequency f_2. A typical example of the emission patterns corresponding to $f_2 = 7.99$ GHz ($f_P = 15$ GHz) is shown in Fig. 12.8(b). Comparing

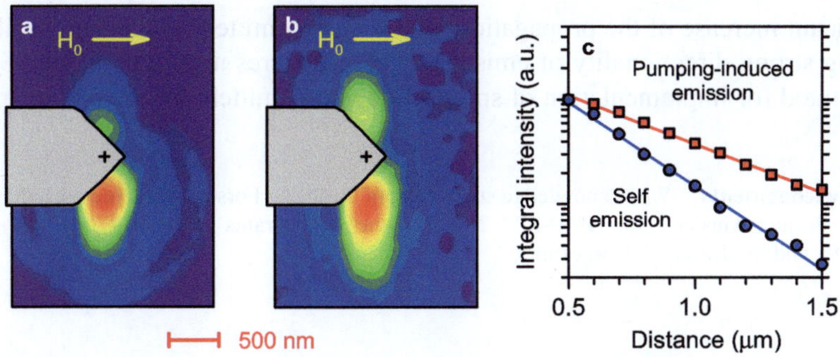

Fig. 12.8 (**a**) and (**b**) Pseudo-color spatial intensity maps of the emitted spin waves, acquired at $I = 5$ mA. (**a**) Emission at the frequency of auto-oscillation, in the absence of the external microwaves. (**b**) Emission at the frequency of the pumping-induced secondary peak, at $f_P = 15$ GHz. (**c**) Logarithmic plots of the integral intensity of the emitted spin waves vs the distance from the nano-contact for the cases shown in panels (**a**) – *circles*, and (**b**) – *squares*

the two maps in Figs. 12.8(a) and (b), one can clearly see that the emission induced by the pumping preserves the main features of the emission at the self-oscillation frequency, but is characterized by a noticeably larger spin-wave propagation length.

Figure 12.8(c) shows the intensity of the emitted spin waves as a function of the distance from the nano-contact, obtained from the emission patterns shown in Figs. 12.8(a) and (b) by integrating along the direction parallel to H_0. These dependencies are well described by an exponential decay, as shown by linear fits on the logarithmic scale. The decay is significantly stronger for spin waves at the frequency of the auto-oscillation compared to the waves at the combinatory frequency. From the data of Fig. 12.8(c), we obtain the values of the decay lengths: $\xi = 540$ nm for the emission at the auto-oscillation frequency $f_1 = 7.15$ GHz, and $\xi = 940$ nm for the pumping-induced emission at $f_2 = 7.99$ GHz.

We associate the observed increase of the propagation length with the increase of the group velocity of the emitted spin waves due to the increase in their frequency [14]. In other words, by transferring the energy generated by the STNO to higher frequencies by using the effect of the nonlinear frequency conversion, one can improve the emission characteristics, and make STNOs better suited for applications as spin-wave emitters.

12.7 Summary

We have experimentally demonstrated the operation of STNOs that are characterized by small working currents, do not require large static magnetic fields for their operation, and support directional steerable emission of spin waves into the surrounding magnetic film. We also showed that the emission characteristics of the STNOs can be controlled by using microwave pumping. This mechanism enables a significant increase of the propagation length of the emitted spin waves, while preserving strong directionality of emission. These features make the studied STNOs well suited for implementation of spin-torque nano-emitters for magnonic applications.

Acknowledgements We acknowledge support from Deutsche Forschungsgemeinschaft, the European Project Master (No. NMP-FP7 212257), the NSF grants DMR-0747609 and ECCS-0967195, and the Research Corporation.

References

1. J.C. Slonczewski, J. Magn. Magn. Mater. **159**, L1–L7 (1996)
2. L. Berger, Phys. Rev. B **54**, 9353–9358 (1996)
3. E.B. Myers, D.C. Ralph, J.A. Katine, R.N. Louie, R.A. Buhrman, Science **285**, 867–870 (1999)
4. J.A. Katine, F.J. Albert, R.A. Buhrman, E.B. Myers, D.C. Ralph, Phys. Rev. Lett. **84**, 3149–3152 (2000)
5. M. Tsoi, A.G.M. Jansen, J. Bass, W.-C. Chiang, V. Tsoi, P. Wyder, Nature **406**, 46–48 (2000)
6. S.I. Kiselev, J.C. Sankey, I.N. Krivorotov, N.C. Emley, R.J. Schoelkopf, R.A. Buhrman, D.C. Ralph, Nature **425**, 380–383 (2003)
7. S. Urazhdin, N.O. Birge, W.P. Pratt Jr., J. Bass, Phys. Rev. Lett. **91**, 146803 (2003)
8. W.H. Rippard, M.R. Pufall, S. Kaka, S.E. Russek, T.J. Silva, Phys. Rev. Lett. **92**, 27201 (2004)
9. I.N. Krivorotov, N.C. Emley, J.C. Sankey, S.I. Kiselev, D.C. Ralph, R.A. Buhrman, Science **307**, 228–231 (2005)
10. S. Kaka, M.R. Pufall, W.H. Rippard, T.J. Silva, S.E. Russek, J.A. Katine, Nature **437**, 389–392 (2005)
11. F.B. Mancoff, N.D. Rizzo, B.N. Engel, S. Tehrani, Nature **437**, 393–395 (2005)
12. A. Slavin, Nat. Nanotechnol. **4**, 479–480 (2009)
13. V.E. Demidov, S. Urazhdin, S.O. Demokritov, Nat. Mater. **9**, 984–988 (2010)
14. V.E. Demidov, S. Urazhdin, V. Tiberkevich, A. Slavin, S.O. Demokritov, Phys. Rev. B **83**, 060406(R) (2011)
15. M. Madami, S. Bonetti, G. Consolo, S. Tacchi, G. Carlotti, G. Gubbiotti, F.B. Mancoff, M.A. Yar, J. Åkerman, Nat. Nanotechnol. **6**, 635–638 (2011)
16. M.R. Pufall, W.H. Rippard, S.E. Russek, S. Kaka, J.A. Katine, Phys. Rev. Lett. **97**, 087206 (2006)
17. A.N. Slavin, V.S. Tiberkevich, Phys. Rev. B **74**, 104401 (2006)
18. S.M. Rezende, F.M. de Aguiar, R.L. Rodriguez-Suarez, A. Azevedo, Phys. Rev. Lett. **98**, 087202 (2007)
19. X. Chen, R.H. Victora, Phys. Rev. B **79**, 180402(R) (2009)
20. V.E. Demidov, S.O. Demokritov, K. Rott, P. Krzysteczko, G. Reiss, Appl. Phys. Lett. **92**, 232503 (2008)
21. S. Neusser, D. Grundler, Adv. Mater. **21**, 2927 (2009)
22. V.V. Kruglyak, S.O. Demokritov, D. Grundler, J. Phys. D, Appl. Phys. **43**, 264001 (2010)
23. B. Lenk, H. Ulrichs, F. Garbs, M. Münzenberg, Phys. Rep. **507**, 107–136 (2011)
24. V.E. Demidov, M.P. Kostylev, K. Rott, P. Krzysteczko, G. Reiss, S.O. Demokritov, Appl. Phys. Lett. **95**, 112509 (2009)
25. V. Vlaminck, M. Bailleul, Phys. Rev. B **81**, 014425 (2010)
26. Y. Au, T. Davison, E. Ahmad, P.S. Keatley, R.J. Hicken, V.V. Kruglyak, Appl. Phys. Lett. **98**, 122506 (2011)
27. V.E. Demidov, M.P. Kostylev, K. Rott, J. Münchenberger, G. Reiss, S.O. Demokritov, Appl. Phys. Lett. **99**, 082507 (2011)

28. V.E. Demidov, S.O. Demokritov, B. Hillebrands, M. Laufenberg, P.P. Freitas, Appl. Phys. Lett. **85**, 2866 (2004)
29. S.O. Demokritov, V.E. Demidov, IEEE Trans. Magn. **44**, 6 (2008)
30. M.A. Hoefer, T.J. Silva, M.D. Stiles, Phys. Rev. B **77**, 144401 (2008)
31. V.E. Demidov, O. Dzyapko, M. Buchmeier, T. Stockhoff, G. Schmitz, G.A. Melkov, S.O. Demokritov, Phys. Rev. Lett. **101**, 257201 (2008)
32. B.A. Kalinikos, IEE Proc. H **127**, 4 (1980)
33. B.A. Kalinikos, A.N. Slavin, J. Phys. C **19**, 7013 (1986)
34. A. Slavin, V. Tiberkevich, Phys. Rev. Lett. **95**, 237201 (2005)
35. W.H. Rippard, M.R. Pufall, S. Kaka, T.J. Silva, S.E. Russek, J.A. Katine, Phys. Rev. Lett. **95**, 067203 (2005)
36. J.C. Sankey, P.M. Braganca, A.G.F. Garcia, I.N. Krivorotov, R.A. Buhrman, D.C. Ralph, Phys. Rev. Lett. **96**, 227601 (2006)
37. B. Georges, J. Grollier, M. Darques, V. Cros, C. Deranlot, B. Marcilhac, G. Faini, A. Fert, Phys. Rev. Lett. **101**, 017201 (2008)
38. S. Urazhdin, P. Tabor, V. Tiberkevich, A. Slavin, Phys. Rev. Lett. **105**, 104101 (2010)
39. S. Urazhdin, V. Tiberkevich, A. Slavin, Phys. Rev. Lett. **105**, 237204 (2010)

Chapter 13
Nano-Contact Spin-Torque Oscillators as Magnonic Building Blocks

Stefano Bonetti and Johan Åkerman

Abstract We describe the possibility of using nano-contact spin-torque oscillators (NC-STOs) as fundamental magnonic building blocks. NC-STOs can act as spin wave generators, manipulators, and detectors, and can hence realize all the fundamental functions necessary for fully integrated magnonic devices, which can be fabricated using available CMOS compatible large-scale spin-torque device production processes. We show in particular how a 200 nm sized nano-contact located on an out-of-plane magnetized permalloy "free" magnetic layer can generate spin waves at $f \approx 15$ GHz that propagate up to 4 μm away from the nano-contact with wavelength $\lambda = 200$–300 nm, decay length $\lambda_r \approx 2$ μm and group velocities $v_g \approx 3$ μm/ns. We propose that the same type of NC-STOs can be used as spin wave manipulators, via control of the local Gilbert damping, and as spin wave detector using the spin torque diode effect.

13.1 Introduction

13.1.1 Magnonics and Magnonic Devices

Magnonics [1] is an emerging and rapidly growing research field, positioned at the intersection of magnetodynamics, spintronics, and microwave electronics, and dealing with magnetic phenomena and device functionalities connected with spin waves

J. Åkerman (✉)
Department of Physics, University of Gothenburg, 412 96 Gothenburg, Sweden
e-mail: johan.akerman@physics.gu.se

J. Åkerman
School of Information and Communication Technology, Materials Physics, KTH – Royal Institute of Technology, Electrum 229, 164 40 Kista-Stockholm, Sweden

S. Bonetti
Stanford Institute for Materials and Energy Science (SIMES), Stanford University, Stanford, CA 94305, USA
e-mail: bonetti@kth.se

S.O. Demokritov, A.N. Slavin (eds.), *Magnonics*, Topics in Applied Physics 125,
DOI 10.1007/978-3-642-30247-3_13,

(SWs). A SW represents a phase-coherent collective precession of the local moments of the magnetic medium [2], and can be considered the magnetic analogue of a sound or a light wave: Magnonics is to SWs what photonics is to light. Magnonic devices show particular promise for novel applications in the GHz to THz frequency ranges where the wavelength of SWs is several orders of magnitude shorter than that of electromagnetic (light) waves and hence offer radically better prospects for miniaturization at these frequencies. To date, magnonics studies have primarily focused on the SW *medium*, often in the form of the so-called Magnonic Crystals (MC; in analogy with Photonic Crystals for light) [3, 4]. The SW spectrum in MCs shows bands of allowed states, alternating with forbidden bandgaps, which are tunable by the applied magnetic field. While the SW medium is obviously central to magnonics, a fully functional and integrated microscopic magnonic device also requires SW generators, SW manipulators, and SW detectors. In the current literature, SW generation is typically performed by macroscopic microwave field generators, which couple to the medium through antenna structures [5]. On the detection side, one either uses inductive antenna pickup of the time-dependent magnetization or sophisticated laboratory based spatially resolved optical techniques such as scanning micro Brillouin light scattering (μ-BLS) [6] or time-resolved scanning Kerr microscopy (TRSKM) [7]. However, for magnonic devices to become a commercial reality, such macroscopic external approaches must be replaced by miniaturized and integrated solutions.

13.1.2 Spin-Transfer Torque

When unpolarized electrons enter a magnetic material and become polarized, they acquire angular momentum. John Slonczewski argued that the reciprocal effect, i.e., the transfer of angular momentum from a spin-polarized current to a magnetic layer may exert a sufficiently large torque on the local magnetization to effectively switch its state [8]. This so-called spin-transfer torque (STT) should also be able to sustain a continuous precession of the magnetization vector [8, 9], which was subsequently confirmed in a number of different sample geometries, including mechanical point contacts [10, 11], lithographically defined nano-contacts [12], and nano-pillars [13]. The actual device is called a Spin-Torque (nano) Oscillator (STO), and consists in its simplest form of a magnetoresistive device, where the free layer is excited into a coherent precessional state which can subsequently be detected as a microwave voltage signal, by virtue of the Giant Magnetoresistance (GMR) or Tunneling Magnetoresistance (TMR) of the device. The precession frequency is tunable both via the direct drive current and through the magnitude and direction of the applied magnetic field.

13.1.3 Scope

In this work, we propose the use of nano-contact spin-torque devices as possibly the ideal candidates for the still-lacking magnonic building blocks. As described

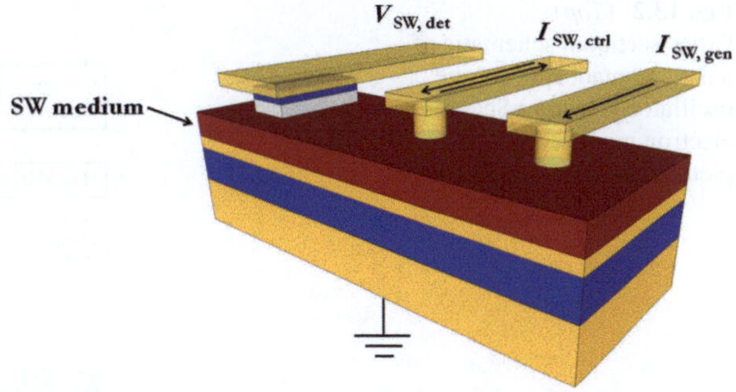

Fig. 13.1 Pictorial schematic of a nano-contact based magnonic circuit

below, the nano-contacts are fabricated on top of the magnonic medium, which is in the form of a GMR trilayer, and the three magnonic functions of SW generation, manipulation, and detection are all three carried out using different aspects of STT. A schematic depiction of such a magnonic circuit is presented in Fig. 13.1.

13.2 Fabrication of Nano-Contact STOs

In the nano-contact geometry, the high-density spin-polarized current necessary to achieve a sizeable steady spin-torque effect, is injected via a nanosized (40–300 nm) contact (NC). The NC is usually located on top of a GMR mesa which extends laterally over several micrometers to allow spin waves to propagate freely, and to be sufficiently damped out before reflecting back from the mesa boundary.

The cross-section of a typical NC structure is depicted in Fig. 13.2. The multilayer thin-film stack comprises [14] a 30 nm Cu base electrode; a 20 nm "fixed" magnetic layer, which for most of the devices discussed here is a CoFe alloy; a 6 nm Cu spacer; a 4.5 nm "free" magnetic layer with low Gilbert damping, typically Permalloy ($Ni_{80}Fe_{20}$); a 5 nm Cu capping layer.

All the layers which form the devices can be deposited by magnetron sputtering. Optical lithography is subsequently used to define mesas of size 26 μm × 8 μm. The NC is patterned by electron-beam lithography to define, on top of each mesa, a central NC and two micrometer-sized openings, which act as ground contacts. The scanning electron microscope (SEM) picture in the lower part of Fig. 13.2 depicts a real structure after these fabrication steps. A final, optical lithography, step is used to define a coplanar waveguide on top of the mesa, which is required for the efficient extraction and detection of the high frequency signal generated by the device [15].

It is important to point out that placing the ground contacts on top of the mesa does not alter the electronic transport in the NC region in any noticeable way: the current flows predominantly orthogonal to the GMR layers. This is achieved thanks to the relatively thick (30 nm), low-resistance, bottom Cu electrode, which acts so as to carry away the electrons flowing from the central NC towards the two lateral micrometer-sized contacts. Placing all the electrodes on the top of the mesa simplifies device fabrication, since a single lithographic step is sufficient for this purpose.

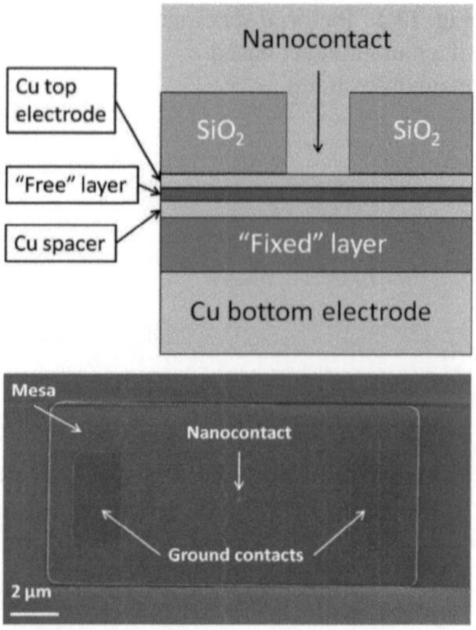

Fig. 13.2 (*Top*) Cross-sectional schematic of a nano-contact spin-torque oscillator. (*Bottom*) Scanning electron microscope (SEM) picture of a fabricated device

13.3 Spin Wave Dynamics in Nano-Contact STOs

13.3.1 Fundamentals

In his pioneering paper [16], Slonczewski predicted that the spin waves excited by the spin-transfer torque of a nano-contact in a perpendicularly magnetized and extended free layer [10] would be exchange-dominated. The spin wave would possess cylindrical symmetry and propagate radially from the nano-contact region, with a wave number k inversely proportional to the nano-contact radius R_c ($k \simeq 1.2/R_c$). Rippard et al. [17] demonstrated that, while Slonczewski's theory correctly describes the frequency and threshold current of spin waves excited in perpendicularly magnetized films, it fails to describe the spin waves excited when the same film is magnetized in the plane. It was later shown theoretically that, in the case of in-plane magnetization, it is also possible to excite a self-localized nonlinear spin wave mode of solitonic character, also known as spin wave "bullet" [18]. The current-induced excitation of the spin wave bullet was then numerically confirmed in several micromagnetic simulations [19–22]. While angular-dependent measurements have been reported on in the literature [23], clear experimental evidence for the spin wave bullet and the characterization of its properties was only recently presented [24].

The key to studying the details of SW dynamics in NC-STOs is the ability to perform measurements in which the direction of the applied magnetic field can be controlled to great precision. The angle of the out-of-plane external angle (θ_e) with respect to the thin film plane is particularly important, since the SW dynamics is critically dependent on it.[1]

[1]Strictly speaking, the *magnetization* angle is the critical parameter. However, since the magnetization angle is a monotonic function of the applied field (at saturation), and since it is easier to experimentally measure the applied field angle, we will refer to the critical field angle.

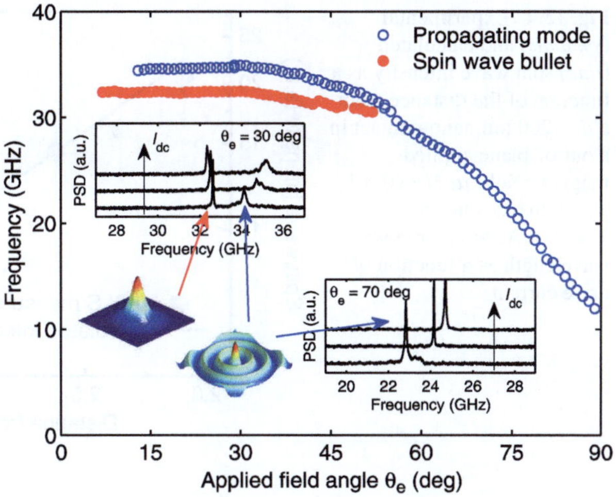

Fig. 13.3 Operating frequency of a nano-contact STO as a function of the applied magnetic field angle. The *two insets* show the power spectral density at two different angles $\theta_e = 30°$ and $\theta_e = 70°$. The calculated mode profiles are also shown

Figure 13.3 illustrates this fact clearly. For fields applied sufficiently far from the film normal, i.e., $\theta_e < 55°$, two distinct and qualitatively different SW modes (shown by the filled and empty symbols) can be excited by the current passing through the NC. These two modes have different frequencies f and opposite signs of the frequency tunability df/dI, as shown by the inset at $\theta_e = 30°$. In contrast, for $\theta_e > 55°$, only one SW modes is excited, with a positive frequency tunability (as illustrated by the inset in Fig. 13.3 for $\theta_e = 70°$). The lower excitation frequency, the negative sign of the frequency tunability and its occurrence only below a certain critical angle θ_c are all characteristics of a spin wave bullet [18, 25]. On the other hand, the higher excitation frequency, the positive sign of the frequency tunability, and its occurrence at all field angles are properties of a propagating spin wave mode [16]. The behavior of the threshold current as a function of applied field angle also confirms these associations [24]. The spatial profiles of the two modes are depicted in Fig. 13.3. The x–y plane in the sketch is the real plane of the sample, while the height and color of the sketch is proportional the magnitude of the out-of-plane component of the magnetization.

13.3.2 Propagating Waves as Magnonic Signals

In the previous subsection, we showed that NC-STOs can excite both localized and propagating SWs. In terms of applications, propagating SWs are of primary interest, since the transmission of a signal requires energy to propagate from a source to a detector. The details of the propagation are also important: What is the wavelength, the characteristic damping length, the group velocity, and the directional profile of the propagating SWs? This constitutes fundamental knowledge required for the design of working magnonic devices.

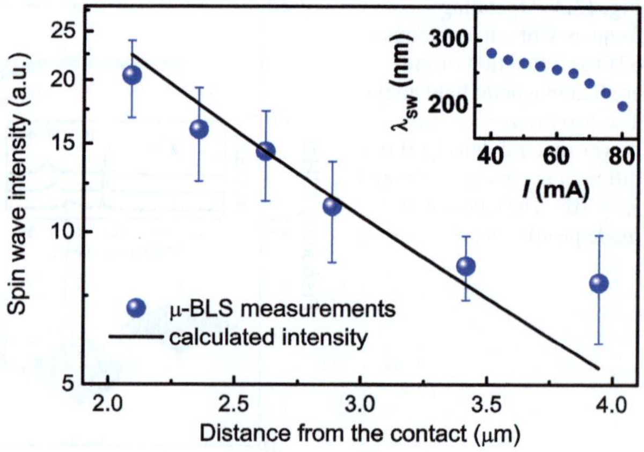

Fig. 13.4 Experimental (*symbols*) and calculated (*line*) spin wave intensity as a function of the distance from a $d = 200$ nm nano-contact in a out-of-plane applied magnetic field $\mu_0 H = 0.6$ T, and with bias current $I = 70$ mA. *Inset*: spin wave wavelength as a function of drive current

The experimental evidence presented for propagating SWs in the previous subsection was only indirect, and direct microscopic evidence for such SW propagation remained elusive until very recently. Following the first direct optical observation of STT generated SWs underneath a nano-contact [26], definitive proof of directional SW propagation came soon thereafter [27]. In both of these works, SWs were detected by a microscope-focused Brillouin Light Scattering (μ-BLS) technique [28], which allows for spatially resolved SW detection.

Madami et al. [27] demonstrated that the SWs emitted by an out-of-plane magnetized NC-STO do indeed have a propagating character. The authors applied the procedure described by Neumann et al. [29] to the μ-BLS technique, which we explain here. If the STT-generated SWs have a well-defined propagation direction in the plane of the free layer, the photons which undergo Stokes and anti-Stokes processes will be scattered at opposite angles with respect to the sample normal. When the full beam is sent to the interferometer, both Stokes and anti-Stokes signals are detected. However, when the authors blinded off half of the beam with a shutter, one of the two peaks disappeared (which peak disappeared depending on which half of the beam had been stopped). This is direct proof that the emitted SWs propagate with a uniquely defined wave-vector direction, away from the NC region. If the SWs had instead been localized, counts would have occurred on both the Stokes and anti-Stokes regions of the spectrum, regardless of which half of the light beam had been blinded off.

The propagation characteristics of the SWs have been quantified by observing the spatial decay of the μ-BLS signal as a function of distance from the NC [27]. The experimental results are shown as symbols in Fig. 13.4. The SW intensity can be modeled as $J(r) = [J(0)/r]\exp(-r/\lambda_r)$, where r is the distance from the NC, $J(0)$ the amplitude at $r = 0$, and λ_r the decay length. In this model, both the energy emission along a cylindrical wavefront ($\sim 1/r$ factor) and the intrinsic damping of the material ($\sim \exp(-r/\lambda_r)$) are considered. $\lambda_r = v_g/(2\alpha\omega)$ represents the decay length related to the SW group velocity v_g, the Gilbert damping parameter α, and the SW angular frequency $\omega = 2\pi f$.

Figure 13.4 also shows the calculated decay in SW intensity, assuming the experimental frequency $f = 15.3$ GHz, $\alpha = 0.008$ (close to the nominal value for permalloy), and using the group velocity $v_g = 3.3$ μm/ns calculated from micromagnetic simulations. It is evident that there is good agreement with the experiment, and one can also extract the decay length $\lambda_r = 2.1$ μm, which is very close to literature values for permalloy [30].

13.4 Nano-Contact-Based Magnonic Building Blocks

13.4.1 Spin Wave Injectors

As described in Sect. 13.3 above, NC-STOs can generate propagating SWs into 5 nm thick NiFe films. They hence represent an obvious candidate for nanoscopic SW injectors in magnonic devices. By orienting the externally applied magnetic field in the film plane, the same type of NC-STO can also generate localized SWs [26]. Such SW injectors offer a remarkable design freedom spanning a very large parameter space in terms of frequency range, modulation rates, SW wave lengths, and material choices.

NC-STOs with in-plane magnetization free layers can oscillate below 5 GHz in low magnetic fields [12], and can be tuned to above 46 GHz if the field is increased to 2.2 T [31]. If the NC is made much smaller, theoretical estimates predict operating frequencies approaching 200 GHz [32]. If the NiFe free layer is replaced by a material with perpendicular magnetic anisotropy (PMA), such as Co/Ni [33] multilayers, operating frequencies as high as 12 GHz can be achieved in zero field, and about 30 GHz in a relatively moderate field of 0.6 T [34]. Using materials with much larger PMA and smaller M_S, such as $Mn_{2.12}Ga$, could potentially raise the zero-field operating frequency to almost 300 GHz [35]. NC-STO-based SW injectors hence have the potential to cover about 2 orders of magnitude in frequency range.

Several NC-STOs can be made to mutually synchronize via the injected SWs [14, 36, 37]. While this increases both the output power and the signal coherence, it also allows for great freedom in tailoring the SW wavefront shape in a magnonic device. For example, if a number of NC-STOs are positioned on a line and made to synchronize, they will emit SWs with a straight wavefront. Alternatively, synchronized NC-STOs on a circle or an arc can be used to focus SW power onto a particular spot.

NC-STOs are readily frequency modulated [38, 39], and have a very high modulation bandwidth, which has been demonstrated up to 3.2 GHz [40]. Since synchronized NC-STOs stay synchronized under modulation [41, 42], it will be possible to frequency modulate SWs both from single NC-STOs and from synchronized lines of NC-STOs. Modulation can also enhance the coherence of the NC-STO signal, and we expect the same enhanced coherence to apply to the STT-generated SWs [43].

As described by Slonczewski [16], the SW wave vector is inversely proportional to the NC radius: $k = 1.2/r_{NC}$. In other words, the SW wavelength λ_{SW} is approximately 2.5 times the NC diameter. Since NC-STOs have been demonstrated with diameters ranging from 40 to 300 nm, there is substantial design freedom in terms of λ_{SW} [44]. It should in principle be possible to vary λ_{SW} from about 90 to 700 nm while keeping the SW frequency constant (see Fig. 2 in [44]), although this has yet to be experimentally confirmed by direct observation of λ_{SW}, using for example μ-BLS.

13.4.2 Spin Wave Manipulators

Spin wave propagation can be controlled and manipulated using, e.g., the Oersted field from a current line on top of the SW medium [45], or by changing the direction and amplitude of an external magnetic field applied to a MC [3]. We argue that it should also be possible to use STT from NC-STOs to control the damping that SWs experience in the region underneath the NC-STO, and through this control of damping it should be possible to manipulate the SW propagation. STT can, for example, be viewed as negative damping, and it is well known that STOs in their sub-threshold regime can enhance thermally excited SW resonances through a local supply of negative damping [46]. When driven with the opposite current polarity, the same NC-STO can provide locally enhanced positive damping and suppress SW propagation.

13.4.3 Spin Wave Detectors

While the examples of NC-STO-based SW generators and manipulators above rely on STT for operation, SW detection can in principle be realized simply through the GMR of the nano-contact region. A passing SW will generate an oscillation in the resistance underneath the nano-contact, which can then be detected. In order to achieve sufficient sensitivity, one would most likely have to use TMR, i.e., the nano-contact would not simply be a metal, but would instead consist of both a tunneling barrier and a fixed magnetic layer, and thus form a complete MTJ on top of the SW medium [47]. While the detection could in principle be done in the microwave regime through a spectrally resolved measurement of the actual frequency of the passing SW, a more convenient detection scheme, removing any need for external microwave equipment, might be to use the spin-torque diode effect, which would generate a direct voltage in proportion to the SW power [48].

13.5 Conclusions

We have described the possibility of using nano-contact spin-torque oscillators as fundamental magnonic building blocks. Such devices consist of a pseudo-spin-valve mesa ($\sim$10 μm in size), on which an electrical nano-contact is fabricated. The relatively high current density ($\sim 10^8$ A/cm^2) that can be pushed through such a nano-contact is capable of inducing, through spin transfer torque, the excitation of spin waves in one of the two ferromagnetic layers constituting the spin valve. Nano-contact spin-torque oscillators can act as spin wave generators, manipulators, and detectors, hence realizing all the fundamental functions necessary for integrated magnonic devices.

Acknowledgements Support from the Swedish Foundation for Strategic Research (SSF), the Swedish Research Council (VR) and the Knut and Alice Wallenberg Foundation is gratefully acknowledged. Stefano Bonetti is a Postdoctoral Fellow supported by a grant from the Swedish Research Council (VR) and the Knut and Alice Wallenberg Foundation. Johan Åkerman is a Royal Swedish Academy of Sciences Research Fellow supported by a grant from the Knut and Alice Wallenberg Foundation. We thank M. Madami, G. Consolo, S. Tacchi, G. Carlotti, G. Gubbiotti, F.B. Mancoff, V. Tiberkevich, and A. Slavin for useful discussions.

References

1. V.V. Kruglyak, S.O. Demokritov, D. Grundler, Magnonics. J. Phys. D, Appl. Phys. **43**(26), 260301 (2010)
2. F. Bloch, Zur theorie des ferromagnetismus. Z. Phys. **61**(3–4), 206–219 (1930)
3. S. Neusser, D. Grundler, Magnonics: spin waves on the nanoscale. Adv. Mater. **21**, 2927–2932 (2009)
4. S.A. Nikitov, P. Tailhades, C.S. Tsai, Spin waves in periodic magnetic structures–magnonic crystals. J. Magn. Magn. Mater. **236**(3), 320–330 (2001)
5. Y.K. Fetisov, C.E. Patton, Microwave bistability in a magnetostatic wave interferometer with external feedback. IEEE Trans. Magn. **35**(2), 1024–1036 (1999)
6. S. Tacchi, M. Madami, G. Gubbiotti, G. Carlotti, A.O. Adeyeye, S. Neusser, B. Botters, D. Grundler, Magnetic normal modes in squared antidot array with circular holes: a combined Brillouin light scattering and broadband ferromagnetic resonance study. IEEE Trans. Magn. **46**(2), 172–178 (2010)
7. M.J. Pechan, C. Yu, R.L. Compton, J.P. Park, P.A. Crowell, Direct measurement of spatially localized ferromagnetic-resonance modes in an antidot lattice (invited). J. Appl. Phys. **97**(10), 10J903 (2005)
8. J. Slonczewski, Current-driven excitation of magnetic multilayers. J. Magn. Magn. Mater. **159**(1–2), L1–L7 (1996)
9. L. Berger, Emission of spin waves by a magnetic multilayer traversed by a current. Phys. Rev. B **54**(13), 9353–9358 (1996)
10. M. Tsoi, A.G.M. Jansen, J. Bass, W.-C. Chiang, M. Seck, V. Tsoi, P. Wyder, Excitation of a magnetic multilayer by an electric current. Phys. Rev. Lett. **80**(19), 4281–4284 (1998)
11. M. Tsoi, A.G. Jansen, J. Bass, W.C. Chiang, V. Tsoi, P. Wyder, Generation and detection of phase-coherent current-driven magnons in magnetic multilayers. Nature **406**(6791), 46–48 (2000)
12. W. Rippard, M. Pufall, S. Kaka, S. Russek, T. Silva, Direct-current induced dynamics in Co90Fe10/Ni80Fe20 point contacts. Phys. Rev. Lett. **92**(2), 027201 (2004)

13. S.I. Kiselev, J.C. Sankey, I.N. Krivorotov, N.C. Emley, R.J. Schoelkopf, R.A. Buhrman, D.C. Ralph, Microwave oscillations of a nanomagnet driven by a spin-polarized current. Nature **425**(6956), 380–383 (2003)
14. F.B. Mancoff, N.D. Rizzo, B.N. Engel, S. Tehrani, Phase-locking in double-point-contact spin-transfer devices. Nature **437**(7057), 393–395 (2005)
15. S. Bonetti, Magnetization dynamics in nano-contact spin torque oscillators. Ph.D. thesis, Kungliga Tekniska Högskolan, The Royal Institute of Technology, Stockholm, Sweden (2010)
16. J. Slonczewski, Excitation of spin waves by an electric current. J. Magn. Magn. Mater. **195**(2), 261–268 (1999)
17. W.H. Rippard, M.R. Pufall, T.J. Silva, Quantitative studies of spin-momentum-transfer-induced excitations in co/cu multilayer films using point-contact spectroscopy. Appl. Phys. Lett. **82**(8), 1260–1262 (2003)
18. A. Slavin, V. Tiberkevich, Spin wave mode excited by spin-polarized current in a magnetic nanocontact is a standing self-localized wave bullet. Phys. Rev. Lett. **95**(23), 237201 (2005)
19. D. Berkov, N. Gorn, Magnetization oscillations induced by a spin-polarized current in a point-contact geometry: mode hopping and nonlinear damping effects. Phys. Rev. B **76**(14), 144414 (2007)
20. D. Berkov, J. Miltat, Spin-torque driven magnetization dynamics: micromagnetic modeling. J. Magn. Magn. Mater. **320**(7), 1238–1259 (2008)
21. G. Consolo, B. Azzerboni, L. Lopez-Diaz, G. Gerhart, E. Bankowski, V. Tiberkevich, A. Slavin, Micromagnetic study of the above-threshold generation regime in a spin-torque oscillator based on a magnetic nanocontact magnetized at an arbitrary angle. Phys. Rev. B **78**(1), 014420 (2008)
22. G. Consolo, B. Azzerboni, G. Gerhart, G. Melkov, V. Tiberkevich, A. Slavin, Excitation of self-localized spin-wave bullets by spin-polarized current in in-plane magnetized magnetic nanocontacts: a micromagnetic study. Phys. Rev. B **76**(14), 144410 (2007)
23. W. Rippard, M. Pufall, S. Kaka, T. Silva, S. Russek, Current-driven microwave dynamics in magnetic point contacts as a function of applied field angle. Phys. Rev. B **70**(10), 100406(R) (2004)
24. S. Bonetti, V. Tiberkevich, G. Consolo, G. Finocchio, P. Muduli, F. Mancoff, A. Slavin, J. Åkerman, Experimental evidence of self-localized and propagating spin wave modes in obliquely magnetized current-driven nanocontacts. Phys. Rev. Lett. **105**(21), 1–4 (2010)
25. G. Gerhart, E. Bankowski, G. Melkov, V. Tiberkevich, A. Slavin, Angular dependence of the microwave-generation threshold in a nanoscale spin-torque oscillator. Phys. Rev. B **76**(2), 024437 (2007)
26. V.E. Demidov, S. Urazhdin, S.O. Demokritov, Direct observation and mapping of spin waves emitted by spin-torque nano-oscillators. Nat. Mater. **9**(12), 984–988 (2010)
27. M. Madami, S. Bonetti, G. Consolo, S. Tacchi, G. Carlotti, G. Gubbiotti, F.B. Mancoff, M.A. Yar, J. Åkerman, Direct observation of a propagating spin wave induced by spin-transfer torque. Nat. Nanotechnol. **6**, 635–638 (2011)
28. V.E. Demidov, S.O. Demokritov, B. Hillebrands, M. Laufenberg, P.P. Freitas, Radiation of spin waves by a single micrometer-sized magnetic element. Appl. Phys. Lett. **85**(14), 2866–2868 (2004)
29. T. Neumann, T. Schneider, A.A. Serga, B. Hillebrands, An electro-optic modulator-assisted wavevector-resolving Brillouin light scattering setup. Rev. Sci. Instrum. **80**(5), 053905 (2009)
30. V.E. Demidov, S.O. Demokritov, K. Rott, P. Krzysteczko, G. Reiss, Self-focusing of spin waves in permalloy microstripes. Appl. Phys. Lett. **91**(25), 252504 (2007)
31. S. Bonetti, P. Muduli, F. Mancoff, J. Åkerman, Spin torque oscillator frequency versus magnetic field angle: the prospect of operation beyond 65 GHz. Appl. Phys. Lett. **94**(10), 102507 (2009)
32. M.A. Hoefer, M.J. Ablowitz, B. Ilan, M.R. Pufall, T.J. Silva, Theory of magnetodynamics induced by spin torque in perpendicularly magnetized thin films. Phys. Rev. Lett. **95**(26), 267206 (2005)

33. W.H. Rippard, A.M. Deac, M.R. Pufall, J.M. Shaw, M.W. Keller, S.E. Russek, C. Serpico, Spin-transfer dynamics in spin valves with out-of-plane magnetized CoNi free layers. Phys. Rev. B **81**(1), 014426 (2010)
34. S.M. Mohseni, S.R. Sani, J. Persson, T.N. Anh Nguyen, S. Chung, Y. Pogoryelov, J. Åkerman, High frequency operation of a spin-torque oscillator at low field. Phys. Status Solidi (RRL) – Rapid Res. Lett. **5**(12), 432–434 (2011)
35. S. Mizukami, F. Wu, A. Sakuma, J. Walowski, D. Watanabe, T. Kubota, X. Zhang, H. Naganuma, M. Oogane, Y. Ando, T. Miyazaki, Long-lived ultrafast spin precession in manganese alloys films with a large perpendicular magnetic anisotropy. Phys. Rev. Lett. **106**(11), 1–4 (2011)
36. S. Kaka, M.R. Pufall, W.H. Rippard, T.J. Silva, S.E. Russek, J.A. Katine, Mutual phase-locking of microwave spin torque nano-oscillators. Nature **437**(7057), 389–392 (2005)
37. M. Pufall, W. Rippard, S. Russek, S. Kaka, J. Katine, Electrical measurement of spin-wave interactions of proximate spin transfer nanooscillators. Phys. Rev. Lett. **97**(8), 087206 (2006)
38. P.K. Muduli, Y. Pogoryelov, S. Bonetti, G. Consolo, F. Mancoff, J. Åkerman, Nonlinear frequency and amplitude modulation of a nanocontact-based spin-torque oscillator. Phys. Rev. B **81**(14), 140408(R) (2010)
39. M.R. Pufall, W.H. Rippard, S. Kaka, T.J. Silva, S.E. Russek, Frequency modulation of spin-transfer oscillators. Appl. Phys. Lett. **86**(8), 082506 (2005)
40. P.K. Muduli, Y. Pogoryelov, Y. Zhou, F. Mancoff, Spin torque oscillators and RF currents modulation, locking, and ringing. Integr. Ferroelectr. **125**, 37–41 (2011)
41. P.K. Muduli, Y. Pogoryelov, F. Mancoff, J. Akerman, Modulation of individual and mutually synchronized nanocontact-based spin torque oscillators. IEEE Trans. Magn. **47**(6), 1575–1579 (2011)
42. Y. Pogoryelov, P.K. Muduli, S. Bonetti, E. Iacocca, F. Mancoff, J. Åkerman, Frequency modulation of spin torque oscillator pairs. Appl. Phys. Lett. **98**(19), 192501 (2011)
43. Y. Pogoryelov, P.K. Muduli, S. Bonetti, F. Mancoff, J. Åkerman, Spin-torque oscillator linewidth narrowing under current modulation. Appl. Phys. Lett. **98**(19), 192506 (2011)
44. F.B. Mancoff, N.D. Rizzo, B.N. Engel, S. Tehrani, Area dependence of high-frequency spin-transfer resonance in giant magnetoresistance contacts up to 300 nm diameter. Appl. Phys. Lett. **88**(11), 112507 (2006)
45. M.P. Kostylev, A.A. Serga, T. Schneider, T. Neumann, B. Leven, B. Hillebrands, R.L. Stamps, Resonant and nonresonant scattering of dipole-dominated spin waves from a region of inhomogeneous magnetic field in a ferromagnetic film. Phys. Rev. B **76**, 184419 (2007)
46. V. Tiberkevich, A. Slavin, J.-V. Kim, Microwave power generated by a spin-torque oscillator in the presence of noise. Appl. Phys. Lett. **91**(19), 192506 (2007)
47. F. Maci, A.D. Kent, F.C. Hoppensteadt, Spin-wave interference patterns created by spin-torque nano-oscillators for memory and computation. Nanotechnology **22**, 095301 (2011)
48. A.A. Tulapurkar, Y. Suzuki, A. Fukushima, H. Kubota, H. Maehara, K. Tsunekawa, D.D. Djayaprawira, N. Watanabe, S. Yuasa, Spin-torque diode effect in magnetic tunnel junctions. Nature **438**(7066), 339–342 (2005)

Part IV
Static and Dynamic Magnonic Crystals

Chapter 14
Spin Waves in Artificial Crystals and Metamaterials Created from Nanopatterned $Ni_{80}Fe_{20}$ Antidot Lattices

Sebastian Neusser, Georg Duerr, Rupert Huber, and Dirk Grundler

Abstract Periodic nanopatterning of metallic ferromagnets allows one to create devices with artificially tailored band structures for spin waves. We discuss our recent experiments performed on two-dimensional periodic lattices of nanoholes created in thin $Ni_{80}Fe_{20}$ films which exhibit man-made allowed minibands and forbidden frequency gaps provoking artificial crystal behavior and metamaterials properties.

14.1 Introduction

Periodically patterned dielectric materials provide the basis of advanced photonic devices which allow one to precisely mold the flow of electromagnetic waves in the optical regime [1]. Depending on the wavelength, metamaterials have been created in mainly two classes. Following [2], they form either bandgap materials which can be described by Bragg reflection and other periodic media concepts (if the wavelength is on the order of the periodicity), or artificial materials which can be described by homogenization and effective-media concepts (if the wavelength is much longer than the periodicity). Bandgap materials include the so-called photonic crystals which exhibit an artificially tailored band structure for electromagnetic waves with allowed minibands and forbidden frequency gaps [1]. For spin waves (magnons), metamaterials are far less explored compared to electromagnetic waves, but relevant building blocks have recently been realized. It has been shown that both the isolating ferrimagnets and metallic ferromagnets exhibit man-made band structures for spin waves if periodically patterned [3, 4]. Going beyond photonics, the band structure has been reprogrammed for one-and-the-same magnonic crystal by making use of different non-volatile magnetic states in the unit cell of the periodic lattice [5–8]. So far much of the work on artificially tailored band structures has been performed on periodic magnetic elements which are separated by air gaps and coupled via dynamic stray fields [6, 9–12]. Though realized rather early in a ferrimagnet [13], metamaterials behavior of the so-called magnetic antidot lattices was

S. Neusser · G. Duerr · R. Huber · D. Grundler (✉)
Lehrstuhl für Physik funktionaler Schichtsysteme, Physik Department E10, Technische Universität München, James-Franck-Str. 1, 85747 Garching b. München, Germany
e-mail: grundler@ph.tum.de

S.O. Demokritov, A.N. Slavin (eds.), *Magnonics*, Topics in Applied Physics 125,
DOI 10.1007/978-3-642-30247-3_14,

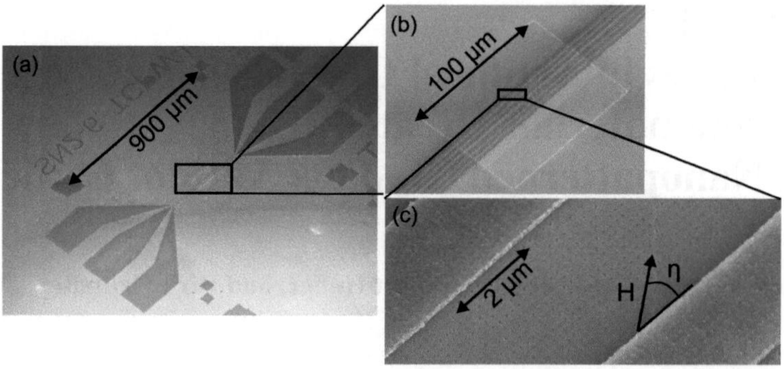

Fig. 14.1 Scanning electron microscopy (SEM) images of (**a**) two collinear coplanar waveguides (CPWs) on a semi-insulating GaAs substrate which run over a (**b**) 100-μm-wide $Ni_{80}Fe_{20}$ mesa in the central region. (**c**) Part of the periodic nanohole array forming the antidot lattice. The holes with a diameter of 120 nm have been etched into the $Ni_{80}Fe_{20}$ film via direct focused ion beam etching. CPWs are made from a trilayer of Cr/Ag/Au and isolated from the metallic mesa by an insulating layer of SiO_2

evidenced only recently in nanopatterned metallic ferromagnets [14–16]. This was surprising since two-dimensional (2D) arrays of nanoholes were a successful concept for photonic crystals [1] and under investigation in magnonics for many years [17–31]. In this contribution, we review our work on antidot lattices which has led to the observation of (i) miniband formation due to coherent coupling of localized modes, (ii) opening of bandgaps due to Bragg reflection, and (iii) an effective-media parameter for spin waves. We expect the findings to promote the research field of magnonics [26, 32, 33]. As the wavelength of spin waves is orders of magnitude smaller if compared to free-space electromagnetic waves at the same GHz frequencies, advancements in magnonics might also lead to novel microwave devices which are miniaturized and operate at the nanoscale.

14.2 Nanofabrication and All-Electrical Spin-Wave Spectroscopy

The magnetic antidot lattices (ADLs) discussed here have been fabricated from 22-nm-thick $Ni_{80}Fe_{20}$ mesas which contain periodic arrays of nanoholes etched by a focused ion beam (Fig. 14.1). The holes have a diameter d of 120 nm and are arranged on square lattices with lattice constants p ranging between 300 and 4000 nm from sample to sample [14]. The metallic ferromagnet is insulated by a thin layer of SiO_2. Then two adjacent collinear coplanar waveguides (CPWs) are integrated to perform all-electrical spin-wave spectroscopy (AESWS) using a vector network analyzer (VNA) connected to the CPWs [29, 34, 35]. The power of the sinusoidal microwave field $\mathbf{h}_{rf}$ at frequency f is small enough to stay in the linear regime [36]. The antidot lattice resides in an in-plane magnetic field $\mathbf{H}$ applied under different angles η with respect to the leads of the CPWs [37]. The widths of the inner conductor lines w_{IC} and gaps of the CPWs define the relevant wave vectors k generated by the inhomogeneous $\mathbf{h}_{rf}$ via the spatially dependent torque $\gamma\mathbf{M}\times\mathbf{h}_{rf}$ where γ ($\mathbf{M}$) is the gyromagnetic ratio (magnetization) [29]. We measure spin-precession-induced voltage signals a, both, in reflection configuration on one of the two adjacent CPWs which we label CPW2, i.e., a_{22}, and in transmission configuration between the two adjacent coplanar wave guides CPW1 and CPW2, i.e., a_{12}. The separation s between inner conductors is varied in the experiments. To explain measured data, we perform micromagnetic simulations using the software package MicroMagus [38]. For simulations at $k=0$, we simulate one unit cell and use 2D periodic boundary conditions. If the excitation field $\mathbf{h}_{rf}$ is assumed to be homogeneous such simulation data are compared to experimental data a_{22} obtained with wide CPWs. Wave vector dependent dispersion relations $f(k)$ have been obtained from simulations considering a long stripe of unit cells with a local time-dependent excitation. The spatial extension of the excitations field h_{rf} is selected based on experimental parameters such as the inner conductor width of the CPWs. Dispersions relations are extracted from fast Fourier transformation of the data in two steps. Further details on the simulations and relevant parameters are given in detail in [29, 39].

14.3 Antidot Lattices in the Short Wavelength Limit: Bandgap Materials

14.3.1 Large-Period Antidot Lattice: Forbidden Frequency Gaps due to Bragg Reflection

In Fig. 14.2, we show spectra obtained at $\mu_0 H = 40$ mT on two ADLs with the same lattice constant $p = 800$ nm, but using two different sets of CPWs exhibiting different widths [39]. Thereby we address different wave vectors k. H is so large that M is close to saturation. In Figs. 14.2(a) and 14.2(c), dark color indicates spin-wave resonances in the signal a_{22} which is measured in reflection configuration on CPW2. To analyze these data further, it is instructive to consider micromagnetic simulations. From Figs. 14.3(a) and 14.3(b), we find that the various resonances seen in a_{22} follow the predicted angular dependencies. Based on Fig. 14.3(b), we label three relevant modes with i, ii, and iii. The remaining discrepancy between experimental and simulated data concerns mode i which we do not observe for the ADL with $p = 800$ nm. Evaluating spatially resolved spin-precession profiles (not shown), mode i is found to be an edge mode [40] residing very close to the edges of the nanoholes (cf. Fig. 14.6(b)). It is localized on a lateral length scale of 100 nm. Close to the edges the demagnetization effect makes the internal field H_{int} to take a minimum value. This is why mode i resides at small eigenfrequencies. We attribute the lack of mode i in the experimental data to inhomogeneous broadening due to unintentional edge roughness and variations in the hole diameters. These parameters vary from hole to hole in

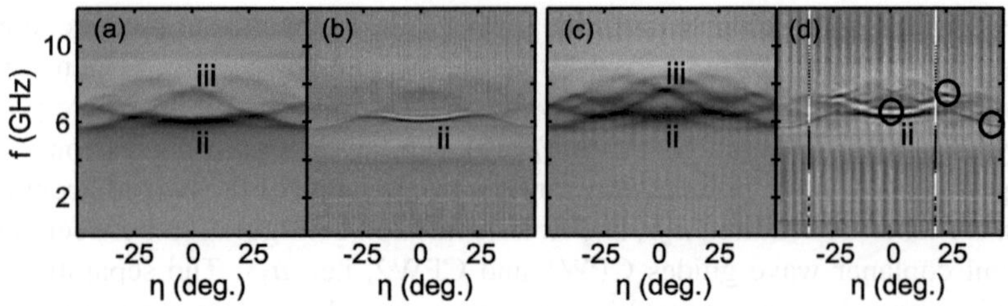

Fig. 14.2 Spectra measured at $\mu_0 H = 40$ mT for two ADLs with identical $p = 800$ nm using different CPWs: imaginary part of (**a**) a_{22} and (**b**) a_{12} using $w_{IC} = 2.0$ μm; imaginary part of (**c**) a_{22} and (**d**) a_{12} using $w_{IC} = 0.34$ μm. For the reflection data in (**a**) and (**c**), *dark color* corresponds to large spin-wave precession amplitude, i.e., a resonance. For the transmission data in (**b**) and (**d**), the *white-black-white* contrast change indicates an accumulated phase shift of spin waves which propagate between CPW1 and CPW2 [29]. Note that a pronounced oscillating contrast in (**b**) and (**d**) occurs only for few modes at specific η [highlighted by *circles* in (**d**)]. From (**a**) to (**c**) [(**b**) to (**d**)] The transferred wave vector k increases. *Vertical stripes* are artifacts

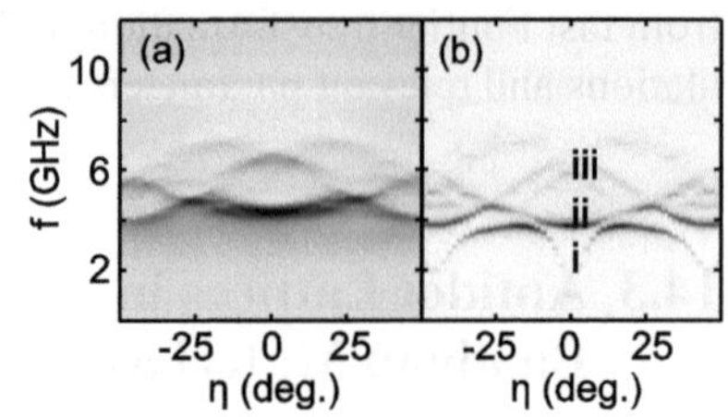

Fig. 14.3 Comparison between (**a**) measured imaginary part of (a_{22}) and (**b**) simulated data for an ADL with $p = 800$ nm at $\mu_0 H = 20$ mT. *Dark color* corresponds to a large spin-wave precession amplitude, i.e., a resonance

the real ADL and change the H_{int} dependent eigenfrequency locally. Since our data are taken over a large amount of nanoholes, a sharp resonance is not resolved. Following [24, 25, 31], modes ii and iii found at larger eigenfrequencies exhibit large spin-precession amplitudes in stripes between rows of nanoholes (cf. Fig. 14.7(b)) and in localized spots between pairs of nanoholes, respectively. The stripes are found to be formed perpendicular to **H** if η equals 0 and 45 degrees.

We observe that mode ii shifts to a larger eigenfrequency when we move in Fig. 14.2 from graph (a) to (c). This increase is not seen for mode iii. The different behavior is attributed to propagating and standing spin waves, respectively, formed in the antidot lattice. Propagating spin waves are dispersive and their eigenfrequency f vary as a function of k. In Figs. 14.2(b) and 14.2(d), we display spectra measured in transmission configuration, i.e., we excite spin waves using $\mathbf{h}_{\text{rf}}$ at the emitter CPW1 and measure the spin-precession induced voltage at the detector CPW2 using the VNA. The separation s amounts to 12.0 μm in (b) and 6.5 μm in (d). An oscillating black–white–black contrast indicates an accumulated phase shift of spin waves which propagate between CPW1 and CPW2 [14, 29, 34, 35]. Note that a pronounced oscillating contrast in (b) and (d) occurs only for few modes at specific η. In particular, mode ii exhibits such a contrast in (b) and (d) substantiating the propagating character suggested by the k dependent eigenfrequency shift extracted from

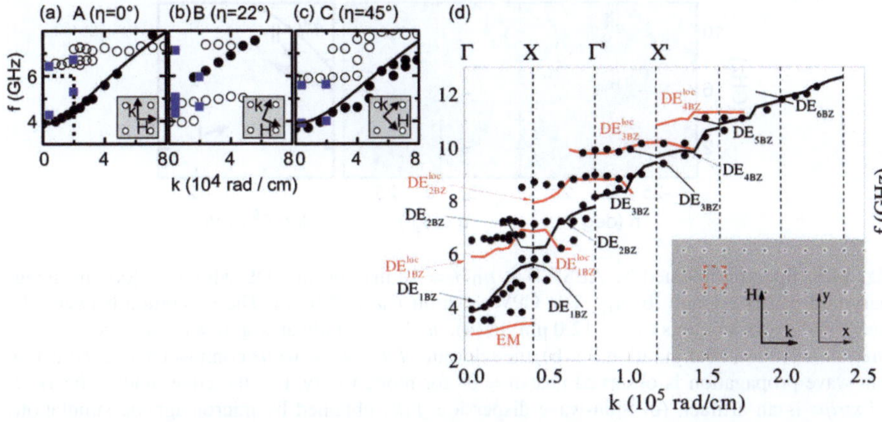

Fig. 14.4 **(a)–(c)** Spin-wave dispersions $f(k)$ for three selected angles η (A, B, and C) obtained by inelastic light scattering (*circles*) at $\mu_0 H = 20$ mT. The wave vector $k = k_\perp$ is always perpendicular to **H**. *Full* (*open*) *circles* highlight the propagating (standing) wave character of different modes. *Squares* refer to the AESWS data using different CPWs. *Full lines* are calculated dispersions for nanowires [41], where $k_\perp$ points along the nanowire and is varied, but the wave vector component $k_\parallel$ parallel to the field exhibits a fixed value: **(a)** $k_\parallel = \pi/600$ nm and **(c)** $k_\parallel = \sqrt{2}\pi/600$ nm. *Insets* illustrate the geometry used for the BLS experiments. From [29]. The *dotted lines* in **(a)** mark the wave-vector regime investigated in detail in Fig. 14.7. **(d)** Experimental data (*solid circles*) from detailed measurements performed on the ADL with $p = 800$ nm using inelastic light scattering at $\eta = 0$. Here the value of the transferred wave vector k extends over several BZs such that allowed minibands and forbidden frequency gaps at BZ boundaries are identified. *Solid lines* are theoretical curves. The *bold black line* highlights the fundamental mode which was indicated by the *solid line* in **(a)**. The *inset* shows an SEM image and relevant parameters. The *square* (*broken line*) illustrates the area of a unit cell of 800 × 800 nm^2. The labels refer to spin-precession profiles discussed in Chap. 15 by G. Gubbiotti et al. Taken from [16]

Figs. 14.2(a) and 14.2(c). The standing spin-wave mode iii does not show an oscillating contrast near $\eta = 0°$ as expected from the same former analysis. The different character of the modes has been evidenced by further experiments using inelastic light scattering on a nominally identical ADL (cf. Figs. 14.4(a)–14.4(c)). Such data have been obtained on thermally excited modes [29]. Propagating modes are highlighted by filled circles and lines. Lines remodel the stripe-like spin-wave modes and are spin-wave dispersions calculated for ferromagnetic nanowires of the appropriate effective width w_{eff} [29, 41]. The data measured on the fundamental mode of the ADL (filled circles) mostly follow the predicted functionality (solid lines) exhibiting a finite slope $v_{\mathrm{g}} = 2\pi\,\mathrm{d}f/\mathrm{d}k$. We observe a group velocity v_{g} of a few km/s over a broad range of k values. Modes with nearly constant eigenfrequency are summarized by open symbols.

Recently, the inelastic light scattering experiment at $\eta = 0°$ was extended to large transferred wave vectors k and a more detailed analysis of the spectra was performed. This provided the data of Fig. 14.4(d) [16]. Here, we display a repeated zone scheme and illustrate Brillouin zone (BZ) boundaries and equivalent Γ points by vertical dashed lines. The qualitative comparison with a more re-

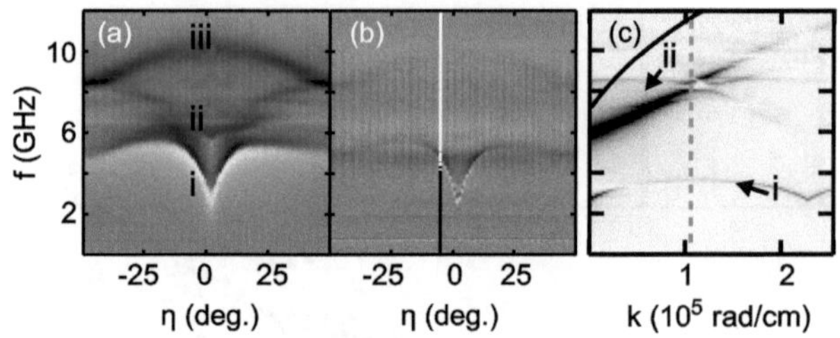

Fig. 14.5 Spectra obtained by AESWS at $\mu_0 H = 40$ mT for an ADL with $p = 300$ nm: imaginary part of (**a**) a_{22} and (**b**) a_{12}. The CPWs exhibit $w_{IC} = 2.0$ μm. The separation between the inner conductors amounts to $s = 12.0$ μm. In (**a**), *dark color* indicates spin-wave resonances measured in reflection configuration. In (**b**), the relevant *white–black–white* contrast change indicating spin-wave propagation is observed near $\eta = 0°$ for mode i only, i.e., the edge mode. The *vertical stripe* is an artifact. (**c**) Spin-wave dispersion $f(k)$ obtained by micromagnetic simulations at $\mu_0 H = 60$ mT and $\eta \approx 0°$ as a function of wave vector k. The formalism is detailed in [14]. *Dark color* indicates eigenmodes. The wave vector k is perpendicular to **H**. The *vertical dashed line* indicates the boundary of the first BZ at π/p. The calculated spin-wave dispersion for the Damon–Eshbach mode ($\mathbf{k} \perp \mathbf{M}$) of the unpatterned $Ni_{80}Fe_{20}$ film is shown for comparison (*solid line*). Adapted from [14]

fined model for spin-wave excitations in ADLs [16] substantiates that the fundamental mode does not follow a monotonous function for all k. Instead it exhibits a forbidden frequency gap at, e.g., $k = \pi/p$, i.e., the boundary of the first BZ in Fig. 14.4(d). The model calculations (solid lines) suggest that forbidden frequency gaps are opened due to Bragg reflection provoked by the periodic patterning of the $Ni_{80}Fe_{20}$ film with holes. The formalism allows one to define the Bloch wave vector. Several minibands are formed from propagating and standing (localized) modes in the ADL. Importantly, the width of the band gaps is found to reflect the amplitude of the spatially oscillating internal field and not directly the contrast of materials parameters between film and nanoholes. This is detailed further in Chap. 15 by G. Gubbiotti et al. In the following section, we address the artificial crystal behavior in ADLs with a small lattice constant.

14.3.2 Short-Period Antidot Lattice: Miniband Formation due to Coherent Coupling of Edge Modes

In Fig. 14.5(a), we depict the spin-wave resonances (dark color) measured on an ADL with $p = 300$ nm at $\mu_0 H = 40$ mT. Again we find various eigenmodes which vary characteristically as a function of the angle η. If compared to the data in Fig. 14.2(a) obtained on an ADL with $p = 800$ nm, the different eigenmodes cover a much larger frequency regime and exhibit more pronounced angular dependencies. This is due to the smaller lattice constant. In particular, we observe a prominent mode residing at about 3 GHz at $\eta = 0°$ in Fig. 14.5(a). The low

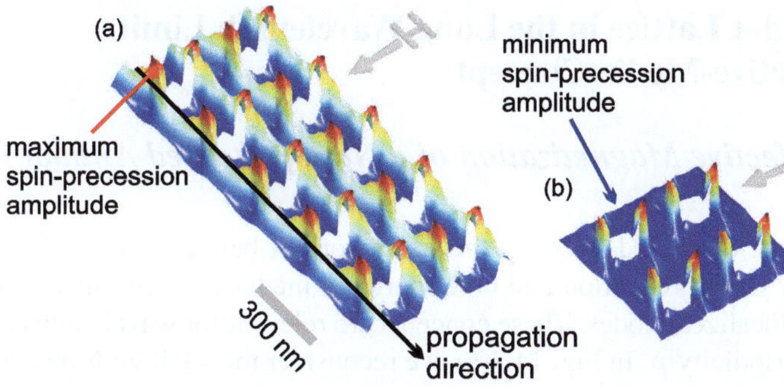

Fig. 14.6 Simulated spin-precession amplitudes of edge modes in an ADL with $p = 300$ nm when a field of $\mu_0 H = 40$ mT is applied as sketched: (**a**) $f = 2.9$ GHz and (**b**) $f = 4.5$ GHz. From (**a**) to (**b**) the field is rotated by 9 degrees. This change in η significantly reduces the edge-mode coupling from hole to hole. Adapted from [14]

eigenfrequency at $\eta = 0°$ and the characteristic field dependence are characteristic for an edge mode (mode i) (cf. Fig. 14.3(b)). The frequency increases to about 5 GHz at $\eta = 20°$. The appearance of the sharp resonance is striking. We attribute this to the coupling of the localized edge modes from hole to hole as illustrated in Fig. 14.6(a) using the spatial mapping of simulated spin-precession amplitudes. The coupling is substantiated by the transmission signal a_{12} depicted in Fig. 14.5(b). Here, a white-black-white contrast change is found at frequencies which indicate spin-wave propagation between CPW1 and CPW2 via edge modes. The simulated dispersion relation $f(k)$ shown in Fig. 14.5(c) evidences that the edge modes indeed form a miniband. The eigenfrequency is an oscillating function of k in the repeated zone scheme. The dipolar coupling through the interconnecting ferromagnetic material is so strong that the spin-wave dispersion $f(k)$ exhibits a rather steep slope v_g for k near the Γ point ($k = 0$). Measured and simulated values for the group velocity v_g agree well and amount to about 6 km/s near $\eta = 0°$ [14]. The formation of both the allowed miniband that supports propagating spin waves and the corresponding forbidden frequency gap is not due to Bragg reflection in the short-period antidot lattice of $p = 300$ nm but coherent coupling of modes which have a localized character for $p = 800$ nm. Note that for mode ii (fundamental mode) we do not resolve a propagation signal in Fig. 14.5(b). This is different from the results obtained on the ADL with $p = 800$ nm shown in Figures 14.2 and 14.4.

Interestingly the propagation signal for mode i is observed only for small values $|\eta|$ in Fig. 14.5(b). Micromagnetic simulations reveal that the coupling between localized edge modes is reduced by tilting the magnetic field away from the high-symmetry direction (cf. Fig. 14.6(b)). As a consequence with increasing angle $|\eta|$, the group velocity v_g is found to decrease significantly. Experimentally we find that the miniband formation is suppressed at $|\eta| \approx 20°$ [14]. This offers further control of artificial crystal behavior in magnonics.

14.4 Antidot Lattice in the Long Wavelength Limit: Effective-Media Concept

14.4.1 Effective Magnetization of a Nanopatterned Antidot Lattice

So far we have discussed ADLs in the framework of bandgap materials which are described by Bragg reflection and other periodic media concepts such as coherent coupling of localized modes. These concepts are relevant for wavelengths on the order of the periodicity p. In Fig. 14.7(a), we reconsider the ADL with $p = 800$ nm of Fig. 14.4, but we now focus on the dispersion relation $f(k)$ in the long wavelength limit, i.e., for small k close to $k = 0$ (cf. dotted frame in Fig. 14.4(a)). Making use of the time- and spatially resolved magneto-optical Kerr effect (MOKE) as outlined in detail in [15], the number of measured data points (open circles) has been increased significantly below $k_\perp = 1.5 \times 10^4$ rad/cm if compared to the data (solid circles) in Fig. 14.4(a). The MOKE experiments have allowed us to measure precisely the dispersion $f(k)$ at spin-wave wavelengths which are far longer than the periodicity $p = 800$ nm. By this means we address the ADL in the framework of an artificial material which is periodically modulated and described by homogenization and effective media concepts.

In Fig. 14.7(a), we compare the experimental data obtained on the ADL (open circles) with different theoretical predictions (lines), a micromagnetic simulation (gray-color coded data), and the dispersion relation $f(k)$ measured on the unpatterned $Ni_{80}Fe_{20}$ film (open squares). Strikingly, the nanowire model (dotted and dash-dotted lines) introduced in Fig. 14.4 fails to describe the data for $k_\perp \le 1.5 \times 10^4$ rad/cm, even if nanowire modes with $n \ge 2$, i.e., an increased number $(n-1)$ of nodal lines perpendicular to **H** (cf. Figs. 14.7(b)–14.7(d)), are considered. Most importantly, the nanowire-based dispersions $f(k)$ suggest a group velocity $v_g = 2\pi \mathrm{d}f/\mathrm{d}k = 0$ for small k whereas the measured data (open circles) exhibit a finite slope v_g even very close to $k = 0$. Simulations performed on the ADL considering the periodic boundary condition in a direction transverse to $k_\perp$ are able to reproduce the measured behavior (gray-color coded data in Fig. 14.7(a)). The behavior with $v_g \neq 0$ resembles the unpatterned-film dispersion (open squares). The dispersion measured on the film is reproduced well by the solid line calculated for a thin $Ni_{80}Fe_{20}$ film assuming the saturation magnetization $M_s = 770$ kA/m [42]. Using the same formalism, the measured ADL dispersion can be reproduced equally well if M_s is replaced by $M_s^* = 600$ kA/m (dashed line), i.e., an effective saturation magnetization. This value is reduced by 22 % if compared to $M_s = 770$ kA/m of the unpatterned film. In this approach, the ADL can be regarded as an effectively continuous magnetic medium for propagating spin waves which experience a renormalized magnetization. Strikingly the effective magnetization $M_s^* = 600$ kA/m is found to be smaller by 22 % whereas the periodic array of holes of diameter 120 nm removes only 1.8 % of the $Ni_{80}Fe_{20}$. The periodic patterning has thus created an artificial material with novel and unexpected properties which cannot be derived from the known properties of the individual sub-components.

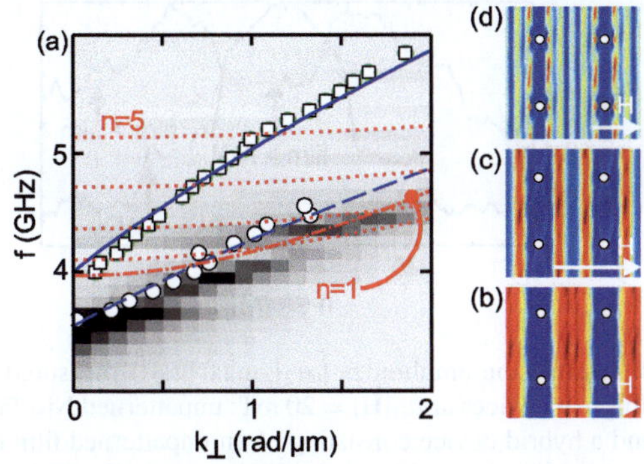

Fig. 14.7 (**a**) Measured and calculated spin-wave dispersions $f(k_\perp)$ for an unpatterned $Ni_{80}Fe_{20}$ film and an ADL with $p = 800$ nm at $\mu_0|\mathbf{H}| = 20$ mT and $\eta = 0$. $k_\perp$ is perpendicular to $\mathbf{H}$. *Open symbols* are data obtained by the magneto-optical Kerr effect experiment of [15] (*squares* – film, *circles* – ADL). The *solid* (*long-dashed*) *line* reflects a calculated plain-film dispersion assuming a saturation magnetization $M_s = 770$ kA/m ($M_s^* = 600$ kA/m). For comparison nanowire dispersions $f(k)$ are depicted: the *dotted and dashed-dotted lines* consider $M_s = 770$ kA/m and quantized wave-vector components $k_\parallel = n\pi/w_{\text{eff}}$ parallel to $\mathbf{H}$ ($w_{\text{eff}} = 600$ nm). The *gray-scale plot* is the spin-wave dispersion as obtained from wave-vector-resolved micromagnetic simulations. *Dark* (*bright*) *color* corresponds to large (small) SW amplitude. Spatial spin-wave profiles are obtained from micromagnetic simulations at $k = 0$: (**b**) $f = 3.7$ GHz ($n = 1$), (**c**) $f = 4.2$ GHz ($n = 2$), and (**d**) $f = 4.4$ GHz ($n = 3$). *Dark* (*bright*) *colors* correspond to small (large) spin-precession amplitude. The direction of $\mathbf{H}$ is indicated. From [15]

In [23–25, 29, 31], it has been argued that the fundamental mode excited in ADLs at $\eta = 0$ resembles confined stripes of large spin-precession amplitudes (cf. Fig. 14.7(b)). The stripes are separated from each other by the rows of nanoholes and periodic inhomogeneities of the internal field H_{int}. These findings motivated the nanowire-based model describing well the overall slope of the dispersion $f(k)$ which was measured in the first BZ (cf. Figs. 14.4(a) and 14.4(c)). Figure 14.7(a) substantiates that the model fails at long wavelengths (small k). The confinement by the rows of nanoholes and periodic inhomogeneities of H_{int} is no longer effective. Quantized wave-vector components $k_\parallel = n\pi/w_{\text{eff}}$ parallel to $\mathbf{H}$ do not play a role for the dispersion close to $k = 0$ (w_{eff} is the effective width of the confined stripe and $n = 1, 2, \ldots$). The data of Fig. 14.7(a) suggest that the stripes of large spin-precession amplitudes of Fig. 14.7(b) are coherently coupled in the long wavelength limit. Thereby the effectively continuous magnetic medium with $M_s^* = 600$ kA/m for the propagating spin waves is formed.

14.4.2 *Transmission of Spin Waves Across the Boundary of an Antidot Lattice*

It is now interesting to discuss the efficiency with which spin waves are injected into an ADL. Using AESWS we have studied the propagation of long-wavelength

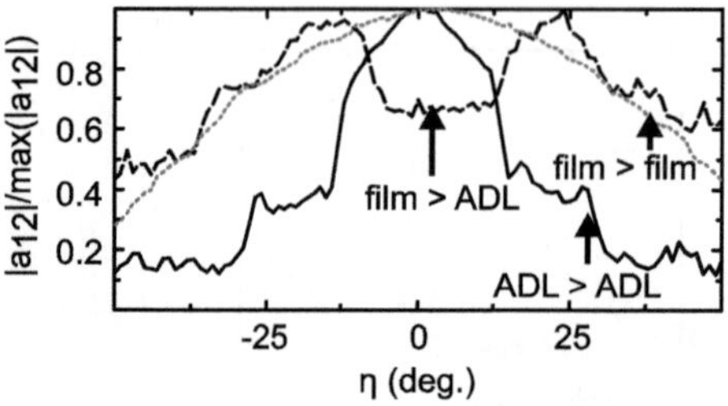

Fig. 14.8 Normalized transmission amplitudes $|a_{12}|/\max(|a_{12}|)$ measured by AESWS between CPW1 and CPW2 on different devices at $\mu_0|\mathbf{H}| = 20$ mT: unpatterned $Ni_{80}Fe_{20}$ film (film > film), ADL (ADL > ADL) and a hybrid device consisting of an unpatterned film connected to an ADL (film > ADL). These labels reflect the positions of CPW1 and CPW2 (CPW1 > CPW2). The width w_{IC} of the CPWs is 4 μm. From angle to angle and device to device the eigenfrequencies vary as we summarize here the signal strengths of the most prominent modes at a given η

spin waves through three different devices: (i) an unpatterned thin film, (ii) an ADL with $p = 800$ nm, and (iii) a hybrid device which consists of a film connected to an ADL being nominally identical to (ii). In Fig. 14.8, we summarize spin-precession induced voltages measured at detector CPWs for different angles η. The separation s between the inner conductors of the CPWs amounts always to 19.5 μm. For the homogeneous thin-film sample (i) the maximum signal a_{12} is observed at $\eta = 0$. At this angle the wave vector $\mathbf{k}$ is perpendicular to $\mathbf{M}$ which itself is parallel to $\mathbf{H}$. Consequently, a Damon–Eshbach-type spin-wave mode propagates between CPW1 and CPW2. For such a mode, the propagation velocity v_g is known to be comparably large leading to a large decay length $l = v_g \tau$ (τ is the relaxation time). The damping along the propagation path s being proportional to $\exp(-s/l)$ is thus small leading to a large a_{12}. An increased angle $|\eta|$ provokes a flattening of the relevant spin-wave dispersion relation $f(k)$ and thereby reduces v_g. This explains the observed behavior that a_{12} decreases with $|\eta|$ [29]. At $\eta = 50°$, a_{12} is reduced to 40 % if compared to $\eta = 0$. For the ADL sample (ii), we observe a functional dependence which is more complex compared with the unpatterned film. However, the overall tendency is the same in that a_{12} decreases with increasing $|\eta|$. At $\eta = 50°$, the signal strength is only about 15 % of the signal a_{12} at $\eta = 0$. The signal suppression with increasing $|\eta|$ is pronounced. A significantly different behavior is observed for sample (iii) which is a combination of samples (i) and (ii). In this sample the spin waves propagate across the interface between the film and the ADL. CPW2 detects the spin waves injected into the ADL. The spin waves propagate within the ADL for 10.7 μm before reaching CPW2. In this sample the maximum signal strength is no longer found at $\eta = 0°$. Instead, we get the maximum a_{12} for $15° < |\eta| < 25°$. At $\eta = 0$ and $50°$, we find almost the same reduced value of about 60 % of the maximum a_{12}. The hybrid device (iii) thus does not behave as the sum of the two sub-systems (i) and (ii). We attribute this discrepancy to the transmission coefficient t of the interface which deviates from one if dispersion relations $f(k)$ on both sides of the interface are different. A detailed analysis shows that t exhibits a pronounced angular and frequency dependence. At a fixed frequency of 4.4 GHz, we find the transmission coefficient t to vary from 0.5 ± 0.1 to 0.8 ± 0.1 when changing η from 0 to 20° [15]. At the same time, t varies roughly from 0.25 to 0.8 for eigenfrequencies f ranging from 4 to 5 GHz at $\eta = 0°$. Convincingly, the variation of t with f is remodeled if we consider the effective-medium concept of Sect. 14.4.1, i.e., the relevant spin-wave dispersion $f(k)$ based on the effective magnetization $M_s^* = 600$ kA/m for the ADL [15]. The metamaterials properties of the ADL thus enter the functionality of the magnonic hybrid device (iii) being a field-controlled semi-transparent mirror for spin waves.

14.5 Conclusions

We have discussed metamaterial properties of magnetic antidot lattices made from thin $Ni_{80}Fe_{20}$ films. Depending on the lattice constant ranging from 300 and 800 nm, we have observed the formation of allowed minibands due to coherent coupling of localized modes and Bragg reflection of propagating spin waves, respectively, in the short-wavelength limit. In the long-wavelength limit, the array of periodic nanoholes is found to behave as an artificial magnetic thin film with an effective saturation magnetization being significantly reduced if compared to the unpatterned material. An external magnetic field allows us to further control the band structure parameters and manipulate the spin-wave propagation. By this means, we vary the coupling strength between localized modes, the damping parameter, and the propagation direction. The field-controlled metamaterial properties of ADLs offer novel perspectives for nanoscale microwave circuits allowing for data transmission and processing with propagating spin waves in the GHz frequency regime.

Acknowledgements We thank C.H. Back, H.G. Bauer, M. Becherer, P. Berberich, D.V. Berkov, B. Botters, F. Brandl, G. Carlotti, L. Giovannini, G. Gubbiotti, M. Krawczyk, V.V. Kruglyak, M. Madami, S. Mamica, F. Montoncello, F. Nizzoli, T. Rapp, D. Schmitt-Landsiedel, T. Schwarze, M.L. Sokolovskyy, S. Tacchi, P. Weiser, G. Woltersdorf, and R. Zivieri for discussions, excellent experimental and theoretical support. The research leading to these results has received funding from the European Community's Seventh Framework Programme (FP7/2007-2013) under Grant No. 228673 (MAGNONICS) and the German Excellence Cluster Nanosystems Initiative Munich (NIM).

References

1. J.D. Joannopoulos, S.G. Johnson, J.N. Winn, R.D. Meade, *Photonic Crystals: Molding the Flow of Light*, 2nd edn. (Princeton Univ. Press, Princeton, 2008)
2. N. Engheta, R.W. Ziolkowski, *Metamaterials: Physics and Engineering Explorations* (Wiley/IEEE Press, New York, 2006)
3. A.A. Serga, A.V. Chumak, B. Hillebrands, J. Phys. D, Appl. Phys. **43**(26), 264002 (2010)
4. G. Gubbiotti, S. Tacchi, M. Madami, G. Carlotti, A.O. Adeyeye, M. Kostylev, J. Phys. D, Appl. Phys. **43**(26), 264003 (2010)
5. J. Topp, D. Heitmann, M.P. Kostylev, D. Grundler, Phys. Rev. Lett. **104**(20), 207205 (2010)

6. S. Tacchi, M. Madami, G. Gubbiotti, G. Carlotti, S. Goolaup, A.O. Adeyeye, N. Singh, M.P. Kostylev, Phys. Rev. B **82**, 184408 (2010)
7. J. Topp, G. Duerr, K. Thurner, D. Grundler, Pure Appl. Chem. **83**, 1989 (2011)
8. J. Topp, S. Mendach, D. Heitmann, M. Kostylev, D. Grundler, Phys. Rev. B **84**, 214413 (2011)
9. G. Gubbiotti, S. Tacchi, G. Carlotti, N. Singh, S. Goolaup, A.O. Adeyeye, M. Kostylev, Appl. Phys. Lett. **90**(9), 092503 (2007)
10. M. Kostylev, P. Schrader, R.L. Stamps, G. Gubbiotti, G. Carlotti, A.O. Adeyeye, S. Goolaup, N. Singh, Appl. Phys. Lett. **92**(13), 132504 (2008)
11. S. Tacchi, M. Madami, G. Gubbiotti, G. Carlotti, H. Tanigawa, T. Ono, M.P. Kostylev, Phys. Rev. B **82**(2), 024401 (2010)
12. S. Tacchi, F. Montoncello, M. Madami, G. Gubbiotti, G. Carlotti, L. Giovannini, R. Zivieri, F. Nizzoli, S. Jain, A.O. Adeyeye, N. Singh, Phys. Rev. Lett. **107**, 127204 (2011)
13. Y. Gulyaev, S. Nikitov, L. Zhivotovskii, A. Klimov, P. Tailhades, L. Presmanes, C. Bonningue, C. Tsai, S. Vysotskii, Y. Filimonov, JETP Lett. **77**(10), 567 (2003)
14. S. Neusser, G. Duerr, S. Tacchi, M. Madami, M.L. Sokolovskyy, G. Gubbiotti, M. Krawczyk, D. Grundler, Phys. Rev. B **84**, 094454 (2011)
15. S. Neusser, H.G. Bauer, G. Duerr, R. Huber, S. Mamica, G. Woltersdorf, M. Krawczyk, C.H. Back, D. Grundler, Phys. Rev. B **84**, 184411 (2011)
16. R. Zivieri, S. Tacchi, F. Montoncello, L. Giovannini, F. Nizzoli, M. Madami, G. Gubbiotti, G. Carlotti, S. Neusser, G. Duerr, D. Grundler, Phys. Rev. B **85**, 012403 (2012)
17. C. Yu, M.J. Pechan, G.J. Mankey, Appl. Phys. Lett. **83**(19), 3948 (2003)
18. C. Yu, M.J. Pechan, W.A. Burgei, G.J. Mankey, J. Appl. Phys. **95**, 6648 (2004)
19. S. McPhail, C.M. Gürtler, J.M. Shilton, N.J. Curson, J.A.C. Bland, Phys. Rev. B **72**(9), 094414 (2005)
20. M.J. Pechan, C. Yu, R.L. Compton, J.P. Park, P.A. Crowell, J. Appl. Phys. **97**(10), 10J903 (2005)
21. M. Yu, L. Malkinski, L. Spinu, W. Zhou, S. Whittenburg, J. Appl. Phys. **101**(9), 09F501 (2007)
22. O.N. Martyanov, V.F. Yudanov, R.N. Lee, S.A. Nepijko, H.J. Elmers, R. Hertel, C.M. Schneider, G. Schönhense, Phys. Rev. B **75**(17), 174429 (2007)
23. M. Kostylev, G. Gubbiotti, G. Carlotti, G. Socino, S. Tacchi, C. Wang, N. Singh, A.O. Adeyeye, R.L. Stamps, J. Appl. Phys. **103**(7), 07C507 (2008)
24. S. Neusser, B. Botters, D. Grundler, Phys. Rev. B **78**, 054406 (2008)
25. S. Neusser, B. Botters, M. Becherer, D. Schmitt-Landsiedel, D. Grundler, Appl. Phys. Lett. **93**(12), 122501 (2008)
26. S. Neusser, D. Grundler, Adv. Mater. **21**(28), 2927 (2009)
27. S. Tacchi, M. Madami, G. Gubbiotti, G. Carlotti, A. Adeyeye, S. Neusser, B. Botters, D. Grundler, IEEE Trans. Magn. **46**(2), 172 (2010)
28. S. Tacchi, M. Madami, G. Gubbiotti, G. Carlotti, A.O. Adeyeye, S. Neusser, B. Botters, D. Grundler, IEEE Trans. Magn. **46**(6), 1440 (2010)
29. S. Neusser, G. Duerr, H.G. Bauer, S. Tacchi, M. Madami, G. Woltersdorf, G. Gubbiotti, C.H. Back, D. Grundler, Phys. Rev. Lett. **105**(6), 067208 (2010)
30. H. Ulrichs, B. Lenk, M. Munzenberg, Appl. Phys. Lett. **97**, 092506 (2010)
31. C.L. Hu, R. Magaraggia, H.Y. Yuan, C.S. Chang, M. Kostylev, D. Tripathy, A.O. Adeyeye, R.L. Stamps, Appl. Phys. Lett. **98**(26), 262508 (2011)
32. V.V. Kruglyak, S.O. Demokritov, D. Grundler, J. Phys. D, Appl. Phys. **43**(26), 264001 (2010)
33. B. Lenk, H. Ulrichs, F. Garbs, M. Munzenberg, Phys. Rep. **507**(4–5), 107 (2011)
34. M. Bailleul, D. Olligs, C. Fermon, S.O. Demokritov, Europhys. Lett. **56**(5), 741 (2001)
35. M. Bailleul, D. Olligs, C. Fermon, Appl. Phys. Lett. **83**(5), 972 (2003)
36. J. Podbielski, D. Heitmann, D. Grundler, Phys. Rev. Lett. **99**(20), 207202 (2007)
37. J. Podbielski, F. Giesen, M. Berginski, N. Hoyer, D. Grundler, Superlattices Microstruct. **37**, 341 (2005)
38. D.V. Berkov, N.L. Gorn, Micromagus – software for micromagnetic simulations (2008). www.micromagus.de

39. S. Neusser, Spin waves in antidot lattices: from quantization to magnonic crystals, Ph.D. thesis, Technische Universität München (2011)
40. J. Jorzick, S.O. Demokritov, B. Hillebrands, M. Bailleul, C. Fermon, K.Y. Guslienko, A.N. Slavin, D.V. Berkov, N.L. Gorn, Phys. Rev. Lett. **88**(4), 047204 (2002)
41. K.Y. Guslienko, S.O. Demokritov, B. Hillebrands, A.N. Slavin, Phys. Rev. B **66**(13), 132402 (2002)
42. B. Kalinikos, A. Slavin, J. Phys. C, Solid State Phys. **19**, 7013 (1986)

Chapter 15
Spin Wave Band Structure in Two-Dimensional Magnonic Crystals

G. Gubbiotti, S. Tacchi, M. Madami, G. Carlotti, R. Zivieri, F. Montoncello, F. Nizzoli, and L. Giovannini

Abstract We present a combined experimental and theoretical study of the spin wave band structure in two-dimensional magnonic crystals consisting of square arrays of either circular permalloy dots (disks) or antidots (holes). The spin wave dispersion has been measured by means of Brillouin light scattering spectroscopy, spanning the wave vector over several Brillouin zones in the reciprocal space of the artificial crystals. The experimental data are satisfactorily interpreted thanks to band structure calculations carried out using the dynamical matrix method. In the case of the array of disks, the frequency dispersion of the different eigenmodes in

G. Gubbiotti (✉)
Sede di Perugia, c/o Dipartimento di Fisica, Istituto Officina dei Materiali del CNR (CNR-IOM), Via A. Pascoli, 06123 Perugia, Italy
e-mail: gubbiotti@fisica.unipg.it

G. Gubbiotti · S. Tacchi · M. Madami · G. Carlotti
CNISM-Unità di Perugia and Dipartimento di Fisica, Università di Perugia, Via Pascoli, 06123 Perugia, Italy

S. Tacchi
e-mail: tacchi@fisica.unipg.it

M. Madami
e-mail: madami@fisica.unipg.it

G. Carlotti
e-mail: carlotti@fisica.unipg.it

R. Zivieri · F. Montoncello · F. Nizzoli · L. Giovannini
Dipartimento di Fisica, Università di Ferrara and CNISM, Via Saragat 1, 44122 Ferrara, Italy

R. Zivieri
e-mail: zivieri@fe.infn.it

F. Montoncello
e-mail: montoncello@fe.infn.it

F. Nizzoli
e-mail: nizzoli@fe.infn.it

L. Giovannini
e-mail: giovannini@fe.infn.it

S.O. Demokritov, A.N. Slavin (eds.), *Magnonics*, Topics in Applied Physics 125,
DOI 10.1007/978-3-642-30247-3_15, © Springer-Verlag Berlin Heidelberg 2013

any direction in the reciprocal space can be explained introducing the concept of a bidimensional effective wave vector. In the case of the antidot array, two families of propagating modes, having extended and localized character, exhibit bandgaps at Brillouin zone boundaries. The bandgap formation is discussed in terms of Bragg reflection as well as of the inhomogeneity of the internal magnetic field experienced by precessing spins.

15.1 Introduction

Magnonic Crystals (MCs) constitute a new class of metamaterials with periodically modulated magnetic properties where collective spin excitations, i.e., spin waves (SWs), can propagate [1–8]. The magnonic density of states is characterized by the presence of frequency ranges where SW propagation is allowed, alternated with forbidden bandgaps. This happens in analogy to the propagation of electromagnetic (elastic) waves in artificial materials with periodically modulated dielectric constant (elastic properties), namely photonic [9] (phononic [10]) crystals. However, the wavelengths of SWs are shorter than those of light in the GHz frequency range, so that MCs offer better prospects for miniaturization of devices in the microwave range of frequencies, with the additional advantage that frequency position and width of the band gap are tunable by the applied magnetic field [11, 12]. This provides an unique opportunity to design and exploit a new generation of spin logic devices, filters and waveguides operating in the GHz frequency range [13, 14], as well as extremely sensitive sensor for detecting magnetic fields [15]. Whatever is the device application, knowledge of the magnonic band structure of a specific MC is preliminary to any desired application.

Up to now, experimental studies have been focused on dense arrays of identical elements (stripes and dots) [16–27] and antidot arrays where circular holes are etched into a continuous ferromagnetic film. A few experimental works have also dealt with bicomponent MCs consisting of stripes of two different ferromagnetic materials (cobalt and permalloy, for example) [28, 29] in direct contact or array of ferromagnetic rods embedded into a matrix of another ferromagnet [30]. All these studies demonstrated the existence of dispersive Bloch waves whose periodicity in the reciprocal space matches the width of the Brillouin zone. Moreover, since in a magnetic system the magnetic order may not coincide with the geometrical one, this can be exploited to obtain reprogrammable (reconfigurable) magnetic ground states and consequently reprogrammable dynamic response of the system. To this respect, it has been demonstrated that by applying a proper magnetic field sequence to an array of stripes it is possible to switch between ferromagnetic (F) and antiferromagnetic (AF) ground states and the consequent SW propagation control [31–33]. Very recently, Ding et al. [34, 35] have showed that MCs represent an excellent model medium with controlled disorder on a crystal lattice while N. Polushkin [36] proposed that a frequency shift of the gap for spin-wave propagation can be obtained by a uniform dc current flowing across the MC.

In this chapter, we review experiments and theoretical calculations of the spin wave band structure in two qualitatively different model systems of 2D MCs. The first system consists of a periodic array of closely-spaced circular permalloy nanomagnets (disks). Here, the discrete frequency levels corresponding to stationary spin waves of each disk become finite-amplitude bands [37, 38], similar to what happens to the electronic bands in solids in the tight-binding approximation [39]. In the present case, the degeneration in energy is removed by dynamical dipolar interaction which couples individual resonances through air gaps [8].

The second system is formed by an array of antidots (ADs), i.e., circular holes etched into a continuous permalloy film. Here the band structure can be considered similar to the nearly-free-electron approximation in solids [39], where forbidden gaps appear in the electronic energy as a result of Bragg diffraction at the Brillouin-zone boundaries. Propagation of SWs in 2D AD arrays of nanopatterned holes etched into a continuous ferromagnetic film has been extensively investigated in the last years because of the possibility to control SW propagation on the nanoscale [40–48].

15.2 Sample Details and Experiment

As stated in the introduction, here we present the results relative to two different kind of samples: a dense array of NiFe disks, fabricated using deep ultraviolet lithography [24], and a NiFe antidot array fabricated using focused ion-beam [48]. The former sample (Sample #1) consists of a large array of 50 nm thick $Ni_{80}Fe_{20}$ circular dots having a diameter $\delta = 600$ nm and arranged in a square matrix with interdot separation of $\Delta = 55$ nm (period $a = \Delta + \delta = 655$ nm). This corresponds to a square first BZ of side $\pi/a = 4.80 \times 10^4$ rad/cm. The second sample (Sample #2) consists of a Permalloy ($Ni_{80}Fe_{20}$, Py) film, 22 nm thick, that was periodically patterned using focused ion-beam into a square lattice with circular holes of diameter $\delta = 120 \pm 30$ nm forming an AD array of periodicity $a = 800$ nm. This corresponds to a square first BZ of side $\pi/a = 3.93 \times 10^4$ rad/cm. Scanning electron microscopy images of the two samples are shown in Fig. 15.1.

Brillouin light scattering experiments were performed in the backscattering geometry by using a Sandercock-type high contrast and high resolution $(3 + 3)$ tandem Fabry–Pérot interferometer. About 200 mW of a solid state laser light operating at $\lambda = 532$ nm wavelength was focused on the sample surface. The sample was mounted on a two-axis goniometer which allows choosing a specified angle of incidence of light (θ) as well as rotating the sample around the surface normal (azimuthal rotation, ϕ), with an accuracy of 1°. By changing θ, it is possible to select the magnitude of the in-plane component $\boldsymbol{q}$ of the SW wavevector $[q = (4\pi/\lambda)\sin\theta]$ entering into the scattering process, while the variation of the azimuthal angle ϕ, measured with respect to the magnetic field direction, corresponds to change the in-plane direction of $\boldsymbol{q}$. For Sample #1, the SW frequency dispersion was studied along the principal high-symmetry directions of the first BZ, i.e., ΓX,

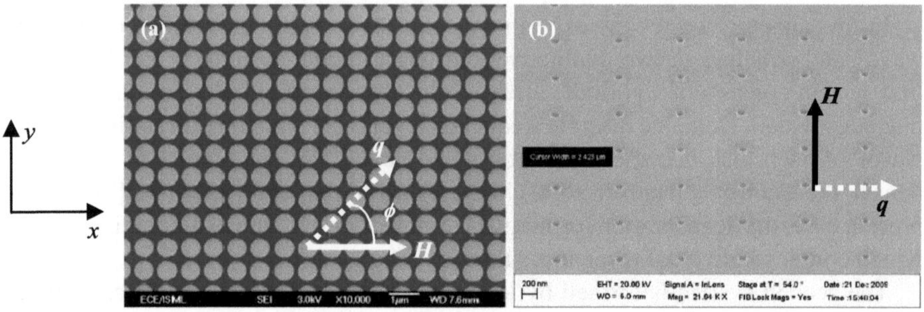

Fig. 15.1 Scanning electron microscopy image of Sample #1 (**a**) and Sample #2 (**b**). The directions of the applied magnetic field and the wavevector are shown by the *arrows*

ΓY, XM, and YM. Note that the first two cases correspond to what in the literature are called Magnetostatic Backward Volume Wave (MSBVW) geometry and Magnetostatic Surface Wave (MSSW) geometry, respectively. The latter is also known as Damon–Eshbach (DE) geometry [49]. A permanent magnet providing a magnetic field of intensity $H = 1.0$ kOe directed along the x-axis (Fig. 15.1) was mounted on the same goniometer, so that the relative orientation between field and the array does not change during the measurements and magnetostatic effects are the same for all the measured dispersion curves. For the AD array, spectra were measured in the DE scattering geometry, where the wavevector $\boldsymbol{q}$ is perpendicular to $\boldsymbol{H}$. We have explored wave vectors up to 2.2×10^5 rad/cm, which corresponds to investigating the reciprocal space up to the 6th Brillouin zone (BZ). A magnetic field of magnitude $H = 200$ Oe, sufficient to saturate the sample, was applied in the sample plane along the y-axis.

15.3 Theoretical Description: Dynamical Matrix Method Applied to 2D MCs

An array of magnetic particles or antidots can be represented by the repetition of a supercell, containing one or more magnetic elements or antidots, repeated at all points of a 2D Bravais lattice. We assume that the ground state is the same for all the supercells of the array (periodic boundary conditions), and that the dynamic magnetization fulfills the Bloch theorem:

$$\delta \boldsymbol{m}_K(\boldsymbol{r} + \boldsymbol{R}) = \delta \boldsymbol{m}_K(\boldsymbol{r}) e^{i\boldsymbol{K}\cdot\boldsymbol{R}} \tag{15.1}$$

where $\boldsymbol{K}$ is a Bloch wavevector and $\boldsymbol{R}$ is a Bravais lattice vector.

The DMM is based on the discretization of the problem's domain, i.e., the subdivision of the magnetic fraction of a supercell into small regular cells ("micromagnetic cells"), so that the magnetization in each cell can be assumed constant [38]. Due to the periodic boundary conditions and the Bloch rule, the number of independent variables in the system is limited to the number N of micromagnetic cells in the supercell, also for an infinite array. The ground state of the system can be calculated with a standard micromagnetic package which minimizes the total energy density due to Zeeman, exchange, and dipolar interaction:

$$E_{\text{Zeeman}} = -M_s \boldsymbol{H} \cdot \sum_{\boldsymbol{R}} \sum_{j=1}^{N} \boldsymbol{m}(\boldsymbol{r}_j + \boldsymbol{R}), \tag{15.2a}$$

$$E_{\text{exch}} = A \sum_{\boldsymbol{R}} \sum_{j=1}^{N} \sum_{n \in \{\mathbf{n.n.}\}} \frac{1}{d_{jn}^2} \left[1 - \boldsymbol{m}(\boldsymbol{r}_j + \boldsymbol{R}) \cdot \boldsymbol{m}(\boldsymbol{r}_n)\right], \tag{15.2b}$$

$$E_{\text{dip}} = \frac{M_s^2}{2} \sum_{\substack{\boldsymbol{R},\boldsymbol{R}' \\ (\boldsymbol{R} \neq \boldsymbol{R}')}} \sum_{k,j=1}^{N} \boldsymbol{m}(\boldsymbol{r}_k + \boldsymbol{R}) \bar{N}\left(\boldsymbol{r}_k + \boldsymbol{R}, \boldsymbol{r}_j + \boldsymbol{R}'\right) \boldsymbol{m}\left(\boldsymbol{r}_j + \boldsymbol{R}'\right). \tag{15.2c}$$

These expressions contain sums running over the micromagnetic cells (k, j) of a given supercell and sums running over the supercells $(\boldsymbol{R})$. As it is customary in numerical approaches, the last sums are calculated by truncation; in particular, we start from the interaction of a supercell with its nearest neighbors, then we sum the terms corresponding to farther and farther elements; when the contributions become negligible with respect to the machine accuracy, the sums are terminated. It turns out that, typically, particles within $R < 3a$, with a the supercell size, contribute to the sums. This is also consistent with experimental results showing that magnetic elements separated by a distance greater than the dot size do not significantly interact [50]. The sum over n includes the nearest neighbors (**n.n.**) of $\boldsymbol{r}_j + \boldsymbol{R}$, A is the exchange stiffness constant, and d_{jn} is the distance between the centers of the jth and the nth micromagnetic cell (see (15.2b)); it can possibly involve micromagnetic cells belonging to different supercells in contact. In (15.2c), $\bar{N}$ is the demagnetizing tensor.

The dynamics of the magnetization is then calculated by expanding the equations of motion to the first order in the magnetization fluctuations, written in terms of fluctuation of the azimuthal and polar angles in each cell $\delta\theta_i \delta\varphi_i$ $(i = 1, 2, \ldots, N)$, respectively. This yield to the following matrix equation

$$\mathrm{A}\mathbf{v} = \lambda \mathrm{B}\mathbf{v} \tag{15.3a}$$

where B is the dynamical matrix, i.e., the Hessian of the energy density calculated at equilibrium, A is given by:

$$\mathrm{A} = \begin{bmatrix} 0 & i\sin\phi_1 & 0 & 0 & \ldots \\ -i\sin\phi_1 & 0 & 0 & 0 & \ldots \\ 0 & 0 & 0 & -i\sin\phi_2 & \ldots \\ 0 & 0 & i\sin\phi_2 & 0 & \ldots \\ \ldots & \ldots & \ldots & \ldots & \ldots \end{bmatrix}, \tag{15.3b}$$

v is the vector of the variables:

$$\mathbf{v} = \begin{bmatrix} \delta\theta_1 \\ \delta\phi_1 \\ \delta\theta_2 \\ \delta\phi_2 \\ \cdots \end{bmatrix}, \tag{15.3c}$$

and $\lambda = \gamma/M_s\omega$ is the eigenvalue with γ the gyromagnetic ratio and ω the angular frequency. Equation (15.3a) has the form of a Hermitian generalized eigenvalue problem that can be efficiently solved by numerical procedures exploiting the symmetries of the matrices. The eigenvalues provide the frequencies of the modes, and the eigenvectors allow calculating the corresponding profiles of the magnetization.

Regarding the classification of collective normal modes in arrays of dots, since modes maintain a symmetry character similar to those found for the isolated dot, they can be labeled in the same way, with reference to the real part of the z-component of the dynamic magnetization. Depending on the number m (n) of nodes perpendicular (parallel) to the direction of the magnetization, the modes are classified as Backward-like modes mBA (Damon–Eshbach-like modes nDE); modes with mixed character $m\text{BA} \times n\text{DE}$ are also found. Moreover, the mode with no nodes is classified as fundamental mode (F), while modes with dynamic magnetization localized at the ends of the particle are labeled nEM, depending on the number of nodes n. For the classification of collective normal modes in arrays of ADs see the next paragraph.

15.4 Results and Discussion

15.4.1 Sample #1: Bidimensional Array of Disks

Two sequences of BLS spectra relative to Sample #1 are reported in Fig. 15.2. The spectra are recorded along different paths within the 2D 1BZ of the artificial crystal, as shown by the red arrows presented in the figure insets.

It can be seen that the spectra exhibit a strong frequency and intensity evolution of peaks when passing from Y to M to X (left figure) as well as from Γ to M. Interestingly, the spectra measured at the M points are characterized by a larger number of peaks with respect to those measured at other k-space points because at this point modes usually not active in Brillouin scattering experiments give rise to an appreciable signal. As an example, at point M, the BLS cross-section of a totally antisymmetric mode like the $1\text{BA} \times 1\text{DE}$ mode, whose spatial profile is shown in Fig. 15.4, becomes appreciable due to a non-vanishing momentum transfer in both X and Y directions.

In Fig. 15.3, the dispersion curves of the most representative modes are reported along the symmetry directions of the 1BZ. A very good agreement between experimental frequencies and calculated curves has been obtained. Comparing the measured spectra to the previously described results of the calculations, we could label

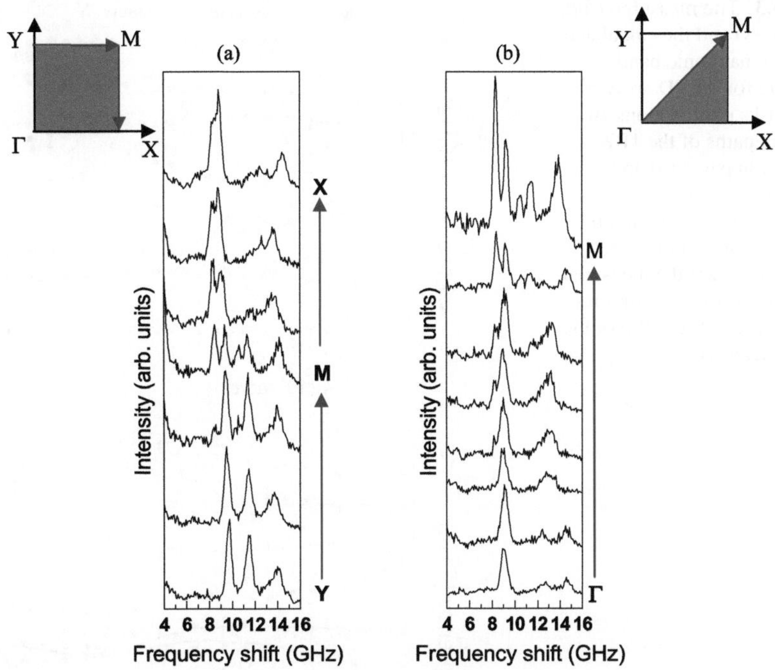

Fig. 15.2 Measured sequences of the anti-Stokes Brillouin spectra taken along the YMX direction (panel (**a**)) and the ΓM direction (panel (**b**)) of the 1BZ

each of the detected modes and confirm the theoretical findings about the modes cross-section. The calculated spatial profiles (inside a single dot) of some representative modes are depicted in Fig. 15.4.

First of all, we discuss the results obtained for the SW dispersion measured along the ΓYMX path. Since the width of each magnonic band is proportional to the mean square dynamic magnetization inside a single dot, the largest dispersion, of about 2 GHz, has been observed for the F mode. For higher order modes, instead, the amplitude of the magnonic band rapidly decreases by increasing the number of oscillations within the single dot. As it can be seen in Fig. 15.3, the measured frequency of the F mode reflects the properties of the dipole–exchange mode of the continuous film (open circles), in the whole first BZ. However, its frequency is significantly downshifted with respect to the spin wave of the continuous film because of the static demagnetizing field in the array along the applied field direction (x-direction). As it has been explained in a previous work [24], the behavior of dispersion curves can be explained introducing an effective wave vector $\boldsymbol{k}^{\mathrm{eff}}$, that includes and replaces the band index, i.e., the mode type, and Bloch wave vector. In the extended zone scheme, $\boldsymbol{k}^{\mathrm{eff}}$ can be defined as:

$$\boldsymbol{k}^{\mathrm{eff}} = \boldsymbol{K} + P(m)\left[m + \frac{1-P(m)}{2}\right]\frac{\pi}{a}\hat{\mathbf{x}} + P(n)\left[n + \frac{1-P(n)}{2}\right]\frac{\pi}{a}\hat{\mathbf{y}} \tag{15.4}$$

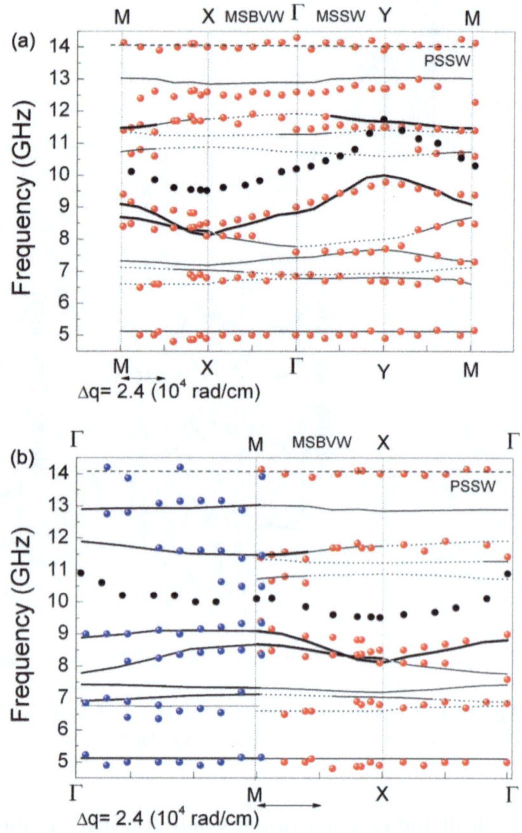

Fig. 15.3 The measured (*full red circles*) and the calculated (*curves*) magnonic band structure for the 2D array of NiFe disks plotted along two different paths of the 1BZ (ΓYMX in panel (**a**) and ΓMX in panel (**b**), respectively) represented by the *red arrows* in Fig. 15.2. *Open circles* are the measured frequency for the continuous (unpatterned) NiFe film of the same thickness

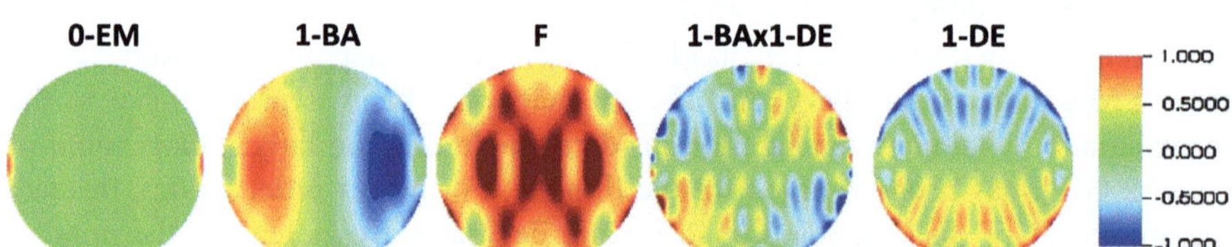

Fig. 15.4 Calculated out-of-plane component of the dynamical magnetization of most relevant modes for a magnetic field of 1.0 kOe applied in the horizontal direction

where $P(i)$ is the parity function ($= +1, -1$ for i even or odd, respectively), $m\mathrm{BA} \times n\mathrm{DE}$ is the mode type, and $\boldsymbol{K}$ is assumed to vary in the reduced BZ. The effective wavevector describes the overall oscillation of the magnetization on the array because it takes into account the oscillation within the dot due to the mode character (second and third terms in (15.4)) and the changing between adjacent dots due to the Bloch wave vector (first term). In disks coupled by dynamic dipolar interaction, the frequency dispersion as a function of $\boldsymbol{k}^{\mathrm{eff}}$ exhibits a behavior analogous to that of the continuous film, where mode frequency increases (decreases), when the modulus of the wavevector increases in a direction perpendicular (parallel) to the applied magnetic field, corresponding to the MSSW (MSBVW) geometry. For example, the F mode at Γ has $\boldsymbol{k}^{\text{eff}} = 0$, while at Y, $|k_x^{\text{eff}}| = 0$ and $|k_y^{\text{eff}}| = K_y = \pi/a$: hence, from Γ to Y, $|k_y^{\text{eff}}|$ increases and this implies a frequency increase. Conversely, from Γ to X, $|k_x^{\text{eff}}|$ increases and this implies a frequency decrease because $\boldsymbol{k}^{\text{eff}}$ is now parallel to $\boldsymbol{H}$. Finally, along the ΓYM path, the frequency of the F mode increases from Γ to Y, while it decreases from Y to M when $\boldsymbol{k}^{\text{eff}}$ is parallel to $\boldsymbol{H}$; correspondingly, the modulus $|\boldsymbol{k}^{\text{eff}}|$ increases from 0 to $\pi\sqrt{2/a}$. The frequency dispersion of the mBA and nDE modes can be explained in the same way. In such a case, however, due to the oscillations of the dynamic magnetization inside the single dot, the modulus of the effective wavevector depends on the number of nodal surfaces.

Looking at the SW dispersion measured along the ΓMX path (panel (b) of Fig. 15.3), it is interesting to notice that the measured and calculated dispersion of F mode along the ΓM high-symmetry direction has less pronounced amplitude with respect to that measured along the XM and YM high-symmetry directions. This is due to a compensation frequency effect associated to the fact that, moving from Γ to M, there is a simultaneous increment of the two in-plane components of the SW wavevector with opposite effects on the mode frequency dispersions (see panel (a) of Fig. 15.3 for comparison). A different behavior is observed for the 1DE and 1BA modes whose frequency oscillation amplitudes are more pronounced with respect to those measured in the XM and YM. In such a case, the increase of wavevector, either in the MSBVW or in the MSSW scattering configuration, has the same effect, i.e., an increase of frequency for 1BA, a decrease of frequency for 1DE: hence, the joint increase along ΓM results in an amplification of those trends.

The physical mechanism which underlines the evolution of the spin-wave frequency of the various modes can also be understood considering the behavior of the dynamic dipolar coupling along both the x and y directions, and taking into account that an increase (decrease) of the dipolar energy induces an increase (decrease) of the spin-wave frequency [8].

15.4.2 Sample #2: Bidimensional Array of Circular Holes

Figure 15.5 shows the measured SW frequencies as a function of the transferred wavevector, together with the calculated dispersion curves. BLS spectra consist of a series of discrete peaks which merge into one for wave vectors larger than 1.5×10^5 rad/cm. Some of the observed modes exhibit a marked dispersive character, while other modes are characterized by an almost constant frequency and are only weakly dispersive. To interpret the rich excitation spectrum of this sample, the first step is the calculation of the equilibrium magnetization state, which was calculated with a micromagnetic code. Typical NiFe magnetic parameters were employed in the calculations: $4\pi M_s = 9.4$ kG, $\gamma/2\pi = 2.95$ GHz/kOe, and $A = 1.3 \times 10^{-6}$ erg/cm with M_s the saturation magnetization, γ the gyromagnetic

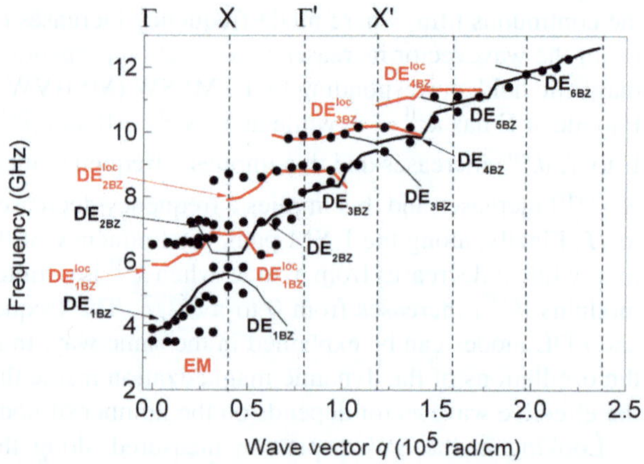

Fig. 15.5 Experimental BLS data (*full circles*) and calculated bands (*full lines*). The main propagating mode is represented by the *thick black curve* in the subsequent BZs. The *thin black curves* are replica of this mode in adjacent BZs. The localized modes are represented by *red curves*. The *dashed vertical lines* separate adjacent BZs

ratio, and A the exchange stiffness constant. The sample was subdivided into cells having size $\Delta_x \times \Delta_y \times \Delta_z = 5\ \text{nm} \times 5\ \text{nm} \times 22\ \text{nm}$. Due to the presence of holes, the calculated ground-state magnetization exhibited small deviations from the direction of $\boldsymbol{H}$ close to the ADs. The experiments were performed with the wavevector directed along the ΓX high-symmetry direction, which corresponds to the Damon–Eshbach scattering geometry for this sample. As seen in Fig. 15.5, the calculated dispersion curves fit the experimental data rather well. According to our calculations, the detected spin-wave modes can be classified taking into account the features of their spatial extension along the x direction.

In fact, the problem can be reduced in a first approximation to an effective 1D problem by singling out the dependence on the Bloch wavevector (along x). However, note that these modes have also less pronounced oscillation along the y direction. From the DMM calculations we can distinguish two distinct families of propagating modes:

1. *Extended modes* having their maximum spin precession amplitude in the horizontal "channels" (along the ΓX direction, parallel to $\boldsymbol{q}$) contained between adjacent rows of holes. These modes are the analogues of the extended surface Damon–Eshbach (DE) mode of a continuous film [49]. However, in this case, a BG appears at each BZ boundary both in measurements and calculations. We indicate the mode with the largest calculated cross-section in the nth BZ as $DE_{n\text{BZ}}$ where $n = 1, 2, \ldots$ takes the role of a "band index". At the center of the first Brillouin zone (1BZ) ($q \approx 0$), the $DE_{1\text{BZ}}$ mode has an almost constant spin precession amplitude in the channels enclosed between rows of holes.

   The frequencies of the calculated modes are shown in the extended zone scheme according to which $\boldsymbol{q} = \boldsymbol{K}$. Note that, unlike the case of array of dots discussed previously (sample #1), for this sample the Bloch wave vector $\boldsymbol{K}$ is defined in the extended zone. For a few modes, the dispersion curves are also repeated in neighboring BZs in order to mark the normal modes of large calculated scattering cross-section. The overall agreement between results of micromagnetic calculations and BLS measurements is very good apart from an overes-

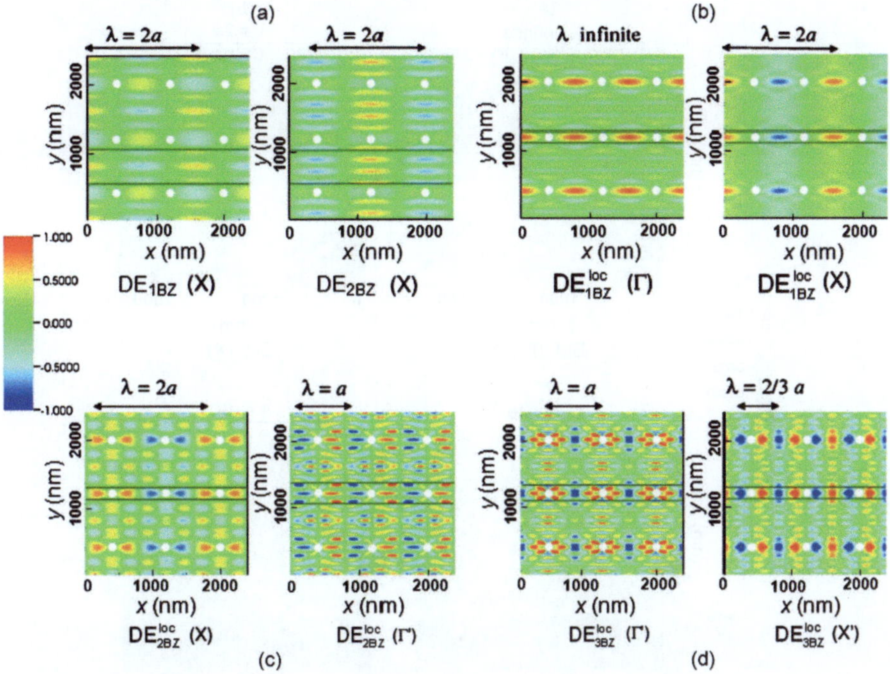

Fig. 15.6 Calculated spatial profiles of representative modes in 3 × 3 primitive cells. The real part of the out-of-plane component of the dynamic magnetization is plotted. For each mode, regions of large precession amplitudes are indicated. The corresponding Bloch wavelengths at different high symmetry points are also shown

timation of the measured frequencies in the vicinity of the border of the 1BZ. The BG between the $\mathrm{DE}_{n\mathrm{BZ}}$ modes decreases with increasing n and eventually vanishes for large values of n. The physical reason underlying the appearance of BGs at BZ boundaries is due to the different spatial localization of the dynamic magnetization as shown in Fig. 15.6. At the edge of the 1BZ the $\mathrm{DE}_{2\mathrm{BZ}}$ and the $\mathrm{DE}_{1\mathrm{BZ}}$ modes behave as two stationary waves. The corresponding spatial profiles differ by a spatial shift of a quarter of the wavelength. In particular, the amplitude of the lowest frequency mode ($\mathrm{DE}_{1\mathrm{BZ}}$) has its maxima between vertical rows of holes in Fig. 15.6(a), while in the horizontal channels the spin precession amplitude can be represented in a form proportional to $\sin[(\pi/a)x]$. Instead, the $\mathrm{DE}_{2\mathrm{BZ}}$ mode has its largest amplitude in between the holes in vertical direction and takes a form proportional to $\cos[(\pi/a)x]$ in horizontal direction.

2. *Localized modes* having amplitudes mainly concentrated along the horizontal rows of holes (see Figs. 15.7(b)–15.7(d)). These modes exhibit a pronounced localized spatial profile and resemble the weak dispersive resonances observed in chains of magnetic dots [21]. They are therefore labeled as $\mathrm{DE}^{\mathrm{loc}}_{n\mathrm{BZ}}$ modes, where label "loc" emphasizes their localized character along rows of ADs. The dispersion of these modes is less pronounced compared to those observed for the extended ones, due to the spatial localization. Note that in such a case the agreement

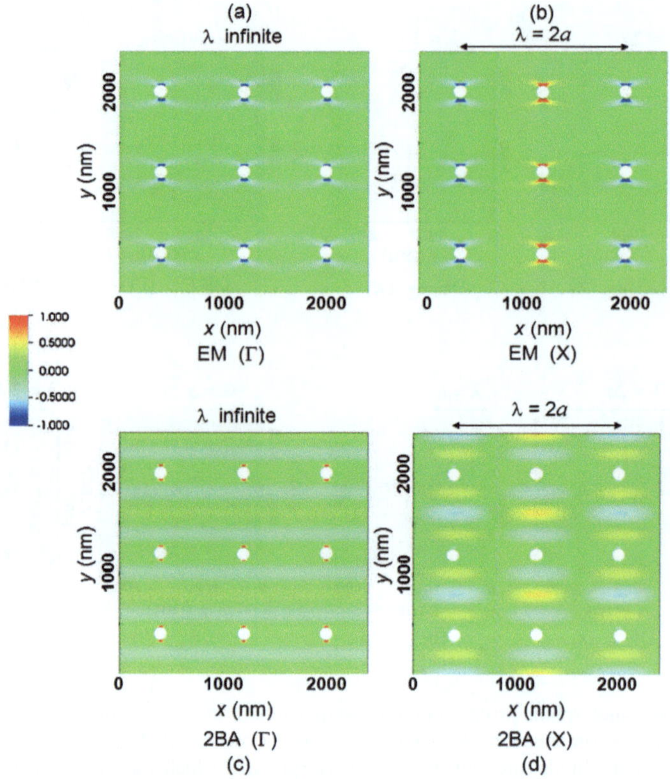

Fig. 15.7 Calculated spatial profiles in 3 × 3 primitive cells of EM at Γ (panel (**a**)) and at X (panel (**b**)) and of 2BA at Γ (panel (**c**)) and at X (panel (**d**)). The real part of the out-of-plane component of the dynamic magnetization is plotted. The corresponding Bloch wavelengths are also shown. A reference frame with the directions of ***H*** and ***q*** is also shown

between the measured and the calculated frequency bandwidth is quite good except for the lowest bands of localized modes. This discrepancy is attributed to possible variation of the hole diameters in the real AD. In Fig. 15.6, also the Bloch wavelength calculated for each mode is indicated at the different symmetry points. This Bloch wavelength is the same both in the channels and in the rows. Finally, in the y-direction the modes appear to be quantized due to the boundary conditions imposed by the presence of the AD rows. Therefore, it is possible to identify a number of nodal planes perpendicularly to the direction of the magnetization and to adopt the nomenclature backward-like modes (mBA) currently used for dots with $m = 1, 2, \ldots$. A spatial profile of the 2BA mode is shown in Figs. 15.7(c) and 15.7(d). Note that the amplitude of this kind of modes is mainly concentrated along the horizontal channels and vanishes in the rows of ADs. In addition, also mixed modes labeled by $m\text{BA} \times n\text{DE}$, namely with both DE and BA character exist. The mBA and $m\text{BA} \times n\text{DE}$ mode dispersions are not

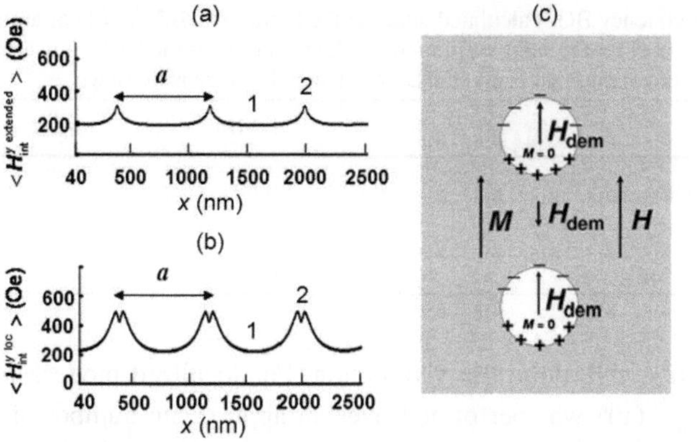

Fig. 15.8 Periodic mean internal field along the x direction relevant for spin wave modes of Fig. 15.2(a) extending through the channels. (**b**) Mean internal field relevant for spin wave modes of Figs. 15.2(b)–15.2(d) localized along rows of holes. In both panels, region 1 (1) is the region corresponding to holes, while region 2 (2) is the region contained between the holes along the x direction. (**c**) Sketch of the AD lattice with two holes (*white*) embedded in the ferromagnetic film (*dark*). The orientations of $\boldsymbol{H}_{\text{dem}}$, $\boldsymbol{H}$, and $\boldsymbol{M}$ are shown. The induced surface "magnetic charges" (positive and negative) are indicated

shown in Fig. 15.5 because their associated calculated cross-section is negligible as also confirmed by BLS measurements.

Eventually, in addition to the above mentioned families of extended and localized modes, we find, at very low frequency, the so-called edge mode (Figs. 15.7(a) and 15.7(b)). This mode exhibits a weakly dispersive behavior, due to its strong spatial localization in the regions adjacent to edges of ADs, where the internal field is highly inhomogeneous. As a final step of our investigation, we analyzed in great detail the behavior of the internal magnetic field in order to investigate opening of BGs for the most representative spin-wave modes of the spectrum. This investigation was carried out in analogy to the well known case of Bragg reflection of electrons in a periodic potential. To achieve this goal, we first estimate the mean internal field acting on precessing spins, namely

$$\langle H_{\text{int}}^{y}(x)\rangle = \frac{1}{N_y}\sum_{N_y} H_{\text{int}}^{y}(x, y), \tag{15.5}$$

by averaging over a relevant number N_y of micromagnetic cells along the y direction using the OOMMF micromagnetic code. In particular, $H_{\text{int}}^{y}(x, y) = H + H_{\text{dem}}^{y}(x, y)$ where H_{dem}^{y} is the y-component of the demagnetizing field $\boldsymbol{H}_{\text{dem}}$. We have also found that the static exchange field contribution is negligible. In order to reproduce the behavior of the internal field of the periodic system in a realistic way, we have simulated a 5×5 AD system, corresponding to 664×664 micromagnetic cells.

The mean internal field $\langle H_{\text{int}}^{\text{extended}}(x)\rangle$ (see Fig. 15.8(a)) experienced by extended modes was calculated by averaging over the total number of micromagnetic cells

Table 15.1 Frequency BGs calculated either at the border of 1BZ ($k = 1$) or at the border of 2BZ ($k = 2$) for the most representative spin modes. Each couple of modes is indicated between *round brackets*. The corresponding observed BLS frequency BGs are also shown

$\Delta\upsilon^k$ (GHz)	DMM	BLS
$\Delta\upsilon^1$ (DE_{2BZ}–DE_{1BZ})	0.60	0.6
$\Delta\upsilon^2$ (DE^{loc}_{2BZ}–DE^{loc}_{1BZ})	1.01	1.6
$\Delta\upsilon^3$ (DE^{loc}_{3BZ}–DE^{loc}_{2BZ})	0.98	0.9

of the primitive cell along the y direction. For localized modes, instead, the averaging $\langle H_{int}^{y\ loc}(x)\rangle$ was performed over an appropriate number of cells in the y direction, corresponding to their region of localization at the border of BZs (see Fig. 15.8(b)). For this special case, the cells number entering in this average were 28. Both $\langle H_{int}^{y\ extended}(x)\rangle$ and $\langle H_{int}^{y\ loc}(x)\rangle$ are inhomogeneous along x. In particular, they both exhibit periodical oscillations corresponding to the periodicity of the system with maxima corresponding to the ADs and minima corresponding to the region between adjacent ADs. This behavior can be understood by taking into account that $\boldsymbol{H}_{dem}$ inside the holes is aligned with $\boldsymbol{H}$. As depicted in Fig. 15.8(c), in fact, this alignment is due to the particular uncompensated distribution of surface "magnetic charges" in the holes which is opposite with respect to that of the magnetic material in which the holes are embedded. Note that, because of its non-local nature, the demagnetizing field does not vanish inside the holes.

The DE_{2BZ} mode exhibits its maximum precession amplitude in region 1 where $\langle H_{int}^{y\ extended}\rangle$ is larger and has thus a larger frequency. Instead, the maximum precession amplitude of the DE_{1BZ} mode is in region 2 where $\langle H_{int}^{y\ extended}\rangle$ is smaller and the mode has thus the smaller frequency. These conclusions can be drawn also for the doublet of modes DE^{loc}_{2BZ} and DE^{loc}_{1BZ}. A comparison between the BG measured by BLS and the values calculated by means of DMM is shown in Table 15.1. The overall agreement is very good apart from an underestimation of the BLS observed BG between the DE^{loc}_{2BZ} and DE^{loc}_{1BZ} modes.

From the above description, it is clear that the occurrence of BGs at the BZ boundaries is caused by the Bragg diffraction for propagating SWs because of the presence of the artificial periodicity of the internal field. Indeed, the inhomogeneity of the internal field gives rise to a counter-propagating "Bragg-reflected" wave. Moreover, the relevant scattering potential for Bragg reflection is provided by the concomitant action of the holes and of the internal field inhomogeneity *between* holes.

This is in contrast with the opening of BGs in photonic and electronic artificial crystals. Indeed, Bragg reflection of light and electrons in corresponding AD lattices and artificial crystals is solely related to the geometry of the system with the holes that can be modeled by means of a scattering potential [51, 52].

It is interesting to make a comparison with electronic bands studied within the nearly-free electron model. Indeed, the eigenfunctions representing frequency modes at the BZs boundaries belonging to nth and $(n+1)$st band ($\sin[(k\pi/a)x]$

and $\cos[(k\pi/a)x]$ with $k = 1, 2, \ldots$, respectively) are interchanged with respect to those of electrons. This can be understood taking into account that the periodic mean internal field has an opposite behavior with respect to the periodic electronic potential. Moreover, contrary to the electronic case, the bandwidth in magnonic AD lattice decreases with increasing Bloch wavevector. Finally, AD behaves not only as waveguide for SWs, but, due to the presence of BGs due to Bragg reflection, also as filter for the frequency of traveling excitations.

15.5 Summary

We have reviewed the experimental and theoretical work about spin wave band structure in two-dimensional magnonic crystals consisting of square arrays of either NiFe disks or holes etched into a continuous NiFe film. Spin wave dispersion has been measured by using Brillouin light scattering along the principal high-symmetry directions of the first Brillouin zone of the artificial crystal. The experimental data have been successfully compared to calculations of the band diagram performed using the dynamical matrix method. With respect to the previous applications of this method to calculate the eigenmodes spectrum of a single nanoobject, in the case of the dense array of disks it has been extended to include the dipolar interaction between the disks. Moreover, periodic boundary conditions have been implemented to reproduce the results obtained in array of antidots. For both investigated 2D Magnonic crystals, frequency bands of finite width and frequency bandgaps have been observed in the experiment and suitably reproduced by the calculations.

Acknowledgements This work was supported by the European Community's Seventh Framework Programme (FP7/2007-2013) under Grant Agreements No. 228673 (MAGNONICS) and No. 233552 (DYNAMAG). The authors acknowledge Prof. A.O. Adeyeye for fabrication of the square array of disks and Prof. D. Grundler, S. Neusser, and G. Duerr for fabrication of the AD sample and cooperation in studying the dynamical properties of the this system.

References

1. J.O. Vasseur, L. Dobrzynski, B. Djafari-Rouhani, H. Puszkarski, Phys. Rev. B **54**, 1043 (1996)
2. J.O. Vasseur, L. Dobrzynski, B. Djafari-Rouhani, H. Puszkarski, Phys. Rev. B **54**, 1043 (1996)
3. V.V. Kruglyak, S.O. Demokritov, D. Grundler, J. Phys. D, Appl. Phys. **43**, 264001 (2010)
4. S.A. Nikitov, P. Tailhades, D.S. Tsai, J. Magn. Magn. Mater. **236**, 320 (2001)
5. Y.V. Gulyaev, S.A. Nikitov, L.V. Zhivotovski, A.A. Klimov, P. Tailhades, L. Presmanes, C. Bonningue, C.S. Tsai, S.L. Vysotskiĭ, Y.A. Filimonov, JETP Lett. **77**, 567 (2003)
6. A.A. Serga, A.V. Chumak, B. Hillebrands, J. Phys. D, Appl. Phys. **42**, 205005 (2009)
7. B. Lenk, H. Ulrichs, F. Garbs, M. Münzenberg, Phys. Rep. **507**, 107 (2011)
8. G. Gubbiotti, S. Tacchi, M. Madami Carlotti, A.O. Adeyeye, M. Kostylev, J. Phys. D, Appl. Phys. **43**, 264003 (2010)
9. E. Yablonovitch, T.J. Gmitter, Phys. Rev. Lett. **63**, 1950 (1989)
10. M.S. Kushwaha, P. Halevi, L. Dobrzynski, B. Djafari-Rouhani, Phys. Rev. Lett. **71**, 2022 (1993)

11. J. Topp, D. Heitmann, M.P. Kostylev, D. Grundler, Phys. Rev. Lett. **104**, 207205 (2010)
12. S. Tacchi, G. Gubbiotti, G. Carlotti, S. Goolaup, A.O. Adeyeye, N. Singh, M.P. Kostylev, Phys. Rev. B **82**, 184408 (2010)
13. G.A. Melkov, V.I. Vasyuchka, A.V. Chumak, V.S. Tiberkevich, A.N. Slavin, J. Appl. Phys. **99**, 08P513 (2006)
14. K.S. Lee, D.-S. Han, S.-K. Kim, Phys. Rev. Lett. **102**, 127202 (2009)
15. M. Inoue, A. Baryshev, H. Takagi, P. Boey Lim, K. Hatafuku, J. Noda, K. Togo, Appl. Phys. Lett. **98**, 132511 (2011)
16. G. Gubbiotti, G. Carlotti, P. Vavassori, M. Kostylev, N. Singh, S. Goolaup, A.O. Adeyeye, Phys. Rev. B **72**, 224413 (2005)
17. G. Gubbiotti, S. Tacchi, G. Carlotti, N. Singh, S. Goolaup, A.O. Adeyeye, M. Kostylev, Appl. Phys. Lett. **90**, 092503 (2007)
18. M. Kostylev, P. Schrader, R.L. Stamps, G. Gubbiotti, G. Carlotti, A.O. Adeyeye, S. Goolaup, N. Singh, Appl. Phys. Lett. **92**, 132504 (2008)
19. M.P. Kostylev, N.A. Sergeeva, in *Magnetic Properties of Laterally Confined Nanometric Structures*, ed. by G. Gubbiotti (Transworld Research Network, Kerala, 2006), pp. 183–207
20. H.T. Nguyen, M.G. Cottam, J. Phys. D, Appl. Phys. **44**, 315001 (2011)
21. R. Zivieri, F. Montoncello, L. Giovannini, F. Nizzoli, S. Tacchi, M. Madami, G. Gubbiotti, G. Carlotti, A.O. Adeyeye, Phys. Rev. B **83**, 054431 (2011)
22. R. Zivieri, F. Montoncello, L. Giovannini, F. Nizzoli, S. Tacchi, M. Madami, G. Gubbiotti, G. Carlotti, A.O. Adeyeye, IEEE Trans. Magn. **47**, 1563 (2011)
23. R. Huber, D. Grundler, Proc. SPIE **8100**, 81000D (2011)
24. S. Tacchi, F. Montoncello, M. Madami, G. Gubbiotti, G. Carlotti, L. Giovannini, R. Zivieri, F. Nizzoli, S. Jain, A.O. Adeyeye, N. Singh, Phys. Rev. Lett. **107**, 127204 (2011)
25. V.V. Kruglyak, P.S. Keatley, A. Neudert, R.J. Hicken, J.R. Childress, J.A. Katine, Phys. Rev. Lett. **104**, 027201 (2010)
26. S. Tacchi, M. Madami, G. Gubbiotti, G. Carlotti, H. Tanigawa, T. Ono, M.P. Kostylev, Phys. Rev. B **82**, 024401 (2010)
27. M. Dvornik, V.V. Kruglyak, Phys. Rev. B **84**, 140405(R) (2011)
28. Z.K. Wang, V.L. Zhang, H.S. Lim, S.C. Ng, M.H. Kuok, S. Jain, A.O. Adeyeye, Appl. Phys. Lett. **94**, 083112 (2009)
29. C.S. Lin, H.S. Lim, Z.K. Wang, S.C. Ng, M.H. Kuok, Appl. Phys. Lett. **98**, 022504 (2011)
30. G. Duerr, M. Madami, S. Neusser, S. Tacchi, G. Gubbiotti, G. Carlotti, D. Grundler, Appl. Phys. Lett. **99**, 202502 (2011)
31. S. Tacchi, M. Madami, G. Gubbiotti, G. Carlotti, S. Goolaup, A.O. Adeyeye, N. Singh, M.P. Kostylev, Phys. Rev. B **82**, 184408 (2010)
32. J. Topp, D. Heitmann, M.P. Kostylev, D. Grundler, Phys. Rev. Lett. **104**, 207205 (2010)
33. J. Topp, S. Mendach, D. Heitmann, M. Kostylev, D. Grundler, Phys. Rev. B **84**, 214413 (2011)
34. J. Ding, M. Kostylev, A.O. Adeyeye, Phys. Rev. Lett. **107**, 047205 (2011)
35. J. Ding, M. Kostylev, A.O. Adeyeye, Phys. Rev. B **84**, 054425 (2011)
36. N.I. Polushkin, Appl. Phys. Lett. **99**, 182502 (2011)
37. G. Gubbiotti, M. Madami, S. Tacchi, G. Carlotti, T. Okuno, J. Appl. Phys. **99**, 08C701 (2006)
38. L. Giovannini, F. Montoncello, F. Nizzoli, Phys. Rev. B **75**, 024416 (2007)
39. H. Ibach, H. Luth, *Solid State Physics* (Springer, New York, 2009)
40. M. Kostylev, G. Gubbiotti, G. Carlotti, G. Socino, S. Tacchi, C. Wang, N. Singh, A.O. Adeyeye, R.L. Stamps, J. Appl. Phys. **103**, 07C507 (2008)
41. S. Neusser, D. Grundler, Adv. Mater. **21**, 2927 (2009)
42. H. Ulrichs, B. Lenk, M. Münzenberg, Appl. Phys. Lett. **97**, 092506 (2010)
43. S. Neusser, G. Duerr, H.G. Bauer, S. Tacchi, M. Madami, G. Woltersdorf, G. Gubbiotti, C.H. Back, D. Grundler, Phys. Rev. Lett. **105**, 067208 (2010)
44. S. Tacchi, M. Madami, G. Gubbiotti, G. Carlotti, A.O. Adeyeye, S. Neusser, B. Botters, D. Grundler, IEEE Trans. Magn. **46**, 1440 (2010)
45. S. Tacchi, M. Madami, G. Gubbiotti, G. Carlotti, A.O. Adeyeye, S. Neusser, B. Botters, D. Grundler, IEEE Trans. Magn. **46**, 172 (2010)

46. C.-L. Hu, R. Magaraggia, H.-Y. Yuan, C.S. Chang, M. Kostylev, D. Tripathy, A.O. Adeyeye, R.L. Stamps, Appl. Phys. Lett. **98**, 262508 (2011)
47. S. Neusser, H.G. Bauer, G. Duerr, R. Huber, S. Mamica, G. Woltersdorf, M. Krawczyk, C.H. Back, D. Grundler, Phys. Rev. B **84**, 184411 (2011)
48. R. Zivieri, S. Tacchi, F. Montoncello, L. Giovannini, F. Nizzoli, M. Madami, G. Gubbiotti, G. Carlotti, S. Neusser, G. Duerr, D. Grundler, Phys. Rev. B **85**, 012403 (2012)
49. R.W. Damon, J.R. Eshbach, J. Phys. Chem. Solids **19**, 308 (1961)
50. J. Jorzick, C. Krämer, S.O. Demokritov, B. Hillebrands, E. Sondergard, M. Bailleul, C. Fermon, U. Memmert, A.N. Müller, A. Kounga, U. Hartmann, E. Tsymbalet, J. Magn. Magn. Mater. **226**, 1835 (2001)
51. J.D. Joannopoulos, S.G. Johnson, J.N. Winn, R.D. Meade, *Photonic Crystals: Molding the Flow of Light*, 2nd edn. (Princeton University Press, Princeton, 2008)
52. C. Flindt, N.A. Mortensen, A.-P. Jauho, Nano Lett. **5**, 2515 (2005)

Chapter 16
Normal Mode Theory for Magnonic Crystal Waveguide

N.Y. Grigoryeva and B.A. Kalinikos

Abstract A general theory for dipole–exchange spin-wave spectrum of magnonic crystal waveguide has been developed. Magnonic crystal waveguide is assumed to be a thin-film ferromagnetic waveguiding medium with periodical modulation of the magnetic parameters along the spin-wave propagation direction. The periodical variation of the magnetic parameters is taken into account in the framework of the Floquet theorem for a differential equation with periodic coefficients. The spin-wave modes' approach, together with the method of tensorial Green's functions, is used to describe propagating spin waves. The exact dispersion relation is obtained in the form of an infinite determinant, which can be easily reduced to a finite one for each particular problem under consideration.

16.1 Introduction

Propagating spin waves in various magnetic nanostructures and particularly in magnonic crystals nowadays attract considerable attention due to their potential application as information carriers in integrated signal processing devices [1–8]. A magnonic crystal waveguide (MCW) may be formed through a periodical variation ("modulation") of any waveguiding parameter of the magnonic structure.

Generally, any dispersion relation for spin waves in a magnonic crystal waveguide may be written as $F(\omega, k; L, w, \mathbf{M}_S, A, p, \mathbf{H}^{\text{int}}, \mathbf{H}^a) = 0$ where L and w are the thickness and width of the magnonic crystal waveguide, respectively, $\mathbf{M}_S$ is the saturation magnetization of the ferromagnetic film, A is the exchange stiffness constant, p is the spin-pinning parameter, $\mathbf{H}^{\text{int}}$ is the constant internal magnetic field, and $\mathbf{H}^a$ is the field of the magnetocrystalline anisotropy. These are the parameters governing the spin-wave dispersion properties. The spatial periodic variation of these parameters (or several of them) and/or their directions produce a magnonic

N.Y. Grigoryeva · B.A. Kalinikos (✉)
Saint-Petersburg Electrotechnical University, Saint-Petersburg 197376, Russia
e-mail: borisk@lamar.colostate.edu

N.Y. Grigoryeva
e-mail: natalygri69@gmail.com

S.O. Demokritov, A.N. Slavin (eds.), *Magnonics*, Topics in Applied Physics 125,
DOI 10.1007/978-3-642-30247-3_16, © Springer-Verlag Berlin Heidelberg 2013

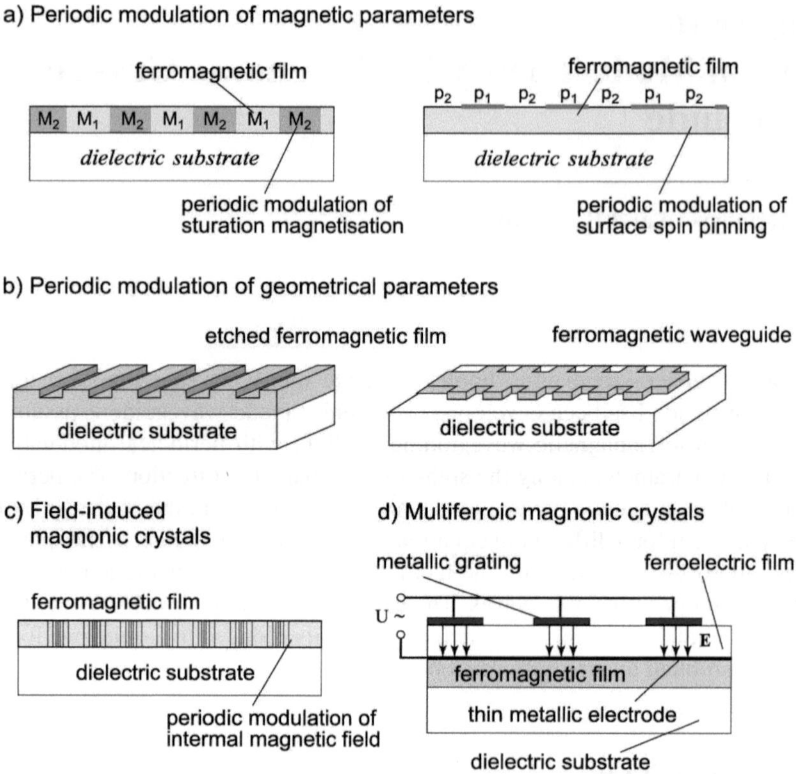

Fig. 16.1 Typical examples of the magnonic crystal waveguides

crystal waveguide. Several typical examples of the magnonic crystal waveguides are shown in Fig. 16.1. Panels (a)–(c) give sketches of the MCWs based on the magnetic films only, while panel (d) gives an example of the MCW based on the ferrite–ferroelectric layered structure (artificial multiferroic heterostructure).

Since the wavelengths of spin waves are shorter than those of light for the same frequencies, magnonic crystal waveguides have tempting prospects for applications in microelectronics and miniaturization of microwave devices [9]. One of the main advantages of the MCW is that the dispersion characteristics of spin waves can be controlled in a wide frequency range through application and variation of the external magnetic and electric field, as well as through choosing the geometry of a waveguiding structure. Electronic tunability of magnonic crystals due to local changing of magnetic and electric fields provides a possibility to control dynamically the dispersion characteristics of the periodic medium during the spin wave propagation [10, 11]. This feature could be essential for the functioning of the devices based on such artificial crystals.

Numerous theoretical and experimental investigations have shown that a large number of spin-wave propagation phenomena interesting both for physics and applications are demonstrated by relatively simple magnetic-film structures [12–17].

This richness is due to the specific combination of the confinement and the so-called "wave intrinsic anisotropy" of the spin-wave dispersion characteristics. The spectrum of spin waves in the magnonic crystals contains regions of reduced group velocity in the vicinity of magnonic bandgaps as in any periodic structure [18]. But in contrast to the photonic and sonic crystals, magnonic crystals have an additional degree of freedom to control the dispersion characteristics which gives rise to new physical effects. Among them are the formation of bright and dark gap solitons [19–22], all-linear time reversal [23], reverse Doppler shift [24], and other effects.

Several methods are used for practical realization of the spatial modulation of the magnetic parameters (or geometric parameters) of magnonic crystals. For example, ion implantation can be applied for spatial modulation of anisotropy [25], and effective gyromagnetic ratio [26]. Also, ion implantation (with lithographic mask or directed ion beam) allows producing MCW with periodically modulated thickness or width. Periodic structures can be also fabricated by lithographic methods, such as e-beam lithography [27] or ultraviolet lithography [28]. Photolithography and wet etching have been employed for the fabrication of one- and two-dimensional YIG based magnonic crystals [22, 29, 30]. The so-called "field-induced magnonic crystals" should also be mentioned. They are realized through the application of the periodically modulated bias magnetic field or electric current to the ferromagnetic layered structures [10, 31]. Such modulations cause a periodic variation of the propagation conditions for spin waves in the waveguiding structure.

A distinct group of the magnonic crystals are artificial multiferroic magnonic crystals or magnonic crystals based on a layered structure containing ferrite and ferroelectric layers. Such magnonic crystals offer a possibility of modulation of their dispersion properties by controlling the electric field applied to the ferroelectric layer [11, 32, 33]. Of principal use is the coupling between electromagnetic and magnetostatic spin waves, due to the broad operating frequency range extending up to the order of gigahertz in hexaferrite structures.

It is obvious that the tendency of miniaturization in microelectronics and nanoelectronics leads to size reduction of the practically used elements. In rather small elements, such as nano-elements, the exchange interaction among spins becomes of the same order as the dipole–dipole one. Thus, an exact spin-wave theory is needed which takes in account both dipole–dipole and exchange interactions as well as periodic modulation of all magnetic waveguiding parameters.

A band structure of all periodic systems (photonic, phononic, and magnonic crystals) can be described by similar theoretical methods [18, 34]. A commonly used analytical tool in this case is the Floquet technique. It is popular because of its conceptual simplicity and applicability to any linear differential equation with periodic coefficients. However, the investigation of the normal waves in the periodically modulated magnetic waveguides has some peculiarities. The main peculiarity is that the spectrum of the ferromagnetic-film waveguide initially has a set of spin-wave modes [35]. As all these modes are involved in a strong mutual dipole–dipole interaction, the so-called "dipole gaps" already appear in the spectrum of the initial ferromagnetic-film structure. In the case these dipole gaps in the initial spectrum appear near the frequency boundaries of the Brillouin zones in MCW, it leads to the additional repulsing (hybridization effects) or to the disappearance of dipole gaps of interacting space harmonics, which are due to the periodicity of the structure. Thus, the physical interpretation of the nature of all dipole gaps in the spin-wave spectrum of the magnonic crystal waveguide is impossible without rigorous calculations. Note that such a situation is in sharp contrast to photonic crystals.

One of the first attempts to study propagation of spin waves in periodic magnetic structures was made by Elachi [36]. Today the number of theoretical and experimental studies on this topic is growing rapidly. Only in the last three years, a lot of papers were published [5–8, 20–24, 27, 30, 33, 37–39]. A lot of new theoretical methods have been developed to describe the infinite or semi-infinite 1D, 2D, and 3D magnonic crystals, e.g., the plane wave method [40], the transfer matrix method [1, 41], the Green function method [42], the spin-wave modes approach [4], the technique using effective magnetic parameters [43], the microscopic approach based on the Heisenberg Hamiltonian [44], and others. At the same time, a general theory with simultaneous treatment of an arbitrary variation of all the magnetic parameters is not yet elaborated.

In the real magnonic crystals, simultaneous periodical modulation of several physical parameters usually takes place. But up to now only one theoretical attempt was made in the work [45] to obtain a semi-analytical solution for the magnonic waveguides in the presence of several spatially periodic perturbations.

In this chapter, a common theory for dipole–exchange spin-wave spectrum of magnonic crystal waveguide is built. In doing so, the previously elaborated method of tensorial Green's functions and spin-wave normal modes expansion technique are used [46]. The theory takes into account a spatial periodic variation of all the parameters determining the dispersion characteristics of propagating spin waves. This model situation is referred to as one-dimensional magnonic crystal waveguide. All types of variation of the magnetic parameters are covered by the presented theory, namely saturation magnetization, external or internal magnetic field, exchange constant, magnetic anisotropy, etc. With some modifications, the same theoretical approach can be applied to the magnonic crystal waveguides with width and thickness modulations.

It is well known that the dispersion relation describes the conditions that should be satisfied for operation of any microwave device utilizing MCW. In this study, we investigate the spin dynamics by examining the dispersion relation for spin waves. The dispersion relation is obtained in the form of an infinite determinant, which can be easily reduced with the required accuracy by means of common perturbation theory. In our theoretical approach, the influence of periodic variation of the magnonic waveguide parameters is incorporated into the initial system of equations using the well-known Floquet theorem.

It should be noted that the previously published theoretical works on the periodic magnetic structures have employed approximations that are too restrictive to be used in many practical cases. Moreover, most of the authors have developed their analysis in the approach, which usually involved no more than one or two modulated parameters and only one main spin-wave thickness mode (single mode regime). The aim of the present work is to determine the properties of a magnonic crystal waveguide by utilizing a rigorous approach which does not employ any of the approximations at the first stage of the analysis and can be generalized to any type of the magnonic crystal waveguide. The solutions are obtained in such a form that they can readily be used for physical interpretation, which is done in terms of individual spin-wave modes and space harmonics. Nevertheless, computer calculations are still necessary to achieve the final results. But in the framework of the presented approach, the precision of calculations can be easily and systematically adjusted to any desired order.

16.2 Landau–Lifshitz Equation of Motion for Space Harmonics in Magnonic Crystal Waveguide

The determination of the spin-wave dispersion relation in homogeneous ferromagnetic materials is a well-established problem widely discussed in the literature (see, e.g., [12–16] and references therein). In periodic magnetic structures, extra factors appear in the dispersion relation. To make our consideration more general, let us consider an arbitrary spatial periodic modulation of all magnetic parameters along the spin-wave propagation direction, i.e., along the ζ-axis (Fig. 16.2). The magnonic crystal waveguide is assumed to have width w in the η-direction, thickness L in the ξ-direction, and is infinite in the ζ-direction. It is magnetized to saturation by a static bias magnetic field $\mathbf{H}^{\text{ext}}(\mathbf{r})$ of an arbitrary direction.

In the classical approach, the spin-wave dispersion relations are determined from the equation of motion for the space- and time-dependent magnetization vector $\mathbf{M}(\mathbf{r}, t)$, referred to as Landau–Lifshitz equation of motion. It reads [47]:

$$\frac{\partial \mathbf{M}(\mathbf{r}, t)}{\partial t} = -|g|\mu_0 [\mathbf{M}(\mathbf{r}, t) \times \mathbf{H}^{\text{eff}}(\mathbf{r}, t)] + \frac{\lambda}{M_0}\left[\mathbf{M}(\mathbf{r}, t) \times \frac{\partial \mathbf{M}(\mathbf{r}, t)}{\partial t}\right], \quad (16.1)$$

where $|g|$ is the gyromagnetic ratio, μ_0 is the permeability of free space, $\mathbf{H}^{\text{eff}}(\mathbf{r}, t)$ is the effective magnetic field, and the last term describes the relaxation effects. As we are not interested in the relaxation effects, in further calculations the last term will be omitted.

The effective magnetic field $\mathbf{H}^{\text{eff}}(\mathbf{r}, t)$ acting on the magnetic moments is introduced here as a sum of several components, each of them can be space- and time-dependent:

$$\mathbf{H}^{\text{eff}}(\mathbf{r}, t) = \mathbf{H}^{\text{int}}(\mathbf{r}) + \mathbf{H}^{\text{ex}}(\mathbf{r}, t) + \mathbf{h}^{d}(\mathbf{r}, t) + \mathbf{H}^{a}(\mathbf{r}, t). \quad (16.2)$$

In the general case, all components can be periodically modulated along the length of the magnonic waveguide (i.e., along the ζ-axis).

The first component of the effective magnetic field $\mathbf{H}^{\text{eff}}$ is the internal magnetic field $\mathbf{H}^{\text{int}}(\mathbf{r})$. It usually consists of the static bias magnetic field $\mathbf{H}^{\text{ext}}(\mathbf{r})$ and the spatially non-uniform static demagnetizing field $\mathbf{H}^{\text{demag}}(\mathbf{r})$. The direction of the internal magnetic field coincides with the direction of the equilibrium orientation of the saturation magnetization $\mathbf{M}_S$ and is determined by angles θ and φ (Fig. 16.2). The angles θ and φ can be obtained from the condition of minimization of the magnetic energy density in static case.

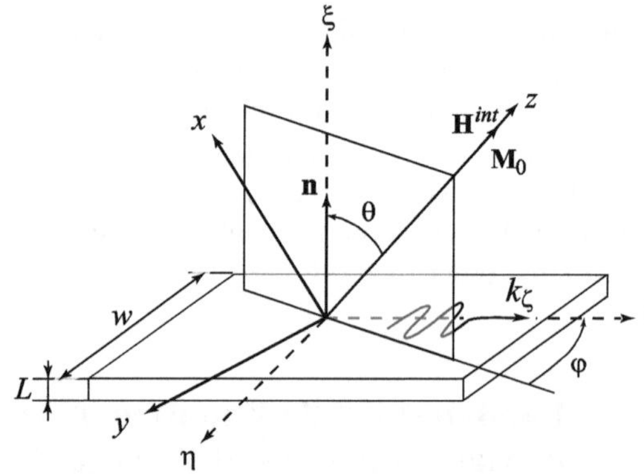

Fig. 16.2 Geometry of the problem

The non-uniform static demagnetizing field $\mathbf{H}^{\text{demag}}(\mathbf{r})$ is due to the shape and magnetocrystalline anisotropy, and in the linear case can be expressed through demagnetizing factors [48, 49]. In some special cases (for example, for width-modulated waveguides [50]) the problem of obtaining the static demagnetization field is so complex that should be solved numerically, via micromagnetic simulations. The external bias magnetic field $\mathbf{H}^{\text{ext}}(\mathbf{r})$ is assumed to be strong enough to enforce parallel alignment of all magnetic moments. If the applied magnetic field $\mathbf{H}^{\text{ext}}(\mathbf{r})$ is supposed to be periodic in space, its periodicity results in a periodic potential with barriers and wells for spin waves inside the waveguide, which cause the Bragg diffraction of spin waves on this magnetic lattice.

Another component of the effective magnetic field is the variable exchange field $\mathbf{H}^{\text{ex}}(\mathbf{r}, t)$. In magnetically inhomogeneous media, both the spatial inhomogeneity of the exchange stiffness constant $A(\mathbf{r})$ and that of the saturation magnetization $M_S(\mathbf{r})$ must be taken into account. Thus, in the case of local direct exchange interaction, exchange field can be expressed as follows [40]:

$$\mathbf{H}^{\text{ex}}(\mathbf{r}, t) = \big(\nabla \cdot \alpha(\mathbf{r}) \nabla\big) \mathbf{M}(\mathbf{r}, t), \tag{16.3}$$

where $\alpha(\mathbf{r}) = 2A(\mathbf{r})/\mu_0 M_S^2(\mathbf{r})$, $\sqrt{\alpha(\mathbf{r})}$ is the exchange length, $A(\mathbf{r})$ is the exchange stiffness constant, and $M_S(\mathbf{r})$ denotes the saturation magnetization.

The next component of the effective magnetic field is the dynamic dipole field $\mathbf{h}^d(\mathbf{r}, t)$ resulting from traveling spin waves. As the dipole field is the result of the non-local electromagnetic interaction between all spins in the magnonic waveguide, Maxwell's equations must be solved to find it. The obtained solutions must satisfy the electrodynamics boundary conditions at all surfaces of the waveguide.

The magnetostatic approximation is usually utilized when solving Maxwell's equations at microwave frequencies. But in some special cases when the effects of retardation cannot be neglected, the solution of the full set of Maxwell's equations is needed. As a result of solving Maxwell's equations, a connection between the variable magnetization and the dipole field should be found. In the framework of the tensorial Green's function formalism considered here, this relation can be obtained in the integral form [35]:

$$\mathbf{h}^d(\mathbf{r}) = \int_{\Omega} \widehat{\mathbf{G}}_{\xi\eta\zeta}(\mathbf{r},\mathbf{r}')\mathbf{m}(\mathbf{r}')\mathbf{r}'. \tag{16.4}$$

It should be noted that the tensorial Green's function $\widehat{\mathbf{G}}_{\xi\eta\zeta}(\mathbf{r},\mathbf{r}')$ is the intrinsic characteristic of the waveguiding medium. It is independent of the magnetic conditions of propagation (e.g., variations of the external magnetic field and exchange constant) as long as the electrodynamics boundary conditions are not changing. Thus, for a regular thin-film waveguide with any variation of magnetic parameters, it will stay unchanged and can be taken from [46]. For the width- and thickness-modulated waveguides, the situation changes considerably. In this case, the periodicity of the dynamic dipole field is due to both the periodicity of the dynamic magnetization and the periodicity of the tensorial Green's function. Moreover, the periodicity of the edges of the waveguide entails changes in the internal static magnetic field and the transverse wave-vector, which also become periodically modulated.

The last component of the effective field is the magnetocrystalline anisotropy field $\mathbf{H}^a(\mathbf{r},t)$. In the linear case, it can be expressed through corresponding demagnetizing factors of different types of anisotropy [51]. The anisotropy field can also be varied along the magnonic waveguide.

Introducing a usual linearization of (16.1) in terms of the small deviation $\mathbf{m}(\mathbf{r},t)$ from the equilibrium position of magnetization $\mathbf{M}_S$, one can split $\mathbf{M}(\mathbf{r},t)$ and $\mathbf{H}(\mathbf{r},t)$ into frequency-independent static parts $\mathbf{M}_S(\mathbf{r})$ and $\mathbf{H}(\mathbf{r})$ and the dynamic parts $\mathbf{h}(\mathbf{r},t)$ and $\mathbf{m}(\mathbf{r},t)$:

$$\begin{aligned} \mathbf{M}(\mathbf{r},t) &= \mathbf{M}_S(\mathbf{r}) + \mathbf{m}(\mathbf{r},t), \quad |\mathbf{m}| \ll |\mathbf{M}_S|, \\ \mathbf{H}(\mathbf{r},t) &= \mathbf{H}(\mathbf{r}) + \mathbf{h}(\mathbf{r},t), \quad |\mathbf{h}| \ll |\mathbf{H}|. \end{aligned} \tag{16.5}$$

Making use of the linearization procedure, we apply relation (16.5) to the equation of motion (16.1) with (16.2)–(16.4). Since $\mathbf{m}(\mathbf{r},t)$ and $\mathbf{h}(\mathbf{r},t)$ are both assumed to be small compared to the static components, we neglect the second-order terms. Thus, for the variable magnetization we arrive to the following equation of motion:

$$\begin{aligned} \frac{\partial \mathbf{m}(\mathbf{r},t)}{\partial t} = &-|g|\mu_0\big\{\big[\mathbf{m}(\mathbf{r},t) \times \big(\mathbf{H}_0(\mathbf{r}) - \widehat{\mathbf{N}}(\mathbf{r})\mathbf{M}_S(\mathbf{r}) - \widehat{\mathbf{N}}^a(\mathbf{r})\mathbf{M}_S(\mathbf{r})\big)\big] \\ &+ \big[\mathbf{M}_S(\mathbf{r}) \times \big((\nabla \cdot \alpha(\mathbf{r})\nabla)\mathbf{m}(\mathbf{r},t)\big) + \mathbf{m}(\mathbf{r},t) \times \big((\nabla \cdot \alpha(\mathbf{r})\nabla)\mathbf{M}_S(\mathbf{r})\big)\big] \\ &+ \big[\mathbf{M}_S(\mathbf{r}) \times \mathbf{h}^d(\mathbf{r},t)\big] - \big[\mathbf{M}_S(\mathbf{r}) \times \big(\widehat{\mathbf{N}}^a(\mathbf{r})\mathbf{m}(\mathbf{r},t)\big)\big]\big\}. \end{aligned} \tag{16.6}$$

Equation (16.6), while written for all components of the vector $\mathbf{m}(\mathbf{r},t)$ in the new coordinate system xyz related to the direction of the equilibrium magnetization (see Fig. 16.2), gives a coupled system of two integro-differential equations [35]. The transition from the coordinate system $\xi\eta\zeta$ to the coordinate system xyz can be done by means of orthogonal transformations of rotation through angles φ and $(\theta - \pi/2)$ (see, e.g., [35]).

We search for a solution of the equation of motion (16.6) written together with the system of Maxwell's equations corresponding to monochromatic spin waves. Thus,

$$\mathbf{m}(\mathbf{r},t) = \mathbf{m}(\mathbf{r})e^{i\omega t}, \qquad \mathbf{h}^d(\mathbf{r},t) = \mathbf{h}^d(\mathbf{r})e^{i\omega t}. \tag{16.7}$$

In order to take into account spatial periodicity of the parameters of the magnonic waveguide, we use the differential method [52] for solving a general eigenvalue problem. This method is based on the Floquet theorem for a differential equation with periodic coefficients. In the framework of this method, a solution for the dynamic magnetization $\mathbf{m}(\mathbf{r})$ and an expression for the dynamic dipole field $\mathbf{h}^d(\mathbf{r})$ can be represented as product of a plane-wave envelope function $\exp(-ik_\zeta\zeta)$ and a corresponding periodic function $\mathbf{f}_k(\mathbf{r}) = \mathbf{f}_k(\xi, \eta, \zeta + m\Lambda)$ [18]. Here Λ is a period of the structure. The periodic functions $\mathbf{f}_k(\mathbf{r})$ may next be expanded in a Fourier series. So, the result is written as:

$$\begin{aligned} \mathbf{m}(\mathbf{r}) &= e^{-ik_\zeta\zeta}\mathbf{m}_k(\mathbf{r}) = \sum_{m=-\infty}^{+\infty} e^{-iK_m\zeta}\mathbf{m}(\xi, \eta; K_m), \\ \mathbf{h}^d(\mathbf{r}) &= e^{-ik_\zeta\zeta}\mathbf{h}_k^d(\mathbf{r}) = \sum_{m=-\infty}^{+\infty} e^{-iK_m\zeta}\mathbf{h}^d(\xi, \eta; K_m), \end{aligned} \tag{16.8}$$

where $K_m = k_\zeta + k_m$, and each $k_m = 2m\pi/\Lambda$ denotes a reciprocal lattice vector. Different components $\mathbf{m}(\xi, \eta; K_m)$ and $\mathbf{h}^d(\xi, \eta; K_m)$ are called space harmonics of the propagating spin wave and dynamic dipole field. The values of $K_m = k_\zeta + 2m\pi/\Lambda$ represent physically the propagation wave numbers of these space harmonics, with k_ζ lying in the first Brillouin zone $0 \le k_\zeta < 2\pi/\Lambda$. Note that the space harmonics do not exist independently. They are parts of a total solution for each spin-wave mode. Each space harmonic has a different wave vector K_m and a different phase velocity $v_m = \omega/K_m$.

The next step is the Fourier transform that maps all space-periodic functions in (16.6), i.e., coefficients of the integro-differential equation, to the reciprocal space. The transformation formulas are as follows:

$$\mathbf{F}(\mathbf{r}) = \sum_{m'=-\infty}^{+\infty} \mathbf{F}_{m'}(\xi, \eta)e^{-ik_{m'}\zeta}. \tag{16.9}$$

Here $\mathbf{F}(\mathbf{r})$ denotes *all* space-periodic functions in (16.6) that should be taken into consideration in each particular case. The Fourier coefficients $\mathbf{F}_{m'}(\xi, \eta)$ for all periodic functions can be found from the inverse transform, which reads:

$$\mathbf{F}_{m'}(\xi, \eta) = \frac{1}{\Lambda}\int_0^\Lambda \mathbf{F}(\mathbf{r})e^{ik_{m'}\zeta}\,d\zeta. \tag{16.10}$$

In the general case, these coefficients can be calculated analytically by integration over the period Λ or determined numerically by means of the fast Fourier transform. For several types of periodicity, the Fourier coefficients are well-known and the examples are given in [53]. For some fast convergent series, the Fourier coefficients may be found, e.g., in [54].

For the clarity of further consideration, we assume that the exchange constant, saturation magnetization, and external magnetic field have no variation in the waveguide cross-section (i.e., along the ξ- and η-axis) and have only periodic variation along the waveguide length (i.e., along the ζ-axis) (Fig. 16.2), and demagnetizing field is assumed to be inhomogeneous also across the width of the waveguide.

Finally, substituting expressions (16.7)–(16.10) into (16.6), we obtain the system of equations for the space harmonics of dynamic magnetization:

$$\begin{aligned}
i\omega m_m^x(\xi,\eta) = -|g|\mu_0 \Bigg\{ & \sum_{m'} \big(H_{m-m'}^{\text{ext}} + H_{m-m'}^{\text{demag}}(\eta) + H_{m-m'}^a\big) m_{m'}^y(\xi,\eta) \\
& - \sum_{m'}\sum_{m''} M_{m-m''}^S \alpha_{m''-m'} \bigg(\frac{\partial^2}{\partial\xi^2} + \frac{\partial^2}{\partial\eta^2} \\
& - (k_{m'} + k_\zeta)(k_{m''} + k_\zeta) \bigg) m_{m'}^y(\xi,\eta) \\
& - \sum_{m'}\sum_{m''} M_{m-m''}^S \alpha_{m''-m'} (k_m - k_{m'})(k_m - k_{m''}) m_{m'}^y(\xi,\eta) \\
& - \sum_{m'} M_{m-m'}^S h_{m'}^y(\xi,\eta) + \sum_{m'}\sum_{m''} M_{m-m''}^S \big[N_{m''-m'}^{ayx} m_{m'}^x(\xi,\eta) \\
& + N_{m''-m'}^{ayy} m_{m'}^y(\xi,\eta) \big] \Bigg\},
\end{aligned} \tag{16.11}$$

$$\begin{aligned}
i\omega m_m^y(\xi,\eta) = |g|\mu_0 \Bigg\{ & \sum_{m'} \big(H_{m-m'}^{\text{ext}} + H_{m-m'}^{\text{demag}}(\eta) + H_{m-m'}^a\big) m_{m'}^x(\xi,\eta) \\
& - \sum_{m'}\sum_{m''} M_{m-m''}^S \alpha_{m''-m'} \bigg(\frac{\partial^2}{\partial\xi^2} + \frac{\partial^2}{\partial\eta^2} \\
& - (k_{m'} + k_\zeta)(k_{m''} + k_\zeta) \bigg) m_{m'}^x(\xi,\eta) \\
& - \sum_{m'}\sum_{m''} M_{m-m''}^S \alpha_{m''-m'} (k_m - k_{m'})(k_m - k_{m''}) m_{m'}^x(\xi,\eta) \\
& - \sum_{m'} M_{m-m'}^S h_{m'}^x(\xi,\eta) + \sum_{m'}\sum_{m''} M_{m-m''}^S \big[N_{m''-m'}^{axx} m_{m'}^x(\xi,\eta) \\
& + N_{m''-m'}^{axy} m_{m'}^y(\xi,\eta) \big] \Bigg\},
\end{aligned}$$

where

$$H_{m'}^{\text{demag}}(\eta) = -\frac{1}{\Lambda} \int_0^\Lambda \widehat{N}_{zz}(\eta,\zeta) M_z^S(\zeta) e^{ik_{m'}\zeta}\, d\zeta$$

and

$$H_{m'}^a = -\frac{1}{\Lambda} \int_0^\Lambda \widehat{N}_{zz}^a(\zeta) M_z^S(\zeta) e^{ik_{m'}\zeta}\, d\zeta.$$

Components of the dipole field $\mathbf{h}_m^d(\xi,\eta)$ in (16.11) can be obtained from solving Maxwell's equations in wavenumber domain. As it was already mentioned above, the tensorial Green's function has different forms for different structures. For a ferromagnetic film it is well-known. More information about other types of tensorial Green's functions can be found in [4, 55, 56].

Once the Green's function in wavenumber domain for the regular waveguide is obtained, it is easy to construct its periodic counterpart. The derivation of $\widehat{\mathbf{G}}_{\xi\eta\zeta}(\xi, \xi', \eta, \eta', k_\zeta)$ is an algebraic calculation, which basically demands the eigenvalues and eigenvectors associated with the partial differential equations under consideration [57]. Thus, when we consider the periodic variation of magnetic parameters along the magnonic waveguide, due to the linearity of the problem (16.5) the dipole field and variable magnetization can be represented as a superposition of corresponding space harmonics (16.8). Therefore, the relation between amplitudes of space harmonics for the dynamic dipole field $\mathbf{h}_m^d(\xi, \eta)$ and variable magnetization $\mathbf{m}_m(\xi, \eta)$ can be defined through the generalized tensorial Green function $\widehat{\mathbf{G}}_{\xi\eta\zeta}(\xi, \xi'; \eta, \eta'; k_\zeta)$ by a simple replacement $k_\zeta \to K_m$.

The Floquet formulation (16.11) of the considered problem is relatively simple. However, the numerical calculations must be involved at the last stage. The structure of the system (16.11) corresponds to an eigenproblem with eigenvalues $i\omega$ and eigenvectors $\mathbf{m}_m(\xi, \eta)$. It should be noted that in the case of an arbitrary periodic structure different types of higher-order interactions can occur. Because of this, the Bragg coupling involves the higher-order space harmonics. In such a case, different order Bragg interactions from different Fourier components can sometimes add destructively, leading to disappearance of some bandgaps even if the Bragg condition is satisfied. Moreover, some of these higher-order space harmonics may modify the nature of the initial guided waves.

In a large number of cases, the periodic perturbation is relatively small. Due to this, only several space harmonics should be taken into account in the vicinity of a stop band. When a finite number of space harmonics is used in the Fourier series (16.9), the system of equations (16.11) becomes finite.

16.3 General Dispersion Relation for Dipole–Exchange Spin Waves in Magnonic Crystal Waveguide

The exact dipole–exchange spin-wave dispersion relation for the magnonic crystal waveguide can be obtained in the framework of spin-wave modes approach developed in our previous works (see [4] and references therein). The spin-wave mode approach is based on the expanding of the solution of the integro-differential equation for variable magnetization (16.6) into an infinite series of normal modes of the linear differential operator. These normal modes form a complete set of vector functions satisfying the exchange boundary conditions. The spin-wave normal mode expansion technique allows one to derive the exact dipole–exchange spin-wave dispersion relation in terms of a vanishing infinite determinant, which gives a clear physical picture of the formation of the dipole–exchange spin-wave spectrum in complicated magnonic systems and reveals the dependence of the propagation characteristics on the parameters of the waveguiding structure. Using the exact dispersion relation, the approximate dispersion relation can be derived by means of the general perturbation theory in each case under consideration.

In the coordinate system xyz (Fig. 16.2), introduced in the previous section when the axis z is parallel to the direction of the saturation magnetization $\mathbf{M}_S$, and due to the linearization procedure (16.5), the vector of variable magnetization becomes two-dimensional:

$$\mathbf{m}(\xi,\eta,\zeta)=\mathbf{e}_x m^x(\xi,\eta,\zeta)+\mathbf{e}_y m^y(\xi,\eta,\zeta)=\begin{pmatrix} m^x(\xi,\eta,\zeta) \\ m^y(\xi,\eta,\zeta)\end{pmatrix}, \tag{16.12}$$

where $\mathbf{e}_x$ and $\mathbf{e}_y$ are the unit vectors. Using (16.12), we rewrite the system of equations for space harmonics (16.11) in the following form:

$$\begin{aligned}\widehat{\mathbf{W}}\mathbf{m}_m(\xi,\eta)&+\sum_{m'}\widehat{\mathbf{T}}_{mm'}\mathbf{m}_{m'}(\xi,\eta)-\sum_{m'}\widehat{\mathbf{F}}_{mm'}\mathbf{m}_{m'}(\xi,\eta)+\sum_{m''}\widehat{\mathbf{N}}_{mm'}\mathbf{m}_{m'}(\xi,\eta)\\ &=|g|\mu_0\sum_{m'}M^S_{m-m'}\int_{L_1}^{L_2}\int_{w_1}^{w_2}\widehat{\mathbf{G}}_{xy}(\xi,\xi',\eta,\eta';K_{m'})\mathbf{m}_{m'}(\xi',\eta')\,d\xi'\,d\eta',\end{aligned} \tag{16.13}$$

where $\widehat{\mathbf{W}}=i\omega\left(\begin{smallmatrix}0 & -1\\ 1 & 0\end{smallmatrix}\right)$ includes frequency-dependent non-diagonal terms,

$$\widehat{\mathbf{T}}_{mm'}=|g|\mu_0\big[H^{\text{ext}}_{m-m'}+H^{\text{demag}}_{m-m'}(\eta)+H^{a}_{m-m'}\big]\begin{pmatrix}1 & 0\\ 0 & 1\end{pmatrix}$$

includes the influence of static magnetic fields,

$$\begin{aligned}\widehat{\mathbf{F}}_{mm'}=|g|\mu_0\sum_{m''}M^S_{m-m''}\alpha_{m''-m'}&\left[\frac{\partial^2}{\partial\xi^2}+\frac{\partial^2}{\partial\eta^2}-(k_{m'}+k_\zeta)(k_{m''}+k_\zeta)\right.\\ &\left.+(k_m-k_{m'})(k_m-k_{m''})\right]\begin{pmatrix}1 & 0\\ 0 & 1\end{pmatrix}\end{aligned}$$

is a linear matrix-differential operator, and

$$\widehat{\mathbf{N}}_{mm'}=|g|\mu_0\sum_{m''}M^S_{m-m''}\begin{pmatrix}N^{axx}_{m''-m'} & N^{axy}_{m''-m'}\\ N^{ayx}_{m''-m'} & N^{ayy}_{m''-m'}\end{pmatrix}$$

represents bulk anisotropy. Here $\widehat{\mathbf{G}}_{xy}(\xi,\xi',\eta,\eta';K_{m'})$ is a two-dimensional part of the tensorial Green's function $\widehat{\mathbf{G}}_{xyz}(\xi,\xi',\eta,\eta';K_{m'})$ in the wavenumber domain. The last is obtained from $\widehat{\mathbf{G}}_{\xi\eta\zeta}(\xi,\xi',\eta,\eta';K_{m'})$ by means of orthogonal transformations of rotation through angles φ and $(\theta-\pi/2)$ [35].

As follows from (16.13), a solution of the integro-differential equation of motion for variable magnetization $\mathbf{m}_m(\xi,\eta)$ cannot be obtained without imposing additional "exchange" boundary conditions, which are due to the exchange energy operator $\widehat{\mathbf{F}}_{mm'}$. Physically, an introduction of the exchange boundary conditions is a phenomenological method to take into account the difference of surface and bulk forces acting on the magnetic moments.

In the framework of our consideration, the linearized version of the mixed exchange boundary conditions initially derived by Rado and Wertman [58] is:

$$\mathbf{M}_S \times \frac{\partial \mathbf{m}(\xi,\eta;K_m)}{\partial n}\bigg|_S + p\big[\big(\mathbf{n}\cdot\mathbf{m}(\xi,\eta;K_m)\big)\mathbf{n}\times\mathbf{M}_S + (\mathbf{n}\cdot\mathbf{M}_S)\mathbf{n}\times\mathbf{m}(\xi,\eta;K_m)\big]\big|_S = 0. \tag{16.14}$$

Here we introduce the so-called spin pinning parameter $p = K_S/A$ [cm^{-1}] characterizing the ratio of surface anisotropy energy and non-uniform exchange energy, which can be different at different surfaces of the MCW. The spin pinning parameter p may be positive and negative depending on the sign of K_S, i.e., depending on whether the easy plane of the magnonic crystal is parallel or perpendicular to the corresponding surface. For the clarity of the description, we omit here the indices of space harmonics and assume $p \neq f(\zeta)$. But in the general case, the spin pinning parameter can also have a periodic variation along the waveguide, causing the variation of the transverse wave number of the spin-wave modes along the propagation direction.

The usually considered limiting cases of the exchange boundary conditions are as follows. There is the case of totally free surface spins which is named as Ament–Rado exchange boundary conditions [59], and also the case of totally pinned surface spins which is named as Kittel's exchange boundary conditions [60]. We emphasize that the applicability of these limiting cases should be estimated in each particular case not only by the value of the material parameter p but also by the rate of variation of the dynamic magnetization in the direction normal to the surface. Below, for generality, we utilize the so-called mixed exchange boundary conditions (16.14).

The solution of the system of the equations for the exchange boundary conditions (16.14), while writing for appropriate variable magnetization components, determines the transverse wave numbers k_n and κ_q of corresponding thickness and width spin-wave modes. The allowed values of the wave numbers k_n could be obtained from (16.14) with normal $\mathbf{n}$ aligned along the ξ-axis and with the boundaries positioned at L_1 and L_2. When necessary, the wave numbers κ_q for the width quantization are obtained from (16.14) with normal $\mathbf{n}$ aligned along the η-axis and with the boundaries positioned at w_1 and w_2 of the MCW.

We solve a set of the integro-differential equations (16.13) by expanding the transverse distribution of variable magnetization $\mathbf{m}(\mathbf{r})$ for each space harmonic m in the infinite series of the orthogonal vector functions:

$$\mathbf{m}_m(\xi,\eta) = M_0^S \sum_{n,q}^{\infty}\big[m_{mnq}^x \mathbf{S}_{nq}^x(\xi,\eta) + m_{mnq}^y \mathbf{S}_{nq}^y(\xi,\eta)\big]. \tag{16.15}$$

Here $\mathbf{S}_{nq}^x(\xi,\eta)$ and $\mathbf{S}_{nq}^y(\xi,\eta)$ are the so-called spin-wave modes; m_{mnq}^x, m_{mnq}^y are the amplitudes of spin-wave modes; M_0^S is the amplitude of zero space harmonic of the saturation magnetization, if the last is varied along the waveguide. The vector functions $\mathbf{S}_{nq}(\xi,\eta)$ are the eigenfunctions of the diagonal matrix-differential operator $\widehat{\mathbf{F}}_{mm'}$, which satisfy the appropriate exchange boundary conditions (16.14). For a regular magnonic waveguide, the variables can be separated and the solution has the following form:

$$\mathbf{S}_{nq}^x(\xi,\eta) = \begin{pmatrix} \Phi_{nq}^x(\xi,\eta) \\ 0 \end{pmatrix}, \qquad \mathbf{S}_{nq}^y(\xi,\eta) = \begin{pmatrix} 0 \\ \Phi_{nq}^y(\xi,\eta) \end{pmatrix}, \tag{16.16}$$

where $n, q = 1, 2, 3, \ldots$. Note that the concrete form of the functions $\Phi^{x,y}_{nq}(\xi, \eta)$ depends on the chosen exchange boundary conditions.

For example, in the case of partially but uniformly pinned surface spins on lateral and normal surfaces of the waveguide, the functions $\Phi^{x}_{nq}(\xi, \eta) = \Phi^{x}_{nq}(\xi, \eta) = \Phi_{nq}(\xi, \eta)$ have the following form:

$$\Phi_{nq}(\xi, \eta) = \Phi_n(\xi)\Phi_q(\eta) = A_n\left[\cos\big(k_n(\xi - L_1)\big) + \frac{p_2}{k_n}\sin\big(k_n(\xi - L_1)\big)\right] \times B_q\left[\cos\big(\kappa_q(\eta - w_1)\big) + \frac{p_4}{\kappa_q}\sin\big(\kappa_q(\eta - w_1)\big)\right]. \quad (16.17)$$

The constants A_n and B_q can be obtained from normalization condition for spin-wave modes, and k_n and κ_q can be found as a solutions of transcendental equation obtained from exchange boundary conditions (16.14). For more details, see [46].

The spin-wave modes, being the eigenfunctions of the Sturm–Liouville problem, form a complete set of the orthogonal functions, which must satisfy the condition of the orthonormality over the interval of their existence Ω:

$$\int_{\Omega} \left[\mathbf{S}^{x}_{nq}(\xi, \eta)\mathbf{S}^{x'*}_{n'q'}(\xi, \eta) + \mathbf{S}^{y}_{nq}(\xi, \eta)\mathbf{S}^{y'*}_{n'q'}(\xi, \eta)\right] d\xi\, d\eta = Lw\delta_{nn'}\delta_{pp'}\delta_{qq'}, \quad (16.18)$$

where δ is the Kronecker delta.

Substituting (16.15) for $\mathbf{m}_m(\xi, \eta; K_m)$ in (16.13) and using the orthonormality condition (16.18), we obtain an infinite system of algebraic equations for the components of the vector spin-wave mode amplitudes $m^{x}_{mnq}(K_m)$ and $m^{y}_{mnq}(K_m)$, which can be written in a simple matrix form:

$$\mathbf{D}^{mm}_{nn;qq}\mathbf{m}_{mnq} + \sum_{n'\neq n \text{ or } q'\neq q}\sum \mathbf{R}^{mm}_{nn';qq'}\mathbf{m}_{mn'q'} + \sum_{m'\neq m}\sum_{n'}\sum_{q'} \mathbf{L}^{mm'}_{nn';qq'}\mathbf{m}_{m'n'q'} = 0. \quad (16.19)$$

Indices n and q number the spin-wave modes, and index m numbers the space harmonics for each spin-wave mode. Here $\mathbf{D}^{mm}_{nn;qq}$, $\mathbf{R}^{mm}_{nn';qq'}$, $\mathbf{L}^{mm'}_{nn';qq'}$ are square matrices 2×2 of the following form:

$$\mathbf{D}^{mm}_{nn;qq} = \begin{pmatrix} \mathrm{H}^{mm}_{nn;qq} + \mathrm{F}^{mm}_{nn;qq} + {}^{xx}\mathrm{N}^{mm}_{nn;qq} - {}^{xx}\mathrm{G}^{mm}_{nn;qq} & -\frac{i\omega}{\omega^0_M} + {}^{xy}\mathrm{N}^{mm}_{nn;qq} - {}^{xy}\mathrm{G}^{mm}_{nn;qq} \\ \frac{i\omega}{\omega^0_M} + {}^{yx}\mathrm{N}^{mm}_{nn;qq} - {}^{yx}\mathrm{G}^{mm}_{nn;qq} & \mathrm{H}^{mm}_{nn;qq} + \mathrm{F}^{mm}_{nn;qq} + {}^{yy}\mathrm{N}^{mm}_{nn;qq} - {}^{yy}\mathrm{G}^{mm}_{nn;qq} \end{pmatrix}, \quad (16.20)$$

$$\mathbf{R}^{mm}_{nn';qq'} = \begin{pmatrix} \mathrm{H}^{mm}_{nn';qq'} - {}^{xx}\mathrm{G}^{mm}_{nn';qq'} & -{}^{xy}\mathrm{G}^{mm}_{nn';qq'} \\ -{}^{yx}\mathrm{G}^{mm}_{nn';qq'} & \mathrm{H}^{mm}_{nn';qq'} - {}^{yy}\mathrm{G}^{mm}_{nn';qq'} \end{pmatrix}, \quad (16.21)$$

$$\mathbf{L}^{mm'}_{nn';qq'} = \begin{pmatrix} \mathrm{H}^{mm'}_{nn';qq'} + \mathrm{F}^{mm'}_{nn';qq'} + {}^{xx}\mathrm{N}^{mm'}_{nn';qq'} - {}^{xx}\mathrm{G}^{mm'}_{nn';qq'} & {}^{xy}\mathrm{N}^{mm'}_{nn';qq'} - {}^{xy}\mathrm{G}^{mm'}_{nn';qq'} \\ {}^{yx}\mathrm{N}^{mm'}_{nn';qq'} - {}^{yx}\mathrm{G}^{mm'}_{nn';qq'} & \mathrm{H}^{mm'}_{nn';qq'} + \mathrm{F}^{mm'}_{nn';qq'} + {}^{yy}\mathrm{N}^{mm'}_{nn';qq'} - {}^{yy}\mathrm{G}^{mm'}_{nn';qq'} \end{pmatrix}. \tag{16.22}$$

Here $\omega^0_M = |g|\mu_0 M^S_0$ and the matrix elements $\mathrm{H}^{mm'}_{nn';qq'}$, $\mathrm{F}^{mm'}_{nn';qq'}$, ${}^{\alpha\beta}\mathrm{N}^{mm'}_{nn';qq'}$ and ${}^{\alpha\beta}\mathrm{G}^{mm'}_{nn';qq'}$ are given by:

$$\begin{aligned} \mathrm{H}^{mm'}_{nn';qq'} &= \delta_{nn'}\delta_{qq'}H^{\text{ext}}_{m-m'} + \delta_{nn'}\delta_{qq'}H^{a}_{m-m'} \\ &\quad + \delta_{nn'}\frac{1}{w}\int_{w_1}^{w_2} H^{\text{demag}}_{m-m'}(\eta)\Phi_q(\eta)\Phi_{q'}(\eta)\,d\eta, \\ \mathrm{F}^{mm'}_{nn';qq'} &= \delta_{nn'}\delta_{qq'}\sum_{m''} M^S_{m-m''}\alpha_{m''-m'}\big[k_n^2 + \kappa_q^2 + (k_{m'}+k_\zeta)(k_{m''}+k_\zeta) \\ &\quad - (k_m - k_{m'})(k_m - k_{m''})\big], \\ {}^{\alpha\beta}\mathrm{N}^{mm'}_{nn';qq'} &= \delta_{nn'}\delta_{qq'}\sum_{m''} M^S_{m-m''}N^{a\alpha\beta}_{m''-m'}, \\ {}^{\alpha\beta}\mathrm{G}^{mm'}_{nn';qq'} &= \frac{M^S_{m-m'}}{wL}\int_{L_1}^{L_2}\int_{L_1}^{L_2}\int_{w_1}^{w_2}\int_{w_1}^{w_2} \Phi_{n'q'}(\xi,\eta)\Phi_{nq}(\xi',\eta') \\ &\quad \times G^{\alpha\beta}_{m'}(\xi,\xi',\eta,\eta';K_{m'})\,d\eta\,d\eta'\,d\xi\,d\xi'. \end{aligned}$$

The physical meaning of the square matrices (16.20)–(16.21) and their matrix elements is straightforward. The matrix $\mathbf{D}^{mm}_{nn;qq}$ gives the spin-wave spectrum of a regular magnetic waveguide in the diagonal approximation (see, e.g., [46]). It includes the contribution of the exchange and dipole–dipole interactions inside the magnetic waveguide, as well as the influence of surface and volume magnetocrystalline anisotropy field. From the condition of vanishing of the determinant of $\mathbf{D}^{mm}_{nn;qq}$, one obtains the spectrum of non-interacting spin-wave modes. The second matrix $\mathbf{R}^{mm}_{nn';qq'}$ takes in account an inter-mode dipole–dipole interaction which gives rise to the dipole gaps in the spin-wave spectrum near the points of the frequency degeneracy. Consequently, the two first terms in (16.19), while taken for $m = 0$, give the dipole–exchange spectrum of the regular magnonic waveguide. The last term in (16.19) takes into account the effect of periodical variation of the magnetic parameters along the magnonic waveguide. This periodical variation causes the appearance of frequency bandgaps in the spectrum at the edges of the Brillouin zones due to the dipole–dipole interaction of different space harmonics. Thus, the matrix element $\mathbf{L}^{mm'}_{nn';qq'}$ describes the dipole–dipole interaction between space harmonics of all spin-wave modes.

Different matrix elements in (16.20)–(16.21) represent the contributions from different interactions to the total energy of spin-wave modes. Namely, the matrix element $\mathrm{H}^{mm'}_{nn';qq'}$ includes the influence of Zeeman energy and its periodic variation on the final spectrum. This matrix element also takes into account the nonuniform character of the demagnetizing field inside a finite-width MCW. $\mathrm{F}^{mm'}_{nn';qq'}$ represents the contribution of the exchange energy and gives rise to the repulsion of the space harmonics, which is due to the variation of the exchange term. The matrix element ${}^{\alpha\beta}\mathrm{N}^{mm'}_{nn';qq'}$ describes the effects of volume magnetocrystalline anisotropy on the spectrum of spin waves in MCW. The element ${}^{\alpha\beta}\mathrm{G}^{mm'}_{nn';qq'}$ describes the influence of the dipole field and the dipole–dipole interaction between the spin-wave modes and space harmonics inside the magnonic crystal waveguide.

We underline that the infinite system (16.19) gives an exact description of the linear processes in an anisotropic magnonic crystal waveguide and enables one to obtain expressions describing the spin-wave spectrum and the transverse distribution of variable magnetization (16.15) in a magnonic crystal waveguide. The condition of vanishing of the determinant of the infinite system (16.19) yields the exact dispersion relation for propagating dipole–exchange spin waves in the anisotropic magnonic crystal waveguide. So, the determination of the exact dispersion law is reduced to the calculation of eigenvalues of the block matrix of the infinite system (16.19).

An inspection of the block matrix in (16.19) reveals that its determinant will be of the Hill's type, so that the eigenvalues in each particular case may be evaluated by a truncation of this determinant. An appropriate theorem on infinite determinants can be found in [61]. After truncation, system (16.19) can be solved by a standard numerical procedures designed for solving hermitian matrix eigenproblems.

The numerical analysis of (16.19) shows that the dipole–exchange spin-wave spectrum of the magnonic crystal waveguide comprises an infinite number of allowed and stop bands. These bands can be numbered according to the numbers of corresponding modes of the spin-wave resonances and numbers of space harmonics.

16.4 Example of Theory Application

The above developed theory was recently used for the theoretical investigation of the field-induced magnonic crystal that was realized through positioning a meander-like structure of the conducting stripes with the electric current near the ferromagnetic-film waveguide (Fig. 16.3). As such an electric current structure produces a periodic magnetic Oersted field in the ferromagnetic waveguide, it can be referred to as MCW. This type of the MCW was experimentally investigated in [10, 23]. A renewed interest to such a type of the magnonic crystals is associated with the possibility to dynamically tune the waveguiding parameters during the spin-wave propagation. As it was already mentioned above, such a dynamical tuning allowed one to obtain several interesting physical effects, namely, all-linear time reversal of pulsed microwave signals, propagated in ferromagnetic film, frequency inversion, etc. [23]. Previously elaborated theories take into account only a sinusoidal form of the Oersted field produced by a conducting array of stripes. However, it is well-known that the Oersted field, produced by a set of current stripes, has a complicated

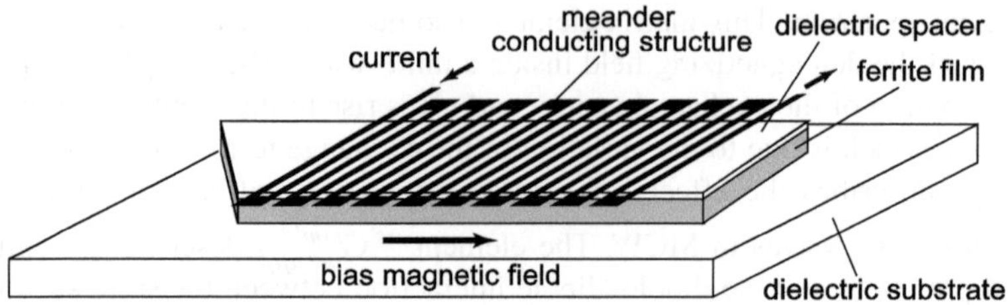

Fig. 16.3 Considered structure of the dynamic magnonic crystal [10]

non-sinusoidal spatial profile. Its Fourier spectrum consists of more than one space harmonics. Therefore, strictly speaking, the first Born approximation is not valid in this case. The application of the above presented theory provides a possibility to calculate the dipole–exchange spin-wave spectrum obtained for a non-sinusoidal periodic variation of the external magnetic field.

To describe the considered model structure, we assume that the spatial distribution of the bias magnetic field has a trapezoidal form:

$$\mathbf{H}_0(\mathbf{r}) = \sum_{m'=-\infty}^{+\infty} \mathbf{H}_{0m'} e^{-ik_{m'}z}, \tag{16.23}$$

where the Fourier coefficients $\mathbf{H}_{0m'}$ for the trapezoidal form of the bias magnetic field modulation can be determined as:

$$\mathbf{H}_{0m'} = i\frac{\delta H \Lambda}{2a\pi^2(m')^2} \sin\left(\frac{2m'a\pi}{\Lambda}\right)\left(1-(-1)^{m'}\right). \tag{16.24}$$

Here δH is the amplitude of the bias magnetic field modulation and a/Λ is a degree of keystoning (at $a/\Lambda \to 0$, we arrive to the meander-type of modulation), and Λ is a period of the structure.

The results of the numerical calculations of the spin-wave spectrum in the field-induced magnonic crystal are presented in Fig. 16.4.

These results sufficiently well describe the experimental data presented in [10]. It should be stressed that the trapezoidal space distribution of the magnetic modulation field utilized in the calculations gives a more realistic result than a sinusoidal distribution. For the same set of parameters, the sinusoidal variation of the external magnetic field produces the first frequency bandgap that is too narrow to describe correctly the first bandgap observed in the experiment [10]. On the other hand, the meander perturbation produces several other bandgaps in the spin-wave spectrum, which were not observed in the cited experiment.

The calculated parameters of the first bandgap are: the central frequency of the bandgap is $\omega_c = 6.516$ GHz and the gap width is $\Delta\omega \approx 13$ MHz. The next three bandgaps are much smaller, and therefore cannot be detected experimentally, which is also in agreement with the cited data [10]. This example shows clearly that even in a very simple case of the bias magnetic field modulation an exact multi-space-harmonic theory is needed to describe the real experiment.

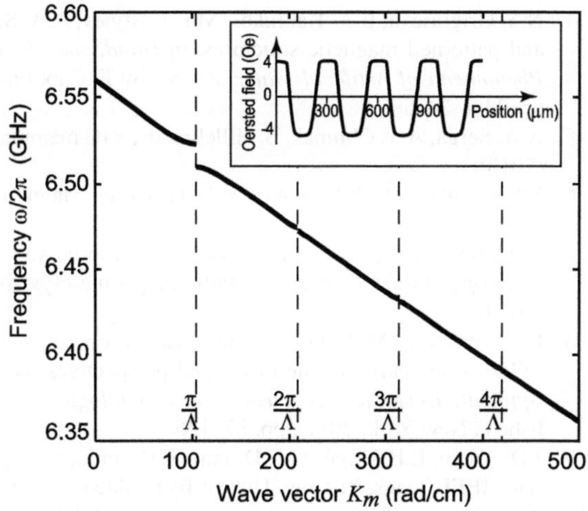

Fig. 16.4 Dispersion characteristic of the dynamic magnonic crystal waveguide. Parameters for the calculations are taken from [10]: $4\pi M_S = 1750$ G, $H_0 = 1625$ Oe, $L = 5$ μm, $\Lambda = 300$ μm, $\delta H = 4$ Oe

16.5 Conclusion

The theory for the spin-wave spectrum in a magnonic crystal waveguide is build. The developed theory takes into account major peculiarities of the dipole–exchange spin-wave propagation in a magnonic crystal waveguide. The arbitrary and simultaneous variation of the main magnetic parameters is considered. The dipole–dipole and exchange interactions are taken into account as well as the magnetocrystalline anisotropy of the ferromagnetic phase. Optionally, the retardation effects can be incorporated in the elaborated theory. The universality of the spin-wave modes approach makes the elaborated theory applicable to a wide class of the magnetic structures such as thin films, finite-width waveguides, layered ferrite–ferroelectric structures, layered structures with metallic screens, multilayered ferrite–dielectric structures, etc. Moreover, using the presented theory for all these structures the dipole–exchange spin-wave spectrum can be obtained in the presence of periodic modulation of several magnetic parameters along the spin-wave propagation direction. The obtained solutions have such a form that they provide a possibility for clear physical interpretation of the particular dispersion properties. An example of application of the developed theory to the calculation of the spin-wave spectrum in a field-induced magnonic crystal waveguide is presented.

References

1. S.A. Nikitov, P. Tailhades, C.S. Tsai, Spin waves in periodic magnetic structures – magnonic crystals. J. Magn. Magn. Mater. **236**(3), 320–330 (2001)
2. H. Puszkarski, M. Krawczyk, Magnonic crystals – the magnetic counterpart of photonic crystals. Solid State Phenom. **94**, 125–134 (2003)
3. V.E. Demidov, S.O. Demokritov, K. Rott, P. Krzysteczko, G. Reiss, Linear and nonlinear spin-wave dynamics in macro- and microscopic magnetic confined structures. J. Phys. D, Appl. Phys. **41**, 164012 (2008)

4. N.Y. Grigorieva, B.A. Kalinikos, M.P. Kostylev, A.A. Stashkevich, Spin waves in multilayered and patterned magnetic structures, in *Handbook of Artificial Materials, Vol. I: Theory and Phenomena of Artificial Materials*, ed. by F. Capolino (Taylor and Francis, Oxford, 2009), pp. 34-1–34-68
5. A.A. Serga, A.V. Chumak, B. Hillebrands, YIG magnonics. J. Phys. D, Appl. Phys. **43**, 264002 (2010)
6. V.V. Kruglyak, S.O. Demokritov, D. Grundler, Magnonics. J. Phys. D, Appl. Phys. **43**, 264001 (2010)
7. G. Gubbiotti, S. Tacchi, M. Madami, G. Carlotti, A.O. Adeyeye, M. Kostylev, Brillouin light scattering studies of planar metallic magnonic crystals. J. Phys. D, Appl. Phys. **43**, 264003 (2010)
8. I.V. Zavislyak, M.A. Popov, Microwave properties and applications of yttrium iron Garnet (YIG) films: current state of art and perspectives, in *Yttrium: Compounds, Production and Applications (Materials Science and Technologies)*, ed. by B.D. Volkerts (Nova Science Publishers, New York, 2011), pp. 87–126
9. J.D. Adam, L.E. Davis, G.F. Dionne, E.F. Schloemann, S.N. Stitzer, Ferrite devices and materials. IEEE Trans. Microw. Theory Tech. **50**(3), 721–737 (2002)
10. A.V. Chumak, T. Neumann, A.A. Serga, B. Hillebrands, M.P. Kostylev, A current-controlled, dynamic magnonic crystal. J. Phys. D, Appl. Phys. **42**, 205005 (2009)
11. N.Y. Grigoryeva, R.A. Sultanov, B.A. Kalinikos, Dual-tunable hybrid wave hexaferrite-ferroelectric millimeter-wave phase shifter. Electron. Lett. **47**(1), 35–36 (2011)
12. M.G. Cottam (ed.), *Linear and Nonlinear Spin Waves in Magnetic Films and Superlattices* (World Scientific, Singapore, 1994)
13. E. Wigen (ed.), *Nonlinear Phenomena and Chaos in Magnetic Materials* (World Scientific, Singapore, 1994)
14. G. Srinivasan, A.N. Slavin (eds.), *High Frequency Processes in Magnetic Materials* (World Scientific, Singapore, 1995)
15. B. Hillebrands, K. Ounadjela (eds.), *Spin Dynamics in Confined Magnetic Structures*, vol. I (Springer, Berlin, 2002)
16. B. Hillebrands, K. Ounadjela (eds.), *Spin Dynamics in Confined Magnetic Structures*, vol. II (Springer, Berlin, 2003)
17. S.O. Demokritov (ed.), *Spin Wave Confinement* (Pan Stanford Publishing Pte. Ltd., Singapore, 2009)
18. L. Brillouin, M. Parodi, *Propagation des Ondes dans les Milieux Periodiques* (Malson Dunod, Paris, 1956)
19. N.-N. Chen, A.N. Slavin, M.G. Cottam, Gap solitons in periodic structures: modulated magnetic thin films. Phys. Rev. B **47**(14), 8667–8671 (1993)
20. A.B. Ustinov, N.Y. Grigorieva, B.A. Kalinikos, Observation of spin-wave envelope solitons in periodic magnetic film structures. JETP Lett. **88**(1), 31–35 (2008)
21. A.B. Ustinov, B.A. Kalinikos, V.E. Demidov, S.O. Demokritov, Formation of gap solitons in ferromagnetic films with a periodic metal grating. Phys. Rev. B **81**, 180406(R) (2010)
22. A.V. Drozdovskii, M.A. Cherkasskii, A.B. Ustinov, N.G. Kovshikov, B.A. Kalinikos, Formation of envelope solitons of spin-wave packets propagating in thin-film magnon crystals. JETP Lett. **91**(1), 16–20 (2010)
23. A.V. Chumak, V.S. Tiberkevich, A.D. Karenowska, A.A. Serga, J.F. Gregg, A.N. Slavin, B. Hillebrands, All-linear time reversal by a dynamic artificial crystal. Nat. Commun. **94**, 141–146 (2010)
24. A.V. Chumak, P. Dhagat, A. Jander, A.A. Serga, B. Hillebrands, Reverse Doppler effect of magnons with negative group velocity scattered from a moving Bragg grating. Phys. Rev. B **81**, 140404(R) (2010)
25. J. McCord, T. Gemming, L. Schultz, J. Fassbender, M.O. Liedke, M. Frommberger, E. Quandt, Magnetic anisotropy and domain patterning of amorphous films by He-ion irradiation. Appl. Phys. Lett. **86**, 162502 (2005)

26. S.G. Reidy, L. Cheng, W.E. Bailey, Dopants for independent control of precessional frequency and damping in $Ni_{81}Fe_{19}$ (50 nm) thin films. Appl. Phys. Lett. **82**, 1254–1257 (2003)
27. A.V. Chumak, P. Pirro, A.A. Serga, M.P. Kostylev, R.L. Stamps, H. Schultheiss, K. Vogt, S.J. Hermsdoerfer, B. Laegel, P.A. Beck, B. Hillebrands, Spin-wave propagation in a microstructured magnonic crystal. Appl. Phys. Lett. **95**, 262508 (2009)
28. J. Wang, A.O. Adeyeye, N. Singh, Magnetostatic interactions in mesoscopic $Ni_{80}Fe_{20}$ ring arrays. Appl. Phys. Lett. **87**, 262508 (2005)
29. S.L. Vysotskii, S.A. Nikitov, Y.A. Filimonov, Magnetostatic waves in 2D periodic structures – magneto-photonic crystals. J. Exp. Theor. Phys. **128**(3), 636–644 (2005)
30. A.V. Chumak, A.A. Serga, S. Wolff, B. Hillebrands, M.P. Kostylev, Design and optimization of one-dimensional ferrite-film based magnonic crystals. J. Appl. Phys. **105**, 083906 (2009)
31. A.V. Voronenko, S.V. Gerus, V.D. Kharitonov, Bragg diffraction of surface magnetostatic waves from magnetic lattices. Russ. Phys. J. **31**, 915–923 (1988)
32. J. Das, B.A. Kalinikos, A.R. Barman, C.E. Patton, Multifunctional dual-tunable low loss ferrite–ferroelectric heterostructures for microwave devices. Appl. Phys. Lett. **91**, 172516 (2007)
33. N.Y. Grigoryeva, B.A. Kalinikos, Dispersion characteristics of dual-tunable 1D magnonic crystal, in *Proc. of Joint European Magnetic Symposia (JEMS 2010)*, Krakow, Poland, 23–28 August (2010), p. 72
34. C. Elachi, Waves in active and passive periodic structures: a review. IEEE Proc. **64**(12), 1666–1698 (1976)
35. B.A. Kalinikos, Dipole-exchange spin-wave spectrum of magnetic films, in *Linear and Nonlinear Spin Waves in Magnetic Films and Superlattices*, ed. by M.G. Cottam (World Scientific, Singapore, 1994), pp. 89–156
36. C. Elachi, Magnetic wave propagation in a periodic medium. IEEE Trans. Magn. **MAG-11**, 36–39 (1975)
37. N.Y. Grigorieva, B.A. Kalinikos, Dispersion characteristics of spin waves in planar periodic structures based on ferromagnetic films. Tech. Phys. **54**(8), 1196–1203 (2009)
38. S.L. Vysotskii, E.N. Beginin, S.A. Nikitov, E.S. Pavlov, Y.A. Filimonov, Effect of ferrite magnonic crystal metallization on Bragg resonances of magnetostatic surface waves. Tech. Phys. Lett. **37**(11), 1024–1026 (2011)
39. K.H. Chi, Y. Zhu, R.W. Mao, J.P. Dolas, C.S. Tsai, An approach for analysis of magnetostatic volume waves in magnonic crystals. J. Appl. Phys. **109**, 07D320 (2011)
40. M. Krawczyk, H. Puszkarski, Plane-wave theory of three-dimensional magnonic crystals. Phys. Rev. B **77**, 054437 (2008)
41. H. Puszkarski, M. Krawczyk, J.-C.S. Lévy, D. Mercier, Ferromagnetic layered composites. Transfer matrix approach. Acta Phys. Pol. A **100**, 195–214 (2001)
42. M.P. Kostylev, A.A. Stashkevich, N.A. Sergeeva, Collective magnetostatic modes on a one-dimensional array of ferromagnetic stripes. Phys. Rev. B **69**, 064408 (2004)
43. V.V. Kruglyak, R.J. Hicken, A.N. Kuchko, V.Y. Gorobets, Spin waves in a periodically layered magnetic nanowire. J. Appl. Phys. **98**, 014304 (2005)
44. L. Dobrzynski, B. Djafari-Rouhani, H. Puszkarski, Theory of bulk and surface magnons in Heisenberg ferromagnetic superlattices. Phys. Rev. B **33**, 3251–3256 (1986)
45. V.V. Kruglyak, A.N. Kuchko, V.I. Finokhin, Spin-wave spectrum of an ideal multilayer magnet upon modulation of all parameters of the Landau–Lifshitz equation. Phys. Solid State **46**(5), 867–871 (2004)
46. B.A. Kalinikos, A.N. Slavin, Theory of dipole–exchange spin wave spectrum for ferromagnetic films with mixed exchange boundary conditions. J. Phys. C, Solid State Phys. **19**, 7013–7033 (1986)
47. A.G. Gurevich, G.A. Melkov, *Magnetization Oscillations and Waves* (CRC Press, Cleveland, 1996)
48. J.A. Osborn, Demagnetizing factors of the general ellipsoid. Phys. Rev. **67**(11–12), 351–357 (1945)

49. R.I. Joseph, E. Schlomann, Demagnetizing field in nonellipsoidal bodies. J. Appl. Phys. **36**(5), 1579–1593 (1965)
50. F. Ciubotaru, A.V. Chumak, A.A. Serga, P. Pirro, N.Y. Grigoryeva, B. Hillebrands, Single band gap microstructured magnonic crystal. J. Phys. D, Appl. Phys. **45**(25), 255002 (2012)
51. B.A. Kalinikos, M.P. Kostylev, N.V. Kozhus, A.N. Slavin, The dipole–exchange spin wave spectrum for anisotropic ferromagnetic films with mixed exchange boundary conditions. J. Phys. Condens. Matter **2**, 9861–9877 (1990)
52. R. Petit, Electromagnetic grating theories: limitations and successes. Nouv. Rev. Opt. **6**(3), 129–135 (1975)
53. Y.V. Gulyaev, S.A. Nikitov, V.P. Plesskii, Reflection of manetostatic waves from periodically corrugated area of the ferrite surface. J. Commun. Technol. Electron. **11**, 2282–2290 (1981)
54. E. Popov, M. Neviére, Grating theory: new equations in Fourier space leading to fast converging results for TM polarization. J. Opt. Soc. Am. A **17**(10), 1773–1784 (2000)
55. V.E. Demidov, B.A. Kalinikos, P. Edenhofer, Dipole–exchange theory of hybrid electromagnetic-spin waves in layered film structures. J. Appl. Phys. **91**(12), 10007–10016 (2002)
56. B.A. Kalinikos, P.A. Kolodin, Excitation of propagating dipole–exchange spin waves in ferromagnetic double-film structure. IEEE Trans. Magn. **28**(5), 3204–3206 (1992)
57. A.R. Baghai-Wadji, Theory and applications of Green's functions. Int. Int. J. High Speed Electron. Syst. **10**(4), 949–1015 (2000)
58. G.T. Rado, J.R. Weertman, Spin-wave resonance in a ferromagnetic metal. J. Phys. Chem. Solids **11**, 315–333 (1959)
59. W.S. Ament, G.T. Rado, Electromagnetic effects of spin wave resonance in ferromagnetic metals. Phys. Rev. **97**(6), 1558–1566 (1955)
60. C. Kittel, Excitation of spin waves in a ferromagnet by a uniform rf field. Phys. Rev. **110**(6), 1295–1297 (1958)
61. L.V. Kantorovich, V.I. Krilov, *Appropriate Methods of Higher Analysis* (Interscience, New York, 1958). Sect. 1.3

Chapter 17
The Dynamic Magnonic Crystal: New Horizons in Artificial Crystal Based Signal Processing

Andrii V. Chumak, Alexy D. Karenowska, Alexander A. Serga, and Burkard Hillebrands

Abstract In this chapter, we describe the development and properties of the first experimental dynamic magnonic crystal devices and highlight certain aspects of the intriguing new physics that they have to offer us. We discuss the significance of the dynamic magnonic crystal both in the context of the furtherance and technological application of magnonics, and in the understanding of *general* wave dynamics in metamaterial systems.

17.1 Introduction

Artificial crystals are wave transmission structures that belong to the category of man-made media known as "metamaterials" [1]. A metamaterial is a synthetic material that has properties derived from an engineered mesoscopic (rather than atomic-scale, or molecular) structuring. The defining feature of an artificial crystal is an "artificial lattice" formed by wavelength-scale modulations in one or more of its material properties.

The transmission spectra of artificial crystals have a wide range of interesting features, the most notable of these are bandgaps: frequency intervals over which wave propagation is prohibited. Bandgap formation in artificial crystal systems occurs as a result of resonant wave-lattice interactions analogous to the atomic-scale

A.V. Chumak · A.A. Serga (✉) · B. Hillebrands
Fachbereich Physik and Forschungszentrum OPTIMAS, Technische Universität Kaiserslautern, 67663 Kaiserslautern, Germany
e-mail: serga@physik.uni-kl.de

B. Hillebrands (✉)
e-mail: hilleb@physik.uni-kl.de

A.V. Chumak
e-mail: chumak@physik.uni-kl.de

A.D. Karenowska
Department of Physics, Clarendon Laboratory, University of Oxford, OX1 3PU Oxford, UK
e-mail: a.karenowska@physics.ox.ac.uk

S.O. Demokritov, A.N. Slavin (eds.), *Magnonics*, Topics in Applied Physics 125,
DOI 10.1007/978-3-642-30247-3_17, © Springer-Verlag Berlin Heidelberg 2013

Bragg scattering encountered in the context of X-ray or electron crystallography of natural crystals [2].

Several types of artificial crystal structure are actively studied, the best known being optical photonic crystals [3, 4] and acoustic phononic crystals [5, 6]. Over the last decade however, a third category of artificial crystal system has stepped into the limelight: the spin-wave embodiment of the artificial crystal, the *magnonic* crystal. This relative newcomer has not only played a pivotal role in the emergence of the field of magnonics, but is more and more widely appreciated [7] as a unique and valuable tool in the study of general wave dynamics in metamaterial systems.

17.2 The State-of-the-Art in Magnonic Crystal Science

Like photonic and phononic crystals, magnonic crystals may be one-, two-, or three-dimensional. However, the understanding and study of two- and three- dimensional structures [8–15] is complicated by the strongly anisotropic dispersion which typifies experimental spin-wave systems.[1] For this reason, the majority of studies to date have focused either on spin-wave propagation in planar magnonic crystals created by periodic patterning or structuring of one-dimensional waveguides [17, 19–37], or on relatively simple two-dimensional crystals formed from magnetic dot or anti-dot lattices [8–13].

Two materials dominate experimental spin-wave studies: monocrystalline yttrium iron garnet (YIG, $Y_3Fe_5O_{12}$) and the polycrystalline metallic alloy Permalloy (Py, $Ni_{81}Fe_{19}$). Yttrium iron garnet is an electrically insulating ferrimagnet, in monocrystalline form its spin-wave damping is uniquely low,[2] allowing propagating spin waves to be observed over centimeter distances [17]. Relatively large-scale magnonic crystals fabricated from monocrystalline YIG films (typical thicknesses in the micron range and lateral dimensions of order several millimeters) provide the perfect model system for the detailed study of the underlying physics of spin-wave propagation in magnonic crystal systems. YIG is perhaps fundamental magnonic crystal science's single most valuable asset [17, 28, 38–48].

The spin-wave free path in thin-film Permalloy is generally of order tens of microns; though substantially larger than in YIG, the material's spin-wave damping is small in comparison with other polycrystalline magnetic metals and alloys. The ease which Py can be deposited and patterned makes it an attractive material for experimental studies of complex geometries and micro- and nanoscale systems [10, 29, 30].

As is evident from the table of contents of this book, there has been a recent explosion of interest in the use of magnons as information carriers in spintronic devices. Capitalizing on elements which convert between information coded into the spin or charge of electrons and magnons, "magnon spintronics" – a concept

[1] For an accessible and enlightening introduction to anisotropic spin-wave dispersion, see [16–18].

[2] Approximately two orders of magnitude lower than all other practical experimental materials.

increasingly widely acknowledged as a field of research in its own right – promises to combine all the advantages of conventional spintronics with those of a uniquely versatile wave-based information platform. Magnonic crystals potentially provide the basis for a wide range of magnon spintronic devices including signal filters, phase shifters, isolators, and – as we shall see later in this chapter – signal processing elements.

As alluded to in the introduction, the field of magnonic crystal science is still in its infancy. Though some studies of periodic spin-wave transmission media were made as early as the 1970s [19], the magnonic crystal concept was not properly recognized until around the turn of the twenty-first century [8]. Whilst it is likely to be some time before they become as well-known as their optical and acoustic cousins, magnonic crystals are rapidly gaining recognition as structures which have much to contribute not only to the study and technological application of spin waves, but also to our general understanding of wave dynamics in metamaterial systems. The magnonic crystal's significance in the context of general wave dynamics derives from the fact that the structures open doors to the experimental study of a range of general wave effects which are extremely difficult to observe elsewhere.

The structure at the focus of this chapter: the *dynamic* magnonic crystal provides a particularly good illustration of the contribution magnonic crystal physics has to make to our fundamental understanding of wave dynamics in metamaterials.

The dynamic magnonic crystal (DMC) is a magnonic crystal which features an artificial lattice that can be switched "off" and "on" on a timescale which is short in comparison with time taken for a spin-wave packet to propagate through it [41]. In the wider context of artificial crystals, this fast-switching functionality is unique. Though it is possible to create photonic crystals with time-varying properties, systems in which substantial (as opposed to subtle) changes to the crystal lattice can be made on optical timescales are, as yet, beyond experimental reach.

In what follows, we describe the development and properties of the first experimental DMC devices, and offer a taste of the intriguing new physics that they have to offer us.

17.3 The Design of Dynamic Magnonic Crystals

A schematic diagram of the dynamic magnonic crystal design at the focus of the work described in this chapter is shown in Fig. 17.1. The DMC comprises a spin-wave waveguide made from a 5 μm thick YIG film grown by liquid phase epitaxy on a gallium gadolinium garnet substrate. A bias magnetic field $B_0 = 160$ mT, applied along the axis of the waveguide supports the propagation of backward volume magnetostatic spin waves[3] (BVMSWs) [16]. The dynamic artificial crystal lattice

[3]Note that backward volume magnetostatic spin waves (BVMSWs) are called so because their dispersion relation $\omega(k)$ has a negative slope, implying that the group and phase velocities of the waves are in opposite directions. We stress, however, that none of the results we describe later in this chapter hinge on this unusual physics; all may be generalized to any waves having a symmetric dispersion function $\omega(k) = \omega(-k)$ propagating in a dynamic artificial crystal structure.

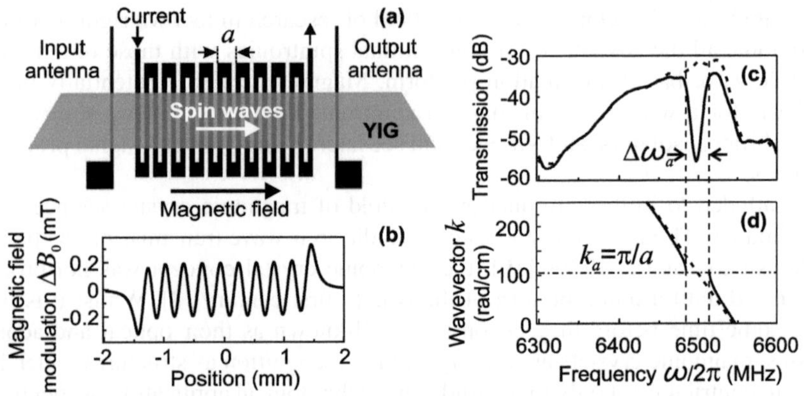

Fig. 17.1 (**a**) The dynamic magnonic crystal (DMC) used in the experiments described in this chapter is fabricated from an yttrium iron garnet spin-wave waveguide with a metallic meander structure having 20 periods of lattice constant $a = 300$ μm (10 shown) positioned $s = 100$ μm from its surface. Two microstrip antennas (input and output) are arranged on the surface of the waveguide and a bias magnetic field B_0 is applied along its length. (**b**) When a current is applied to the meander structure, the magnetic field it produces spatially modulates the waveguide's static magnetic bias field (calculated field distribution shown, park-to-peak amplitude ΔB_0) and the DMC is "on". (**c**) Spin wave transmission characteristics of the DMC in the "off" state (no current applied to the meander structure, *dotted line*) and in the "on" state (static current of 1 A in the meander structure, *solid line*). In the latter case, there is a bandgap corresponding to the wavevectors $\pm k_a = \pm\pi/a$ centered at $\omega_a = 2\pi \cdot 6500$ MHz of -3 dB width approximately 30 MHz. (**d**) Calculated spin-wave dispersion relationship for the DMC system in the "off" (*dotted line*) and "on" (*solid line*) states

is created using a metallic "meander" structure: a planar array of parallel copper strips connected together alternately at their top and bottom edges and positioned close to the surface of the waveguide (Fig. 17.1(a)) [41–43]. If a current I is passed through the meander structure, the magnetic field it produces gives rise to a small spatially periodic modulation (amplitude ΔB_0) of the YIG waveguide's static bias magnetic field B_0. Since the dispersion of spin waves in the YIG is strongly dependent on their local magnetic environment, this modulation defines an artificial crystal lattice (Fig. 17.1(b)). The meander structure comprises 40 series-connected conducting strips of width 75 μm and thickness 10 μm (perpendicular to the plane of the substrate), spaced by 75 μm [41]. This geometry gives an artificial crystal of total length 5.92 mm, having $N = 20$ periods of lattice constant $a = 300$ μm. The structure was patterned using photolithography; a high thermal conductivity substrate was chosen (aluminium nitride) so as to minimize heating of the waveguide.

We note that, in principle, as an alternative to the meander type current carrying structure, a grating-like arrangement with conducting wires arranged in parallel might also be employed. However, the meander configuration is preferable for two main reasons: Firstly, the effective per-unit-current field modulation it produces is larger, since the in-plane components of the magnetic field modulation produced by adjacent stripes oppose. Secondly, the magnetic field averaged over the structure is zero, independently of the magnitude of the current flowing through it.

An $s = 100$ μm thick SiO_2 spacer layer is positioned between the waveguide and the meander structure [41]. The thickness of this spacer layer is chosen so as to minimize direct electromagnetic interaction between the conducting wires and propagating spin waves in the YIG, whilst still maintaining sufficient field modulation to define the crystal lattice. If the conducting surface of the meander structure is placed in direct contact with the YIG, the electromagnetic interaction is strong: the conducting wires act as secondary antennas and have so profound an effect on the transmission properties of the waveguide that the system does not function as an artificial crystal [39, 40]. The interaction versus modulation compromise can be made successfully since the GHz dipole field from spin waves propagating in the YIG reduces exponentially with distance d from its surface, whilst the field modulation reduces as $1/d$, i.e., much more slowly.

Two microstrip antennas, 8 mm apart, are positioned approximately 1 mm from each end of the dynamic magnonic crystal: an input antenna to launch a spin-wave signal $S_{\text{in}}(t)$ into the DMC, and an output antenna to receive the signal $S_s(t)$ transmitted through it. (Note that the input antenna can also be used as a receiver to detect any signal $S_r(r)$ reflected, or traveling "back" from the DMC.)

Figure 17.1(c) shows the frequency dependence of spin-wave transmission through the DMC (for constant spin-wave input power). In the absence of a current flowing in the meander structure, the DMC is "off" and its transmission properties are those of a *uniform* spin-wave waveguide (dashed line): there is a transmission maximum close to $k = 0$ (the ferromagnetic resonance frequency of the film) and transmission reduces with decreasing frequency (increasing k) due to a fall in the excitation and detection efficiencies of the antennas [17]. When a current is applied to the meander structure, the DMC is "on" and a magnonic bandgap centered at the DMC's "Bragg" resonance frequency $\omega_a/2\pi = 6500$ MHz appears, corresponding to the Bragg wavevectors $\pm k_a = \pm\pi/a \simeq \pm 105$ cm^{-1}. The bandgap creates a deep trough in the experimental transmission characteristic; with a current of 1 A in the meander structure, the -3 dB width of this trough is approximately $\Delta\omega_a/2\pi = 30$ MHz (solid line). In Fig. 17.1(d), calculated spin-wave dispersion curves for the DMC in the "off" (dashed line) and "on" states (solid line) are shown.

17.4 The Engineering of Dynamic Magnonic Crystal Characteristics

Both the width of the bandgap and the efficiency with which waves having frequencies lying in the gap are excluded from the DMC are very sensitive to the strength of the magnetic field modulation ΔB_0, and therefore the current applied to the meander structure (see [41]). The larger the current in the meander structure, the deeper the magnetic field modulation which defines the artificial lattice, and thus the stronger its coupling to spin-wave excitations in the waveguide. The wave-lattice coupling is the principal factor in determining the bandgap width: the stronger the coupling, the wider the gap [43].

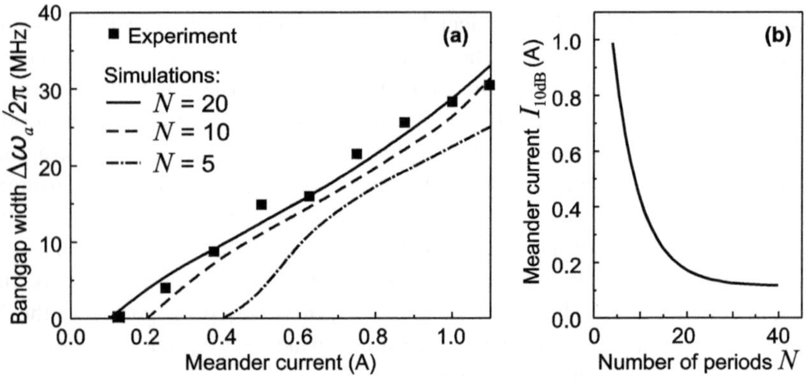

Fig. 17.2 (**a**) Comparison of experimentally measured -10 dB bandgap width of an $N = 20$ period dynamic magnonic crystal with the results of numerical simulations (*solid line*). Data are plotted as a function of the current applied to the meander structure. (**b**) Calculated meander current required to reduce transmission through the DMC in the vicinity of the bandgap by a factor of 10 (i.e., to suppress transmission by 10 dB)

The dependence of the edge-to-edge width of the bandgap (measured -10 dB down relative to the transmission through the system with the DMC "off") on the meander current is plotted in Fig. 17.2(a) (square symbols). It can be seen that the bandgap width is tunable over the range 0–30 MHz by applying a meander current between zero and 1.1 A. On account of this tuning capability, the DMC potentially finds application in a wide range of microwave devices including adjustable filters and isolators. It should be noted that the currents required to operate micro-scale DMCs would be several orders of magnitude smaller than those used to control our proof-of-concept structure.

In a series of investigations detailed in [41], we have performed numerical simulations of the operational characteristics of DMCs having different numbers of periods N. Our calculations are based on a numerical solution of an integral expression for the microwave-frequency dipole field of propagating spin waves obtained from the Landau–Lifshitz equation. The solid line in Fig. 17.2(a), calculated for a DMC of $N = 20$ periods, is in excellent agreement with our experimental data. The bandgap width is found to decrease as the number of DMC periods reduces. Above a minimum value of meander current (which depends on the number of crystal periods), the rejection band width increases monotonically with current. Once the current is above a certain threshold (also dependent on the number of crystal periods) a regime of linear current-width dependence is reached.

Figure 17.2(b) shows how the meander current required to achieve a bandgap of depth 10 dB ($I_{10\mathrm{dB}}$) varies with the number of crystal periods N. As we might intuitively expect, the lower N, the higher the current required. At low currents, the dependence of $I_{10\mathrm{dB}}$ on the number of crystal periods can be approximated by $N \sim 1/I$. This result can be related to the formula derived in [41] for the Fourier transform of the magnetic field which defines the magnonic crystal lattice. The fraction of the incident spin-wave energy reflected from the DMC in the vicinity of the bandgap is proportional to the Fourier amplitude at the spatial frequency $2\pi/a$ which is, in turn, dependent on the applied current and the number of periods N, and decreases exponentially with increasing spacing s between the meander structure and the waveguide. It follows that if the current is lowered, the reflection efficiency of the magnonic crystal may be maintained either by increasing the number of periods so that $N \cdot I = const$, or reducing the spacing s between the meander structure and the waveguide so that $\exp(-k_a s) \cdot I = const$ for $k_a = 2\pi/a$ (the spin-wave wavenumber at the center of the DMC bandgap).

17.5 A Signal See-Saw: Oscillatory Inter-modal Energy Exchange in the Dynamic Magnonic Crystal

Thus far, we have focused solely on the transmission properties of our proof-of-concept dynamic magnonic crystal when it is held continuously either in the "off" or the "on" state. We shall now explore the interesting physics which becomes accessible if the crystal is *switched* from "off" to "on" whilst a spin-wave packet is *inside*.

As described above, if a spin wave with wavevector $k_s \approx +k_a = +\pi/a$ is incident on the DMC whilst it is in the "on" state, it is reflected. However, if the crystal is *switched* from "off" to "on" whilst such a wave *is inside*, the situation is quite different [43]. In this fast-switching case, the incident wave becomes confined to the artificial crystal and coupled to a secondary wave having wavevector $k_r = k_s \pm 2k_a$ via the artificial crystal lattice. The coupling is pronounced when the frequencies of the incident and secondary waves $\omega_s = \omega(k_s)$ and $\omega_r = \omega(k_r)$ are close to each other, i.e., for $k_s \approx k_a$, and therefore *counter-propagating* waves $k_r = k_s - 2k_a \approx -k_a$. The dynamics of the complex amplitudes b_s and b_r of the interacting incident and secondary waves can be modeled by analogy with a system of coupled harmonic oscillators:

$$\frac{db_s}{dt} = i\omega_s b_s + i\Omega_a b_r, \tag{17.1a}$$

$$\frac{db_r}{dt} = i\omega_r b_r + i\Omega_a b_s, \tag{17.1b}$$

where Ω_a is a coupling constant (with units of radians per second) which is numerically equal to the half-width of the DMC bandgap. The frequencies of the oscillatory solutions of (17.1a), (17.1b), are:

$$\omega_{1,2} = \overline{\omega} \pm \sqrt{\Omega_a^2 + \delta\omega^2} = \overline{\omega} \pm \Omega, \tag{17.2}$$

where $\delta\omega = (\omega_s - \omega_r)/2$ is the difference in frequency between the incident and secondary waves, and $\overline{\omega} = (\omega_s + \omega_r)/2$ is their average frequency. Equation (17.2) describes the dispersion relation for a magnonic crystal in the "on" state in the steady-state limit with a bandgap of width $2\Omega_a$ centered at the resonance frequency $\omega_s = \omega_r = \omega_a$ (Fig. 17.1(d)).

In a fast-switching experiment, at the moment when the crystal is switched "on" ($t = 0$), the amplitude of the incident wave is maximum ($b_s = 1$), whilst the coupled secondary wave is absent ($b_r = 0$). Using these initial conditions in conjunction with (17.1a), (17.1b), we can solve for the time-dependence of the power $P_{s,r} = |b_{s,r}|^2$ of each wave:

$$P_s(t) = \cos^2(\Omega t) + \frac{\delta\omega^2}{\Omega^2}\sin^2(\Omega t), \tag{17.3a}$$

$$P_r(t) = \frac{\Omega_a^2}{\Omega^2}\sin^2(\Omega t). \tag{17.3b}$$

The equations above describe *oscillatory* energy exchange at frequency Ω (see (17.2)) between confined incident and secondary spin-wave modes. The maximum power fraction transferred to the secondary mode $F_{\max} = \Omega_a^2/\Omega^2 = \Omega_a^2/(\Omega_a^2 + \delta\omega^2)$ is approximately unity for waves having frequencies which lie within the bandgap ($|\delta\omega| < \Omega_a$) and is achieved after a characteristic energy exchange time

$$T = \frac{\pi}{2\Omega} = \frac{\pi}{2\sqrt{\Omega_a^2 + \delta\omega^2}}, \tag{17.4}$$

or odd integer multiplies thereof (see Fig. 17.3).

If the DMC is switched "off", the coupling is eliminated and the waves, which have powers given by the instantaneous solutions of (17.3a), (17.3b), travel in opposite directions out of it. The oscillatory energy exchange process may be investigated experimentally by measuring the ratio of the powers of these propagating waves for different coupling durations τ:

$$\kappa(\tau) = \frac{P_r(\tau)}{P_s(\tau)} = \frac{\Omega_a^2\tan^2(\Omega\tau)}{\Omega^2 + \delta\omega^2\tan^2(\Omega\tau)}, \tag{17.5}$$

which – bar a constant factor – is independent of their damping. Since the process of oscillatory energy exchange between the incident and secondary waves is efficient only over a relatively narrow range of wavevectors $k_s \approx k_a$, the dispersion relation can be approximated by a first order Taylor expansion, $\omega_s = \omega(k_s) = \omega_a + v_g(k_s - k_a)$, $\omega_r = \omega(k_r) = \omega_a - v_g(k_r + k_a)$, where $v_g = d\omega(k)/dk$ is the spin-wave group velocity [44] at $k = k_a$, and we assume a symmetric dispersion law $\omega(k) = \omega(-k)$:

$$\omega_r = 2\omega_a - \omega_s, \tag{17.6}$$

i.e., the incident and secondary mode frequencies are related by *inversion* of their detuning from the bandgap center frequency ω_a.

The mode coupling mechanism was investigated experimentally using a current pulse $I_0(t)$ of duration τ to switch the DMC "on" whilst an incident spin-wave signal packet of center frequency ω_s and duration 200 ns (excited by a microwave

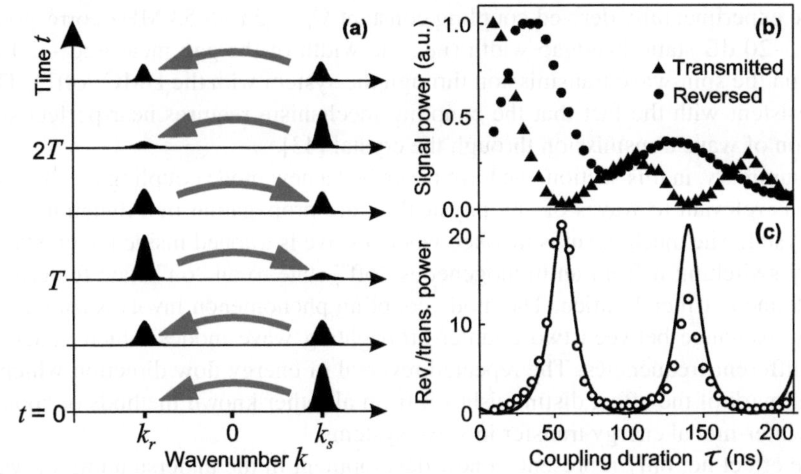

Fig. 17.3 (**a**) Schematic diagram illustrating the process of oscillatory energy exchange in the dynamic magnonic crystal system. If the DMC is switched "on" at $t = 0$ whilst an incident mode $k_s \approx k_a$ is inside, this mode is coupled to a secondary mode $k_r = k_s - 2k_a$. The heights of the peaks centered at k_s and k_r represent the power in the incident and secondary modes. Near complete energy exchange occurs every T seconds for as long as the DMC remains "on". (**b**) Experimental reversed (P_r, *circular symbols*) and transmitted (P_s, *triangular symbols*) signal powers as a function of the coupling duration τ. (**c**) Ratio of reversed to transmitted signal powers as a function of the coupling duration τ, the *solid line* is theory, the *open circular symbols* are points corresponding to the experimental data of (**b**)

current applied to the input antenna) was contained within it [43]. As soon as the DMC is switched "on" at $t = 0$, a transfer of energy from the incident k_s mode into the secondary k_r mode initiates (see Fig. 17.3(a)), both modes being confined to the crystal. At $t = \tau$ when the crystal is switched "off" and the coupling deactivated, the k_s mode propagates to the output antenna, producing a "transmitted" signal of power $S_s \propto P_s(\tau)$, whilst the k_r mode travels to the input antenna, originating a "reversed" signal of power $S_r \propto P_r(\tau)$.

The antiphase variation of the transmitted and reversed signal powers with the coupling time τ predicted by (17.1a), (17.1b) is confirmed by the experimental data of Fig. 17.3(b) for $\omega_s = \omega_r = \omega_a$. Figure 17.3(c) shows the ratio of the reversed to transmitted signals of Fig. 17.3(b) (open circular symbols) and the corresponding theory ($\kappa = P_r/P_s$, solid line). For the theoretical curve, we assume $\Omega_a = 2\pi \cdot 5.53$ MHz, a small detuning $\delta\omega = 0.03$ %, and an effective coupling duration $\tau_{\mathrm{eff}} = \tau - 2\Delta\tau$ where $2\Delta\tau = 12.5$ ns to account for the effect of the rise and fall times of the current pulse applied to the meander structure. The curve is also scaled to match the experimental and theoretical maximum ratios (absolute values being dependent on the efficiency of the antennas and losses). Note that the reduction in the signal amplitudes of Fig. 17.3(b) with increasing τ occurs as a result of damping, while the difference in the heights of the two peaks in Fig. 17.3(c) is a result of the finite spectral width of the input spin-wave signal.

The experimentally derived coupling constant $\Omega_a = 2\pi \cdot 5.53$ MHz corresponds to the -20 dB static bandgap width (i.e., the width of the gap measured -20 dB down on the spin-wave transmission through the system with the DMC "off"). This is consistent with the fact that the coupling mechanism requires near-perfect suppression of wave transmission through the crystal [43].

In summary, in this section, we have described a new mode coupling mechanism, which is relevant to waves of any physical nature propagating in a dynamic artificial crystal. The mechanism is invoked when a wave is trapped inside the crystal by rapidly switching it from an homogeneous "off" state to an "on" state featuring a well defined artificial lattice. The mode coupling phenomenon involves *oscillatory* energy exchange between two counter-propagating wave modes which generally have different frequencies. The repeated reversal of energy flow direction which is the hallmark of the effect distinguishes it from all other known methods of spontaneous inter-modal energy transfer in wave systems.

The effect not only represents a new development in the understanding of wave dynamics in metamaterial systems, but potentially finds technological application in the context of magnonic and magnon spintronic information processing. Moving forward from our large-scale proof-of-concept YIG based DMC, the development of micro-scale structures (based, for example, on Permalloy waveguides) having significantly smaller operating currents (below 1 mA) is – using modern fabrication techniques – not only a realistic near-field vision, but an exciting next step towards the development of the first DMC based information processing devices.

17.6 Frequency Conversion and Time Reversal in the Dynamic Magnonic Crystal

In this, the penultimate section of the chapter, we demonstrate that the mode coupling mechanism described in Sect. 17.5 has further hidden depth: We show that the oscillatory energy exchange phenomenon can be used to perform frequency conversion and time reversal of propagating signals, enhancing further still its status as an interesting and potentially technologically significant fundamental result [42].

The ability to perform time reversal of propagating signals and wavepackets *directly* (i.e., without recourse to signal storage) opens doors to a wide range of powerful information processing functionalities. For this reason, the development and understanding of time-reversal techniques has long been a prominent focus of research activity [49–52].

Time-reversing a signal is equivalent to inverting its spectrum about a reference frequency. This equivalence is best appreciated by considering the Fourier domain description of an arbitrary signal envelope $S(t) = \sum_{-\infty}^{\infty} S_m e^{\mathrm{i}(\omega_\mathrm{R}-\omega_m)t}$, where ω_R is a reference frequency which may take any value (including zero), and the sum is over all of the spectral components ω_m. The time-reversed envelope $S(-t)$ is then described by $S(-t) = \sum_{-\infty}^{\infty} S_m e^{-\mathrm{i}(\omega_\mathrm{R}-\omega_m)t}$, from which it becomes clear that the transformation $t \to -t$ is indistinguishable from the transformation

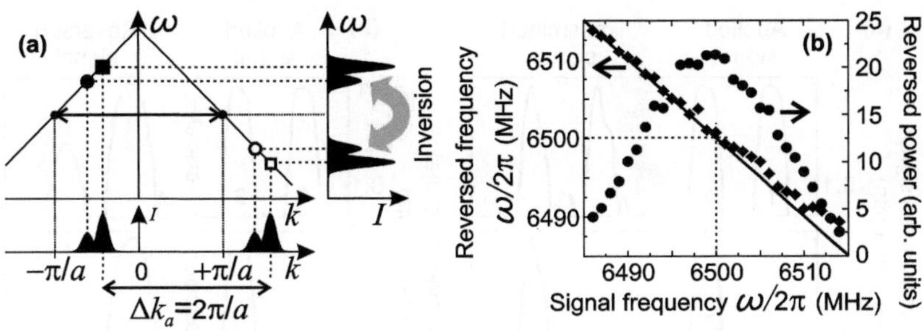

Fig. 17.4 (**a**) Schematic diagram illustrating frequency inversion in the dynamic magnonic crystal system. The *diagonal lines* represent the spin-wave dispersion curve. The applied and reversed signals are counterpropagating. Applied (incident) signals are shown here with positive wavevectors; reversed waves, negative wavevectors. The *small filled dots* indicate the bandgap center frequency ω_a which corresponds to the wavevectors $\pm k_a = \pm\pi/a$. The *open circular and square symbols* illustrate two spectral components of an incident signal waveform. When the DMC is switched "on", the artificial crystal lattice couples these components with corresponding components of a reversed waveform (*filled circular and square symbols*). The difference between the wavevectors of the incident and reversed waves is fixed by the lattice constant a of the DMC so that the k-spectrum of the reversed waveform is uniformly shifted to the left by $\Delta k_a = 2\pi/a = 2k_a$ (*lower panel*) and the reversed waves have frequencies related to those of the incident signals by *inversion* of their detuning from the bandgap center frequency (*right panel*). (**b**) Experimental demonstration of frequency conversion in the dynamic magnonic crystal system. When the carrier frequency of the incident spin-wave signal packet is detuned from the resonance value $\omega_a = 2\pi \cdot 6500$ MHz, the frequency of the reflected signal is shifted (*left ordinate axis, diamonds*). The *solid line* is the theory of (17.6). The efficiency of the inter-modal energy exchange process (*right ordinate axis, circular symbols*) is maximum at the bandgap center and decreases symmetrically with detuning

$(\omega_R - \omega_m) \to -(\omega_R - \omega_m)$, i.e., an "inversion" of the spectrum about ω_R. Returning to the discussion of the previous section of the chapter, we see that the mode coupling mechanism in the DMC provides the means to perform just such an operation: the frequencies of the incident and secondary spin-wave modes are related by an inversion of their detuning $|\delta\omega|$ from the bandgap center frequency ω_a (see (17.6)).

Figure 17.4(a) is a schematic illustration of the frequency inversion process in the DMC. The diagonal lines represent the spin-wave dispersion curve for incident (positive k) and secondary (negative k) waves for $|k| \sim |k_a|$. The small filled dots indicate the wavevectors $\pm k_a$ corresponding to the bandgap center frequency ω_a. The open circular and square symbols represent two spectral components (modes) of an incident signal waveform assumed to be inside the crystal at $t = 0$ when the DMC is switched "on". The filled symbols mark the corresponding secondary (reversed) modes of wavevectors $k_r = k_s - 2k_a$ to which the incident modes are coupled. The difference between the wavevectors of the incident and reversed waves is determined by the lattice constant of the DMC so that the spectrum of the reversed waveform is uniformly shifted to the left by $\Delta k_a = 2\pi/a = 2k_a$. Since the dispersion relationship is symmetrical about $k = 0$, this uniform shift in k-space implies that the frequencies of the incident and reversed waves are related by an inversion of their detuning about ω_a [42].

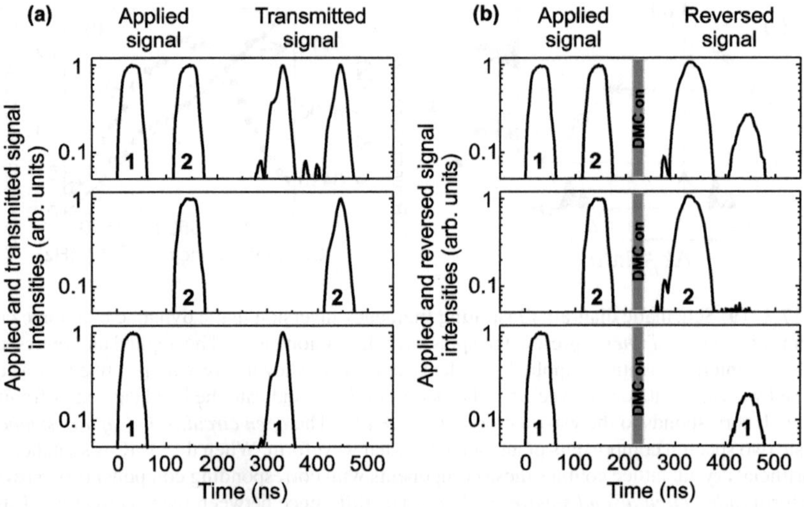

Fig. 17.5 Experimental demonstration of dynamic artificial crystal based time reversal. A train of two pulses, each of width 70 ns, spacing 40 ns and carrier frequency $\omega_s = \omega_a$ was applied to the input antenna. The applied signals shown are the envelopes of those supplied by the microwave generator. (**a**) If the magnonic crystal remains "off" throughout the experiment, transmitted signals appear after a delay of approximately 300 ns (*upper panel*). When the first of the two applied pulses is removed the first transmitted pulse disappears (*middle panel*) and likewise for the second (*lower panel*). (**b**) Reversed spin-wave signals obtained after briefly switching the DMC "on" whilst the spin-wave pulse train is inside. The *shaded area* indicates the time interval of DMC activation. When two pulses are applied, two corresponding reversed pulses are observed (*upper frame*). When the first of the two is removed, the second reflected pulse is absent (*middle panel*) and vice versa (*lower panel*), confirming time reversal

The data of Fig. 17.4(b) provide experimental confirmation of the frequency inversion process in the DMC. Plotted are measured frequencies $\omega(k_r)$ of reversed signals for input spin-wave signal packets of duration 200 ns and frequency $\omega(k_s)$ for $\tau = T$ (see (17.4)). The results are a near-perfect fit to the theory of (17.6) (solid line). The circular symbols (right ordinate axis) indicate the corresponding reversed signal powers; the efficiency of the energy exchange process is maximum at $\omega_s = \omega_a = \omega_r$ and decreases symmetrically with increasing $|\delta\omega|$.

In order to observe time reversal experimentally in the DMC system, an input spin-wave signal comprising two identical spin-wave pulses of carrier frequency $\omega_a = 2\pi \cdot 6500$ MHz was employed [42].

Panel (a) of Fig. 17.5 shows – for reference – the time profiles of applied (input) and transmitted signals detected in the absence of DMC activation (i.e., for the case that the DMC remained "off" for the duration of the experiment). If the first of the two applied pulses is removed, the first transmitted pulse disappears. Similarly, the removal of the second applied pulse leads to the disappearance of the second transmitted envelope. The reversed signals displayed in panel (b) of Fig. 17.5 were obtained by applying the train to the input antenna and allowing the resulting spin wave packets to propagate to the center of the experimental system before switching the DMC "on" via the application of the current pulse (duration $\tau = T$, see (17.4)) to the meander structure. When both spin-wave pulses are applied, two corresponding reversed pulses are observed (upper frame). Here however, unlike the transmitted signals, the removal of the first applied pulse results in the disappearance of the *second* reversed signal envelope. The corresponding situation, when the second applied pulse is absent – the disappearance of the first reversed signal – is illustrated in the lower frame. This behavior provides direct evidence of time reversal.

Note that the relative intensity of the two reversed signals in Fig. 17.5(b) is consistent with the fact that the second reversed signal has a longer effective propagation distance in the YIG than the first, and is therefore attenuated more significantly.

17.7 Concluding Comments

In this chapter, we have progressed from a technical description of the design of the first experimental dynamic magnonic crystals, to a discussion of some of the interesting new physics that these structures allow us to observe. We have placed emphasis on the significance of DMCs as structures with much to contribute both to the fields of spin-wave dynamics and magnonics and, more generally, to our fundamental understanding of wave dynamics in metamaterial systems [7].

The experimental work described in Sects. 17.5 and 17.6 serves to confirm that non-stationary artificial crystal systems provide a means to realize complex spectral transformations such as time reversal which until now have only been demonstrated through the exploitation of non-linear wave media. In the particular context of spin-wave dynamics, our experimental results directly demonstrate that dynamically controlled magnonic crystal systems open doors to sophisticated spin-information processing functionality which might be exploited in future magnonic and spintronic devices [42, 43].

Acknowledgements Financial support from the Deutsche Forschungsgemeinschaft (Grant No. SE 1771/1-2). Dynamic magnonic crystals were fabricated with the technical assistance of the Nano-Structuring Center, TU Kaiserslautern. A.D.K. is grateful for the support of Magdalen College, Oxford.

References

1. D.R. Smith, J.B. Pendry, M.C.K. Wiltshire, Science **305**, 788 (2004)
2. W.L. Bragg, Proc. Camb. Philol. Soc. **17**, 43 (1912)
3. J.D. Joannopoulos, P.R. Villeneuve, F. Fan, Nature **386**, 143 (1997)
4. J. Bravo-Abad, M. Soljacĉić, Nature **6**, 799 (2007)
5. J.O. Vasseur, P.A. Deymier, B. Chenni, B. Djafari-Rouhani, L. Dobrzynski, D. Prevost, Phys. Rev. Lett. **86**, 3012 (2001)
6. T. Miyashita, Meas. Sci. Technol. **16**, R47 (2005)
7. Y. Sivan, J.B. Pendry, Opt. Express **19**, 14502 (2011)
8. S.A. Nikitov, P. Tailhades, C.S. Tsai, J. Magn. Magn. Mater. **236**, 320 (2001)

9. Y.V. Gulyaev, S.A. Nikitov, L.V. Zhivotovskii, A.A. Klimov, P. Tailhades, L. Presmanes, C. Bonningue, C.S. Tsai, S.L. Vysotskii, Y.A. Filimonov, JETP Lett. **77**, 567 (2003)
10. G. Gubbiotti, S. Tacchi, M. Madami, G. Carlotti, A.O. Adeyeye, M.P. Kostylev, J. Phys. D **43**, 264003 (2010)
11. S. Neusser, G. Duerr, H.G. Bauer, S. Tacchi, M. Madami, G. Woltersdorf, G. Gubbiotti, C.H. Back, D. Grundler, Phys. Rev. Lett. **105**, 067208 (2010)
12. B. Lenk, H. Ulrichs, F. Garbs, M. Muenzenberg, Phys. Rep. **507**, 107 (2011)
13. C.-L. Hu, R. Magaraggia, H.-Y. Yuan, C.S. Chang, M. Kostylev, D. Tripathy, A.O. Adeyeye, R.L. Stamps, Appl. Phys. Lett. **98**, 262508 (2011)
14. M. Krawczyk, H. Puszkarski, Phys. Rev. B **77**, 054437 (2008)
15. M. Krawczyk, J. Klos, M.L. Sokolovskyy, S. Mamica, J. Appl. Phys. **108**, 093909 (2010)
16. A.G. Gurevich, G.A. Melkov, *Magnetization Oscillations and Waves* (CRC, New York, 1996)
17. A.A. Serga, A.V. Chumak, B. Hillebrands, J. Phys. D, Appl. Phys. **43**, 264002 (2010)
18. T. Schneider, A.A. Serga, A.V. Chumak, C.W. Sandweg, S. Trudel, S. Wolff, M.P. Kostylev, V.S. Tiberkevich, A.N. Slavin, B. Hillebrands, Phys. Rev. Lett. **104**, 197203 (2010)
19. K.W. Reed, J.M. Owens, R.L. Carter, Circuits Syst. Signal Process. **4**, 157 (1985)
20. P.A. Kolodin, B. Hillebrands, J. Magn. Magn. Mater. **161**, 199 (1996)
21. C. Mathieu, J. Jorzick, A. Frank, S.O. Demokritov, A.N. Slavin, B. Hillebrands, B. Bartenlian, C. Chappert, D. Decanini, F. Rousseaux, E. Cambril, Phys. Rev. Lett. **81**, 3968 (1998)
22. V.V. Kruglyak, A.N. Kuchko, J. Magn. Magn. Mater. **272**, 302 (2004)
23. M.P. Kostylev, A.A. Stashkevich, N.A. Sergeeva, Phys. Rev. B **69**, 064408 (2004)
24. G. Gubbiotti, S. Tacchi, G. Carlotti, N. Singh, S. Goolaup, A.O. Adeyeye, M. Kostylev, Appl. Phys. Lett. **90**, 092503 (2007)
25. A.V. Chumak, A.A. Serga, B. Hillebrands, M.P. Kostylev, Appl. Phys. Lett. **93**, 022508 (2008)
26. M.E. Dokukin, K. Togo, M. Inoue, J. Magn. Soc. Japan **32**, 103 (2008)
27. M. Kostylev, P. Schrader, R.L. Stamps, G. Gubbiotti, G. Carlotti, A.O. Adeyeye, S. Goolaup, N. Singh, Appl. Phys. Lett. **92**, 132504 (2008)
28. A.V. Chumak, A.A. Serga, S. Wolff, B. Hillebrands, M.P. Kostylev, Appl. Phys. Lett. **94**, 172511 (2009)
29. K.S. Lee, D.S. Han, S.K. Kim, Phys. Rev. Lett. **102**, 127202 (2009)
30. A.V. Chumak, P. Pirro, A.A. Serga, M.P. Kostylev, R.L. Stamps, H. Schultheiss, K. Vogt, S.J. Hermsdoerfer, B. Laegel, P.A. Beck, B. Hillebrands, Appl. Phys. Lett. **95**, 262508 (2009)
31. Z.K. Wang, V.L. Zhang, H.S. Lim, S.C. Ng, M.H. Kuok, S. Jain, A.O. Adeyeye, Appl. Phys. Lett. **94**, 083112 (2009)
32. A.V. Chumak, A.A. Serga, S. Wolff, B. Hillebrands, M.P. Kostylev, J. Appl. Phys. **105**, 083906 (2009)
33. A.D. Karenowska, A.V. Chumak, A.A. Serga, J.F. Gregg, B. Hillebrands, Appl. Phys. Lett. **96**, 082505 (2010)
34. V.S. Tkachenko, V.V. Kruglyak, A.N. Kuchko, Phys. Rev. B **81**, 024425 (2010)
35. F.S. Ma, H.S. Lim, Z.K. Wang, S.N. Piramanayagam, S.C. Ng, M.H. Kuok, Appl. Phys. Lett. **98**, 153107 (2011)
36. J. Ding, S. Jain, A.O. Adeyeye, J. Appl. Phys. **109**, 07D301 (2011)
37. H. Al-Wahsha, A. Akjouj, B. Djafari-Rouhani, L. Dobrzynski, Surf. Sci. Rep. **66**, 29 (2011)
38. A.N. Myasoedov, Y.K. Fetisov, Sov. Phys. Tech. Phys. **34**, 666 (1989)
39. Y.K. Fetisov, N.V. Ostrovskaya, A.F. Popkov, J. Appl. Phys. **79**, 5730 (1996)
40. Y.K. Fetisov, J. Commun. Technol. Electron. **10**, 1171 (2004)
41. A.V. Chumak, T. Neumann, A.A. Serga, B. Hillebrands, M.P. Kostylev, J. Phys. D **42**, 205005 (2009)
42. A.V. Chumak, V.S. Tiberkevich, A.D. Karenowska, A.A. Serga, J.F. Gregg, A.N. Slavin, B. Hillebrands, Nat. Commun. **1**, 141 (2010)
43. A.D. Karenowska, V.S. Tiberkevich, A.V. Chumak, A.A. Serga, J.F. Gregg, A.N. Slavin, B. Hillebrands, Phys. Rev. Lett. **108**, 015505 (2012)
44. A.V. Chumak, P. Dhagat, A. Jander, A.A. Serga, B. Hillebrands, Phys. Rev. B **81**, 140404 (2010)

45. N.-N. Chen, A.N. Slavin, M.G. Cottam, IEEE Trans. Magn. **28**, 3306 (1992)
46. N.-N. Chen, A.N. Slavin, M.G. Cottam, Phys. Rev. B **47**, 8667 (1993)
47. A.B. Ustinov, N.Y. Grigor'eva, B.A. Kalinikos, JETP Lett. **88**, 31 (2008)
48. A.B. Ustinov, B.A. Kalinikos, V.E. Demidov, S.O. Demokritov, Phys. Rev. B **81**, 180406 (2010)
49. Y.B. Zeldovich, N.F. Pilipetsky, V.V. Shkunov, *Principles of Phase Conjugation* (Springer, New York, 1985)
50. A. Korpel, M. Chatterjee, Proc. IEEE **69**, 1539 (1981)
51. M.F. Yanik, S. Fan, Phys. Rev. Lett. **93**, 173903 (2004)
52. G.A. Melkov, A.A. Serga, V.S. Tiberkevich, A.N. Oliynyk, A.N. Slavin, Phys. Rev. Lett. **84**, 3438 (2000)

Index

S.O. Demokritov, A.N. Slavin (eds.), *Magnonics*, Topics in Applied Physics 125,
DOI 10.1007/978-3-642-30247-3, © Springer-Verlag Berlin Heidelberg 2013

MIX
Papier aus verantwortungsvollen Quellen
Paper from responsible sources
FSC® C105338

If you have any concerns about our products,
you can contact us on
ProductSafety@springernature.com

In case Publisher is established outside the EU,
the EU authorized representative is:
Springer Nature Customer Service Center GmbH
Europaplatz 3, 69115 Heidelberg, Germany

Printed by Libri Plureos GmbH
in Hamburg, Germany